CATALYST SEPARATION, RECOVERY AND RECYCLING

Catalysis by Metal Complexes

Volume 30

The titles published in this series are listed at the end of this volume.

CATALYST SEPARATION, RECOVERY AND RECYCLING

Chemistry and Process Design

Edited by

DAVID J. COLE-HAMILTON

EaStCHEM, School of Chemistry, University of St. Andrews, St. Andrews, Fife, Scotland

and

ROBERT P. TOOZE

Sasol Technology (UK) Ltd., St. Andrews, Fife, Scotland

Springer

A C.I.P. Catalogue record for this book is available from the Library of Congress.

ISBN-10 94-007-8852-5 (HB)
ISBN-13 978-94-007-8852-7 (HB)
ISBN-10 1-4020-4087-3 (e-book)
ISBN-13 978-1-4020-4087-0 (e-book)

Published by Springer,
P.O. Box 17, 3300 AA Dordrecht, The Netherlands.

www.springer.com

Printed on acid-free paper

TABLE OF CONTENTS

CHAPTER 1 HOMOGENEOUS CATALYSIS – ADVANTAGES AND PROBLEMS **1**
1.1 Catalysis **1**
1.2 Catalyst Stability **4**
1.2.1 THERMALLY INDUCED DECOMPOSITION 4
1.2.2 CHEMICALLY INDUCED DECOMPOSITION 5
1.2.3 PHYSICAL LOSS FROM THE PROCESS 6
1.3 Layout of the Book **6**
1.4 References **8**

CHAPTER 2 CLASSICAL HOMOGENEOUS CATALYST SEPARATION TECHNOLOGY **9**
2.1.1 Coverage of Chapter **9**
2.2 General Process Considerations **9**
2.3 Everything is a Reactor **10**
2.4 Overview of Separation Technologies **10**
2.4.1 TRADITIONAL COBALT WITH CATALYST DECOMPOSITION 10
2.4.2 UNION CARBIDE-DAVY GAS RECYCLE PROCESS 11
2.4.3 LIQUID RECYCLE 12
2.4.4 BIPHASIC SYSTEMS; WATER-ORGANIC 14
2.4.5 INDUCED PHASE SEPARATION 14
2.4.6 NON-AQUEOUS PHASE SEPARATION 15
2.4.6.1 NAPS Using a Non-Polar Catalyst 16
2.4.6.2 NAPS Using a Polar Catalyst 17
2.4.6.3 Ligand Structure and Solubility Properties 17
2.5 Hypothetical processes - How Might the Product be Separated from the Catalyst? **18**
2.5.1 PROPENE HYDROFORMYLATION 19
2.5.2 1-OCTENE HYDROFORMYLATION 20
2.5.3 ALLYL ALCOHOL 20
2.5.4 METHOXYVINYLNAPHTHALENE 21
2.5.5 SEPARATION TECHNOLOGY FOR LESS STABLE CATALYSTS 22
2.5.5.1 Mitsubishi TPPO/TPP Separation 22
2.5.5.2 Organic Polymer for Catalyst Stabilization 22
2.6 Real-World Complications **22**
2.6.1 ORGANOPHOSPHORUS LIGAND DEGRADATIONS 23

2.6.1.1 Oxidation 23
2.6.1.2 Alkyldiarylphosphine Formation 23
2.6.1.3 Ligand Scrambling 24
2.6.1.4 Phosphine Reactions with Conjugated Systems 24
2.6.1.5 Phosphite Oxidation 24
2.6.1.6 Simple Phosphite Hydrolysis 25
2.6.1.7 Poisoning Phosphite Formation 25
2.6.1.8 Aldehyde Acid Formation 25
2.6.1.9 Acidity Control 26
2.6.2 SEPARATING BYPRODUCTS FROM REACTANTS OR PRODUCTS 27
2.6.2.1 Alkene Hydrogenation 27
2.6.2.2 Alkene Isomerization 27
2.6.2.3 Aldehyde Dimerization and Trimerization 27
2.6.2.4 Formation of Conjugated Carbonyls 28
2.6.3 INTRINSIC CATALYST DEACTIVATION 28
2.7 Further Separation Challenges 29
2.7.1 RECOVERY OF METAL VALUES FROM A SPENT CATALYST 29
2.7.1.1 Catalyst Containment and Capture Technologies 30
2.8 Concluding Remarks 35
2.9 References 36

CHAPTER 3 SUPPORTED CATALYSTS 39
Immobilisation of Tailor-made Homogeneous Catalysts 39
3.1 Introduction 39
3.2 Short Historical Overview 40
3.3 Polystyrene Supported Catalysts 41
3.4 Silica Supported Catalyst 44
3.5 Catalysis in Interphases 53
3.6 Ordered Mesoporous Support 58
3.7 Non-covalently Supported Catalysts 60
3.8 Supported Aqueous Phase Catalysis 63
3.9 Process Design [71] 65
3.10 Concluding Remarks 68
3.11 References 69

CHAPTER 4 SEPARATION BY SIZE-EXCLUSION FILTRATION 73
Homogeneous Catalysts Applied in Membrane Reactors 73
4.1 Introduction 73
4.2 Reactors 74

4.2.1 DEAD-END FILTRATION REACTORS 75
4.2.2 CROSS-FLOW FILTRATION REACTORS 76
4.3 Membranes 78
4.3.1 CLASSIFICATION OF FILTRATION TYPES 78
4.3.2 MEMBRANE MATERIALS 79
4.4 Dendrimer Supported Catalysts 80
4.4.1 KHARASCH ADDITION REACTION 81
4.4.2 ALLYLIC SUBSTITUTION REACTIONS 82
4.4.3 HYDROVINYLATION REACTION 86
4.4.4 HYDROGENATION REACTION 88
4.4.5 MICHAEL ADDITION REACTION 89
4.5 Dendritic Effects 90
4.6 Unmodified or Non-dendritic Catalysts 94
4.6.1 HYDROGENATION 95
4.6.2 PHASE TRANSFER CATALYSIS 97
4.7 Soluble Polymer Supported Catalysts 98
4.8 Concluding Remarks 102
4.9 References 102

CHAPTER 5 BIPHASIC SYSTEMS: WATER – ORGANIC 105
5.1 Introduction 105
5.2 Immobilization with the Help of Liquid Supports 106
5.2.1 GENERAL 106
5.2.2 BIPHASIC SYSTEMS 107
5.2.3 AQUEOUS BIPHASIC CATALYSIS 108
5.2.3.1 Water as a Solvent 108
5.2.3.2 Aqueous-phase Catalysis as a Unit Operation 110
5.2.4 EXAMPLES OF AQUEOUS BIPHASIC CATALYSIS 114
5.2.4.1 Hydroformylation (Ruhrchemie/Rhône-Poulenc[RCH/RP] process) 114
5.2.4.2 Other Industrially Used Aqueous-biphasic Processes 116
5.2.4.3 Short Overview of Other Reaction 118
5.2.5 OTHER PROPOSALS FOR WATER - BIPHASIC SYSTEMS 119
5.2.6 INTERLUDE - BIPHASIC SYSTEMS: ORGANIC-ORGANIC 123
5.3 Recycle and Recovery of Aqueous Catalysts 124
5.3.1 RECYCLING 126
5.3.2 RECOVERY 128
5.3.3 ECONOMICS OF THE PROCESS 132
5.3.4 ENVIRONMENTAL ASPECTS 132
5.4 Concluding Remarks 134
5.5 References 135

CHAPTER 6 FLUOROUS BIPHASIC CATALYSIS 145
6.1 Introduction 145

6.2 Alkene Hydrogenation 148
6.3 Alkene Hydrosilation 151
6.4 Alkene Hydroboration 151
6.5 Alkene Hydroformylation 152
6.6 Alkene Epoxidation 158
6.7 Other Oxidation Reactions 161
6.8 Allylic Alkylation 163
6.9 Heck, Stille, Suzuki , Sonagashira and Related Coupling Reactions 164
6.10 Asymmetric Alkylation of Aldehydes 166
6.11 Miscellaneous Catalytic Reactions 169
6.12 Fluorous Catalysis Without Fluorous Solvents 170
6.13 Continuous Processing 171
6.14 Process Synthesis for the Fluorous Biphasic Hydroformylation of 1-Octene 175
6.15 Conclusions 178
6.16 Acknowledgement 179
6.17 References 179

CHAPTER 7 CATALYST RECYCLING USING IONIC LIQUIDS 183
7.1 Introduction 183
7.1.1 INTRODUCTION TO IONIC LIQUIDS 183
7.1.2 INTRODUCTION TO TRANSITION METAL CATALYSIS IN IONIC LIQUIDS 187
7.1.3 MULTIPHASIC CATALYSIS WITH IONIC LIQUIDS – ENGINEERING ASPECTS 189
7.2 Liquid-liquid Biphasic, Rh-catalysed Hydroformylation Using Ionic Liquids 192
7.3 Rhodium Catalysed Hydroformylation Using Supported Ionic Liquid Phase SILP) Catalysis 201
7.3.1 SUPPORTED IONIC LIQUIDS BY CHEMICAL BONDS 203
7.3.2 SUPPORTED IONIC LIQUIDS BY IMPREGNATION 204
7.4 Costs And Economics 206
7.5 Conclusions 209
7.6 References 210

CHAPTER 8 SUPERCRITICAL FLUIDS 215
Compressed Gases as Mobile Phase and Catalyst Support 215
8.1 Introduction to supercritical fluids 215
8.2 Applications of $scCO_2$ in Catalyst Immobilisation 217
8.2.1 CO_2 AS THE ONLY MASS SEPARATING AGENT 217
8.2.2 BIPHASIC SYSTEMS CONSISTING OF CO_2 AND LIQUID PHASES ... 223
8.2.2.1 Water as the Liquid Phase 223
8.2.2.2 Poly(ethyleneglycol) (PEG) as the Liquid Phase 225
8.2.2.3 Ionic Liquids as the Liquid Phase 225
8.2.3 BIPHASIC SYSTEMS CONSISTING OF CO_2 AND SOLID PHASES 230
8.2.3.1 Inorganic Supports 230
8.2.3.2 Organic Polymer Supports 231

8.3 Economic Evaluation and Summary 232
8.3.1 POTENTIAL FOR SCALE-UP 232
8.4 Summary 234
8.5 References 234

CHAPTER 9 AREAS FOR FURTHER RESEARCH 237
9.1 Introduction 237
9.2 Conventional Separation Methods (See Chapter 2) 239
9.3 Catalysts on Insoluble Supports (Chapter 3) 240
9.4 Catalysts on Soluble Supports (Chapter 4) 241
9.5 Aqueous Biphasic Catalysis (Chapter 5) 242
9.6 Fluorous Biphasic Catalysis (Chapter 6) 243
9.7 Reactions Involving Ionic Liquids (Chaoter 7) 244
9.8 Reactions Using Supercritical Fluids (Chapter 8) 245
9.9 Conclusions 247
9.10 References 247

CHAPTER 1 HOMOGENEOUS CATALYSIS – ADVANTAGES AND PROBLEMS

D. J. COLE-HAMILTON[a] AND R. P. TOOZE[b]
[a]*EaStCHEM, School of Chemistry, Univesity of St. Andrews, St. Andrews, Fife, KY19 9ST, Scotland*
[b]*Sasol Technology UK, Ltd., Purdie Building, North Haugh, St. Andrews, Fife, KY19 9ST, Scotland*

1.1 Catalysis

Catalysts speed up chemical reactions but can be recovered unchanged at the end of the reaction. They can also direct the reaction towards a specific product and allow Chemistry to be carried out at lower temperatures and pressures with higher selectivity towards the desired product. As a result they are used very extensively in the Chemical Industry. Chris Adams, writing for The North American Catalysis Society estimates that "35% of global GDP depends on catalysis, although this excludes the emergent genetic business. Confining the analysis to the chemicals industry, with global sales of perhaps $\$1.5 \times 10^{12}$ the proportion of processes using catalysts is 80% and increasing. The catalyst market itself is US$\$10^{10}$, so that catalysis costs are much less than 1% of the sales revenue from the products which they help create. Small wonder that the catalyst market is increasing at 5% per annum" [1]

TABLE 1.1 Comparison of homogeneous and heterogeneous catalysts

	Heterogeneous	Homogeneous
Catalyst form	Solid, often metal or metal oxide	Metal complex
Mode of use	Fixed bed or slurry	Dissolved in reaction medium
Solvent	Usually not required	Usually required – can be product or byproduct
Selectivity	Usually poor	Can be tuned
Stability	Stable to high temperature	Often decompose < 100°C
Recyclability	Easy	Can be very difficult
Special reactions	Haber process, exhaust clean up etc.	Hydroformylation of alkenes, methanol carbonylation, asymmetric synthesis etc

There are two kinds of catalysts. Heterogeneous catalysts are insoluble in the medium in which the reaction is taking place so that reactions of gaseous or liquid reagents occur at the surface, whilst homogeneous catalysts are dissolved in the reaction medium and hence all catalytic sites are available for reaction. Some of the properties of catalysts are collected in Table 1.1, where heterogeneous and homogeneous catalysts are compared.

D. J. Cole-Hamilton and R. P. Tooze (eds.), Catalyst Separation, Recovery and Recycling, 1–8.

Heterogeneous catalysts are generally metals or metal oxides and they tend to give rather unselective reactions. They are very stable towards heat and pressure, so can be used at high temperature. Only the surface atoms are available for reaction. Homogeneous catalysts, on the other hand are usually complexes, which consist of a metal centre surrounded by a set of organic ligands. The ligands impart solubility and stability to the metal complex and can be used to tune the selectivity of a particular catalyst towards the synthesis of a particular desirable product. By varying the size, shape and electronic properties of the ligands, the site at which the substrate binds can be constrained in such a way that only one of a large number of possible products can be produced. As an example, Figure 1.1 shows a range of products that might be produced from a mixture containing an alkene, carbon monoxide, hydrogen and an alcohol. All of the products have their uses, but it is a triumph of homogeneous catalysis that any one of the products can now be made with > 90 % selectivity by careful selection of the metal centre, ligands, reaction conditions and in some cases substrate [2]

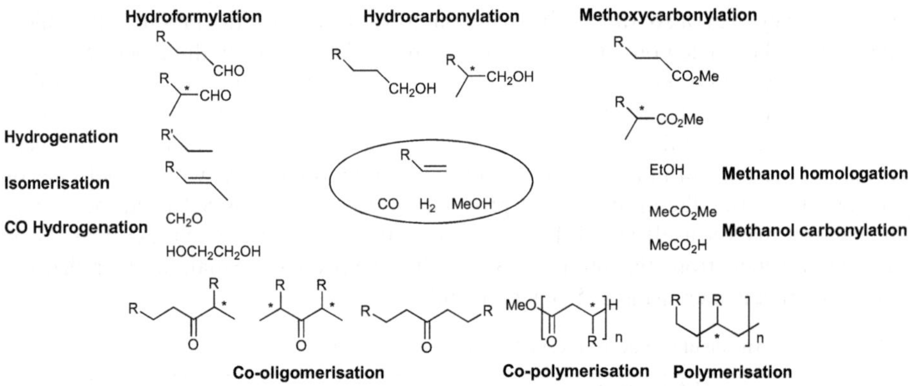

Figure 1.1. Some of the products that can form from an alkene, carbon momoxide, hydrogen and methanol. The asterisks represent asymmetric centres in chiral molecules

Various different kinds of selectivity are represented in Figure 1.1. These include:

- Chemoselectivity, the production of one product type such as alcohols rather than aldehydes
- Regioselectivity, the production of a linear ester rather then one with a branching methyl group
- Stereoselectivity, the production of one enantiomer of a chiral compound (chiral products are marked with an asterisk in Figure 1.1)

In general, heterogeneous catalysts do not show the selectivity shown by chiral catalysts, although current research on surface modifiers has shown that even enantioselective reactions, albeit for a restricted range of substrates is becoming possible [3, 4]

Despite this selectivity advantage of homogeneous catalysts, almost all of the industrial catalytic processes use heterogeneous catalysts, because of their one major advantage, their ease of separation form the reaction product. Being insoluble in the reaction medium, heterogeneous catalysts can often be used as fixed beds over which the substrates flow continuously in the liquid or gaseous form. This means that the catalyst can be contained within the reactor at all times. Not only does this mean that the separation of the products from the catalyst is built into the process, but also, the catalyst is always kept under the conditions of temperature, pressure, contact with the substrate and products, for which it has been optimised.

For homogeneous catalysts, which are dissolved in the reaction medium containing the substrates, products and dissolved gases, the separation can be extremely energy intensive and time consuming. Only rarely, when the product can be evaporated under the reaction conditions, can homogeneous catalytic reactions be carried out under continuous flow conditions, where the substrates are introduced continuously into the reactor whilst the products are continuously removed. More often, commercial processes are carried out under what we shall refer to as batch continuous conditions. Part of the liquid catalytic solution containing the product(s), unreacted substrates and catalyst is removed continuously from the reactor to a separator, which is usually a distillation system operating at lower pressure than the reactor. The products and unreacted substrates are then separated from the catalyst and lower boiling byproducts by fractional distillation before the fraction containing the catalyst is returned to the reactor. Since the separation is carried out under conditions that are far removed from those for which the catalyst has been optimised, there is a danger that the catalyst may precipitate, thus clogging pipework or, worse still, decompose in the recyling loop.

In general homogeneous catalysis has only been commercialised when there is no heterogenous catalyst that is capable of promoting the desired reaction or when selectivity to a higher added value product is possible using a homogeneous catalyst. Creative chemists and process engineers have then joined forces to provide a cost effective solution to the separation problem.

Increasing environmental concerns and dwindling supplies of raw materials and energy sources mean that there is now a significant pressure to introduce cleaner processing in the chemical and pharmaceutical industries.

Ideally reactions should have as many as possible of the following properties:

- Use renewable feedstocks
- Make a single product
- Have 100 % atom efficiency (all the atoms in the starting materials end up in the products, although expulsion of water might be acceptable)
- Operate under mild conditions (preferably ambient temperature and pressure)
- Be tolerant of the presence of air
- Produce no waste or other byproducts (these are often quantified using the *E*-Factor, which is the mass (kg) of waste produced per kg of product formed. For a fuller discussion see Chapter 5, Section 5.3.4)
- Have a very long-lived catalyst if, as is very likely, one is required
- Have a simple separation method for the catalyst from the products
- Operate under continuous flow conditions

The catalysts that come closest to meeting these requirements are enzymes, but in general, the products of biological process are not separated from the medium in which they are formed, rather being used in situ. The demand for high selectivity and low environmental impact suggests that there will be a desire to commercialise more processes using homogenous catalysis and hence the problem of separating, recovering and/or recycling the catalysts must be addressed, perhaps using innovative solutions [5, 6]

There are, then, three critical requirements of any catalyst if it is to be exploited on a commercial scale; these are activity, selectivity and stability. It has been widely demonstrated and generally accepted that homogeneous catalysts are superior to their heterogeneous counterparts in terms of both activity (certainly under mild reaction conditions) and selectivity (the classical example is chiral catalysis).

The fatal weakness that has prevented the predicted proliferation of homogeneous catalysts is lack of stability. This further illustrates that all of the above criteria need to be fulfilled.

1.2 Catalyst Stability

Catalyst stability can be defined in terms of turnover number (TON). A textbook definition of this is:

$$\text{TON} = \frac{\text{mol of feed consumed before activity ceases}}{\text{mol of catalyst utilised}}$$

In reality the limiting case it not complete loss of activity but rather reduction of activity below a critical threshold determined by the economics of any given process and reactor design.

The TON can be reduced in a number of ways

- Thermally induced decomposition.
- Chemically induced decomposition, of which two further categories can be considered namely substrate induced decomposition and poisoning by impurities or products.
- Physical loss from the process.

These will now be considered briefly in turn

1.2.1 THERMALLY INDUCED DECOMPOSITION

As mentioned above one of the fundamental attributes ascribed to homogeneous catalysts is superior activity at low temperature. However, even within classes of such catalysts, improvements in catalyst activity can be made allowing operation at lower temperatures, thus reducing or avoiding completely this mode of catalyst decay. One such example can found in recent advances in palladium catalysed ethene carbonylation (Equation 1.1).

$$CH_2{=}CH_2 + CO + ROH \longrightarrow H\text{-}\{CH_2CH_2C(O)\}_n\text{-}OR$$

Equation 1.1. Ethene carbonylation leading to alkyl (R) propionates (n = 1) or to ethene carbon monoxide copolymers (n is large)

This reaction to give methyl propanoate can be catalysed by a combination of palladium acetate, triphenylphosphine and methanesulphonic acid [7]. However in order to obtain acceptable rates the reaction must be carried out at over 100 °C. At this temperature catalyst life is short due to a variety of side reactions [8, 9] and very expensive palladium is lost. Advances in catalyst design have identified alternative phosphine ligands such as 1,2-*bis*(ditertiarybutylphosphinomethylbenzene) that can give much increased activity even at 30° lower temperatures and thus avoid thermal decomposition [10]. Whilst this and other examples show that increased understanding can lead to improvements in catalyst design such that reactor operating conditions can be changed so as to avoid decomposition, no such activity enhancement will impact on the separation process which is governed by the physical properties of all products and reactants. An example here would be the hydroformylation of alkenes. Scientific advances have resulted in phosphine modified rhodium catalysts that exhibit much greater activity than for example unmodified cobalt catalysts, but when this technology is applied to higher molecular weight alkenes (C10+) the separation of products from catalyst in such a way as to avoid extensive decomposition of expensive catalyst is a formidable technological challenge. Different possible separation methodologies for long chain aldehydes formed by hydroformylation reactions are explored in more detail in the subsequent chapters of this book.

1.2.2 CHEMICALLY INDUCED DECOMPOSITION

No catalyst has an infinite lifetime. The accepted view of a catalytic cycle is that it proceeds via a series of reactive species, be they transient transition state type structures or relatively more stable intermediates. Reaction of such intermediates with either excess ligand or substrate can give rise to very stable complexes that are kinetically incompetent of sustaining catalysis. The textbook example of this is triphenylphosphine modified rhodium hydroformylation, where a plot of activity versus ligand:metal ratio shows the classical "volcano plot" whereby activity reaches a peak at a certain ratio but then falls off rapidly in the presence of excess phosphine, see Figure 1.2 [11]. On occasion these over ligated complexes are materials that can be identified in solution or perhaps more tellingly isolated from catalytic reactions. Such reactions can often be reversed by removal of the excess reagent. Such processes are not considered in the context of this book as resulting in loss of overall turnover number.

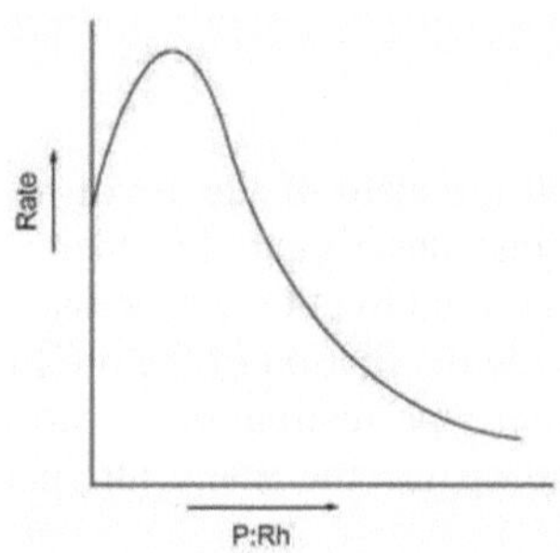

Figure 1.2. Typical plot of the effect of rate on the P:Rh ratio for a Rh/PPh_3 hydroformylation catalyst. The exact position of the maximum depends on [Rh], p_{CO} and T

A more serious but potentially soluble problem is that of poisoning by impurities in the feed to a catalytic reaction. Homogeneous catalysts are again believed to be more susceptible to this mode of decomposition than their heterogeneous counterparts. This problem may be solved by the development of more robust catalysts, but a more usual solution is feedstock purification. An example where this purification is carried out *in situ* involves the addition of large quantities of aluminium reagents (such as methylaluminoxane, MAO) to polymerisation and oligomerisation reactions. The ratio of aluminium to transition metal can be very high (>500) and part of the purpose of this large excess is believed to be removal of oxygenates from alkene feeds [12].

1.2.3 PHYSICAL LOSS FROM THE PROCESS

The loss of expensive catalyst from the reactor system can be fatal for any process. Physical loss involves the removal of active catalyst from the closed loop of the process. This can include the plating out of metal or oxides on the internal surfaces of the manufacturing plant, failure to recover potentially active catalyst from purge streams and the decomposition of active catalyst by the process of product recovery. The first two can be alleviated to some extent by improvements in catalyst or process design, the last is an intrinsic problem for all manufacturing operations and is the subject of this book.

Catalysts are traditionally designed and optimised based on their performance in the reactor and not for their ability to withstand traditional separation processes. However, on taking any system from the laboratory to the pilot plant and beyond, this need to isolate product whilst efficiently recovering the catalyst often becomes the most important single issue. The best option is selection of a product isolation method that maintains the integrity of the catalyst and requires no further treatment of the catalyst prior to reintroduction into the reactor, or leaves the catalyst in the reactor at all times.

A compromise solution can be that, although a catalyst may not be in its active form in the separation unit, it can be recovered and regenerated easily at the production facility. A final option is that spent catalyst can be recovered, concentrated and returned to the original supplier for reprocessing. Whilst this is an expensive and inelegant option, it remains the most pragmatic solution until technologies described in this book reach maturity.

1.3 Layout of the Book

In this book, we report on the state of the art of methods for catalyst separation recovery and recycling, not just describing the chemistry, but also discussing the process design that would be required to put the processes into practice.

Conventional processes involving distillation of the product directly from the reactor or batch continuous operation where the distillation is carried out in a separate chamber (Chapter 2) provide the backdrop for the many alternative processes that are being discussed.

These alternative processes can be divided into two main categories, those that involve insoluble (Chapter 3) or soluble (Chapter 4) supports coupled with continuous flow operation or filtration on the macro – nano scale, and those in which the catalyst is immobilised in a separate phase from the product. These chapters are introduced by a discussion of aqueous biphasic systems (Chapter 5), which have already been commercialised. Other chapters then discuss newer approaches involving fluorous solvents (Chapter 6), ionic liquids (Chapter 7) and supercritical fluids (Chapter 8).
No attempt is made to provide comprehensive coverage of all the work carried out in these different media, but rather to give a flavour of the kind of systems for which the different approaches may be appropriate. In all the chapters, a more detailed discussion of the rhodium catalysed hydroformylation of 1-octene to nonanal, as a representative example of the synthesis of a long chain aldehyde with relatively low volatility, is provided [13, 14]. This reaction has been chosen because:

- It is a reaction demonstrating 100 % atom economy
- It is a reaction which uses both gas and liquid substrates
- The rate of the reaction is crucial for successful commercialisation
- There are important issues relating to chemoselectivity (aldehydes or alcohols may be the products and alkene isomerisation is a competing side reaction, which must be reduced to a minimum) and regioselectivity (linear aldehyde is much preferred over branched)
- It is a commercially important reaction as a step in the synthesis of nonanol, an important plasticizer alcohol. Other long chain alcohols, derived from product aldehydes by hydrogenation are used as the basis of soaps and detergents,
- Currently the reaction is carried out using cobalt based catalysts with severe penalties in terms of harsh operating conditions (80 bar CO/H_2, 200°C). In addition, substantial loss of substrate (*ca.* 10%) to hydrogenation makes the overall selectivity to the linear alcohol *ca.* 80% [15]. Rhodium based systems are capable of giving higher selectivities (>90%) to the desired linear aldeyde product under milder conditions (20 bar, 100°C) [13]
- The reaction has been studied using all of the different possible separation methods and represents a system where advantages and disadvantages of the various processes can be compared [5]
- Despite the very attractive properties of the rhodium-based system, no commercial plants used it because the low stability of the catalyst meant that the catalyst separation problem prevented commercialisation. Very recently, this situation has changed with the introduction of rhodium-based plant by Sasol in South Africa which uses technology developed by Kvaerner Process Technology (now Davy Process Technology). This batch continuous plant produces medium - long chain aldehydes and the separation is carriedout by low pressure distillation [16-18]

In the final Chapter of the book (Chapter 9), all the different processes are compared with a discussion of the various areas where further research will be required to improve the new processes to a point where they may be commercially attractive.

1.4 References

[1] C. J. Adams, *The North American Catalyst Society*, http://www.nacatsoc.org/edu_info.asp?edu_infoID=1.
[2] 'Applied Homogenous Catalysis with Organometallic Compounds', ed. B. Cornils and W. A. Herrmann, VCH, Weinheim, 1996.
[3] Q. H. Xia, H. Q. Ge, C. P. Ye, Z. M. Liu, and K. X. Su, *Chem. Rev.*, 2005, **105**, 1603.
[4] T. Burgi and A. Baiker, *Accounts Chem. Res.*, 2004, **37**, 909.
[5] D. J. Cole-Hamilton, *Science*, 2003, **299**, 1702.
[6] C. C. Tzschucke, C. Markert, W. Bannwarth, S. Roller, A. Hebel, and R. Haag, *Angew. Chem.-Int. Edit.*, 2002, **41**, 3964.
[7] E. Drent, *Eur. Pat.*, 1984, 0106.
[8] R. P. Tooze, K. Whiston, A. P. Malyan, M. J. Taylor, and N. W. Wilson, *J. Chem. Soc.-Dalton Trans.*, 2000, 3441.
[9] W. G. Reman, G. B. J. Deboer, S. A. J. Van Langen, and A. Nahuijsen, *Eur. Pat.*, 1991, 0411.
[10] W. Clegg, G. R. Eastham, M. R. J. Elsegood, R. P. Tooze, X. L. Wang, and K. Whiston, *Chem. Commun.*, 1999, 1877.
[11] K. L. Oliver and F. B. Booth, *Hudrocarbon Process.*, 1970, **49**, 112.
[12] E. Y. X. Chen and T. J. Marks, *Chem. Rev.*, 2000, **100**, 1391.
[13] 'Rhodium catalysed hydroformylation', ed. P. N. W. M. Van Leeuwen and C. Claver, Kluwer, Dordrecht, 2000.
[14] C. D. Frohling and C. W. Kohlpaintner, in 'Applied Homogeneous Catalysis with Organometallic Compounds', ed. W. A. Herrmann, VCH, Weinheim, 1996.
[15] B. Cornils, in 'New synthesis with Carbon Monoxide', ed. J. Falbe, Springer Verlag, Berlin, 1980.
[16] J. A. Banister and G. E. Harrison, *US Patent*, 2004, 0186323.
[17] *Chem. Eng. News.*, 1999, **77**, 19.
[18] *Chem. Eng. News.*, 2004, **82**, 29.

CHAPTER 2 CLASSICAL HOMOGENEOUS CATALYST SEPARATION TECHNOLOGY

DAVID R. BRYANT
Intellectual Property and Technology, Inc.
1201 Shady Way
South Charleston, WV 25309-2419
USA
drbryant@charter.net

2.1 Coverage of Chapter

When considering a separation technique for a homogeneous catalytic process, one must realize that catalyst/product/byproduct separation is an integral part of the entire process. The selection and design of the separation technology goes hand-in-hand with catalyst design, often in an iterative fashion. That is, a catalyst is selected and tested in a continuous unit, with recycle of streams, to discover if there are problems that will necessitate redesign of the catalyst. Redesign is more often the fact than the exception.

The objective of this chapter is to detail considerations that must be addressed in order to successfully marry a catalyst technology with catalyst/product separation technology. The focus of this chapter is hydroformylation, but the general principles should apply to many homogeneous precious-metal catalyzed processes.

2.2 General Process Considerations

There are four principal factors that are paramount in selecting the best separation technique. They are the energy required for the separation, the capital required for the equipment used in the separation, the efficiency/effectiveness of the separation, and the vitality of the catalyst after the separation. General process considerations include:

- Transitions of any type including temperature, pressure or phase changes should be minimized.
- Cooling below 40 degrees Celsius becomes more expensive (river water cannot be used).
- Vacuum below 20 mm Hg is challenging.
- Byproduct formation should be minimized. Single product processes are better. A distillation column, or other step, will be required for each material in the mixture.

D. J. Cole-Hamilton and R. P. Tooze (eds.), Catalyst Separation, Recovery and Recycling, 9–37.

- Everything feasible should be recycled so as to minimize waste.
- Pressures should be kept below 35 bar, at least below 100 bar, to minimize costs and because most process design experience is here.
- The use of rotating equipment such as compressors or centrifuges should be minimized to minimize maintenance costs.
- Corrosive materials, particularly chloride, should be avoided.
- Batch operations should be avoided.
- The handling of solids should be avoided.

2.3 Everything is a Reactor

This may be a good time to introduce a very simple principal of process chemistry, but one that is not widely recognized. It is taught in chemical engineering that the only things in chemistry that matter are temperature and concentration. Every other variable can be reduced to these two. For example, time is simply a reflection of changing concentration.

Now a corollary: since every piece of process equipment has associated with it temperature and concentration, all pieces of process equipment are reactors. Stated differently, everything is a reactor.

There is a tendency to think that once the catalyst is removed from the reactor, all chemistry ceases. Chemistry is occurring throughout the process, and that is why separation of products cannot be viewed in isolation from the process that made them.

2.4 Overview of Separation Technologies

2.4.1 TRADITIONAL COBALT WITH CATALYST DECOMPOSITION

Traditional cobalt hydroformylation separations will not be covered in detail since they have been described in many excellent references.[1] A key factor in understanding cobalt hydroformylation conditions and cobalt/product separations is to recognize that cobalt is a relatively unreactive catalyst that requires high temperatures to achieve commercially viable rates. Cobalt carbonyls have limited thermal stability. By using high (200 bar) partial pressures of syn gas (CO/H_2), thermal stability is achieved during hydroformylation. However, to separate the cobalt catalyst from the hydroformylation products the pressure must be reduced. Separation is achieved by decomposing the catalyst in a step referred to as decobalting. There have been a variety of techniques disclosed for achieving this goal.[2]

A major advance in homogeneous catalysis was the introduction of a trialkylphosphine to supplement the role of carbon monoxide in catalyst stabilization.[3] A ligand modifier such as trialkylphosphine serves three principal roles in a homogeneous catalytic process. It stabilizes the metal, it influences the reaction rate, and it influences process selectivity.

In cobalt hydroformylation, the trialkylphosphine provides a more thermally stable catalyst so that decobalting is not required. Its influence on reaction rate is not a desirable one in that the TOF (turnover frequency) of the cobalt is reduced with the consequence that higher operating temperatures are needed to achieve commercial rates. Finally, the trialkylphosphine significantly alters process selectivity. Rather than making mainly aldehydes, as is the case with unmodified cobalt, the principal product in phosphine-modified cobalt catalysis is the corresponding alcohol. For many alkenes this is not undesirable since higher molecular weight aldehydes would probably be reduced to alcohol in subsequent processing steps. For butanal, however, the circumstance is different.

2.4.2 UNION CARBIDE-DAVY GAS RECYCLE PROCESS

Butyl alcohol is not the principal use of butanal obtained by propene hydroformylation. Rather its major market is 2-ethylhexanol that is prepared via aldol condensation followed by hydrogenation.[4] Thus formation of alcohols when aldehydes are desired is not only a direct efficiency loss, but also the alcohol impurity will form hemiacetals and acetals that complicate refining and lead to increased operating costs.

A breakthrough in hydroformylation was achieved with the introduction of a triarylphosphine-modified, in particular triphenylphosphine-modified, rhodium catalyst.[5] This innovation provided simultaneous improvements in catalyst stability, reaction rate and process selectivity. Additionally, products could be separated from catalyst under hydroformylation conditions. One variant is described as Gas Recycle (Figure 2.1) since the products are isolated from the catalyst by vaporization with a large recycle of the reactant gases.[6] The recycle gas is chilled to condense butanals.

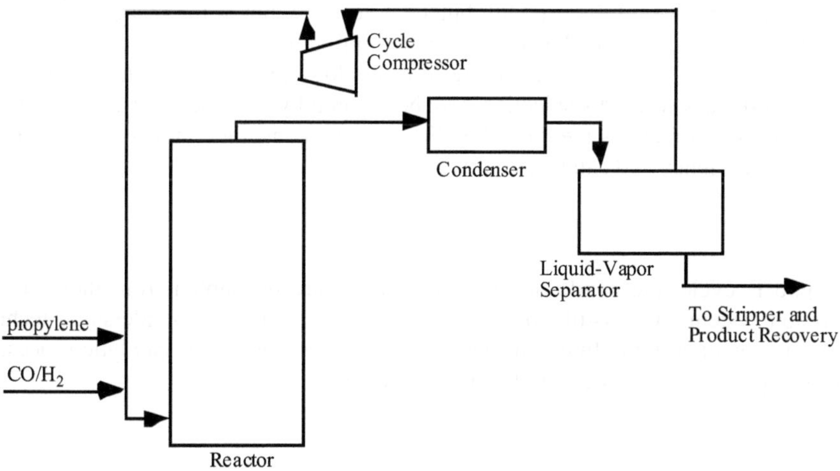

Figure 2.1. Gas Recycle Hydroformylation Process

In the practice of gas recycle hydroformylation [7], rhodium complex and triphenylphosphine are dissolved in a suitable solvent. The reactor is pressurized with the reactants, carbon monoxide, hydrogen and propene. The entering gas is passed through the catalyst solution and becomes saturated with aldehyde. Gas exiting the reactor is chilled to condense butanal, and the reactant gases are compressed and returned to the reactor.

One advantage of Gas Recycle operation is that the catalyst remains in the reactor and is thus always working. This reduces the inventory of the expensive precious metal. Another advantage is that part of the heat of reaction is used to vaporize aldehyde products. A downside is the energy consumption of the recycle compressor. Another downside is that the large gas flows through the catalyst solution expand its volume such that a greater reactor volume is required resulting in increased capital cost.

Gas Recycle is a relatively simple operation. Rather than being circulated through a variety of pipes, pumps and columns, the catalyst remains in one place. A key control variable is maintaining a constant liquid level in the reactor. This is not as simple as it might first seem because in addition to butanal isomers forming, butanal condensation products including dimers and trimers also form to give what are collectively termed "heavies".

Heavies formation is accelerated by a variety of materials.[8] Successful Gas Recycle operation depends on keeping the catalyst solution as pristine as possible to limit heavies formation since in Gas Recycle there is no independent way to remove heavies. There are a single set of conditions for product formation, product removal and by-product (heavies) removal. A key to successful operation is identifying conditions under which the heavies can be removed essentially at their rate of formation. A downside of Gas Recycle is that it may be difficult to recover from upsets in operation, which result in the catalyst solution containing a disproportionate amount of heavies.

Gas Recycle technology has been licensed worldwide by Union Carbide-Davy for the hydroformylation of propene.[9] It has also been operated by Union Carbide for ethene hydroformylation. Its use with butene is feasible, but at the margin of operability. Liquid Recycle, described below, is a better option for butene.

In spite of its limitations, Gas Recycle technology remains a viable option in certain circumstances where its selection may be favored by plant scale or capacity. Other keys to the decision are as mentioned earlier: energy consumption, capital investment, separation efficiency and catalyst vitality.

2.4.3 LIQUID RECYCLE

In Liquid Recycle, the conditions for the reaction are decoupled from those for the separation system.[10] Distillation is a widely practiced and well-understood technology, so it is generally the first consideration for any homogeneous catalytic process. A typical Liquid Recycle system is shown in Figure 2.2.

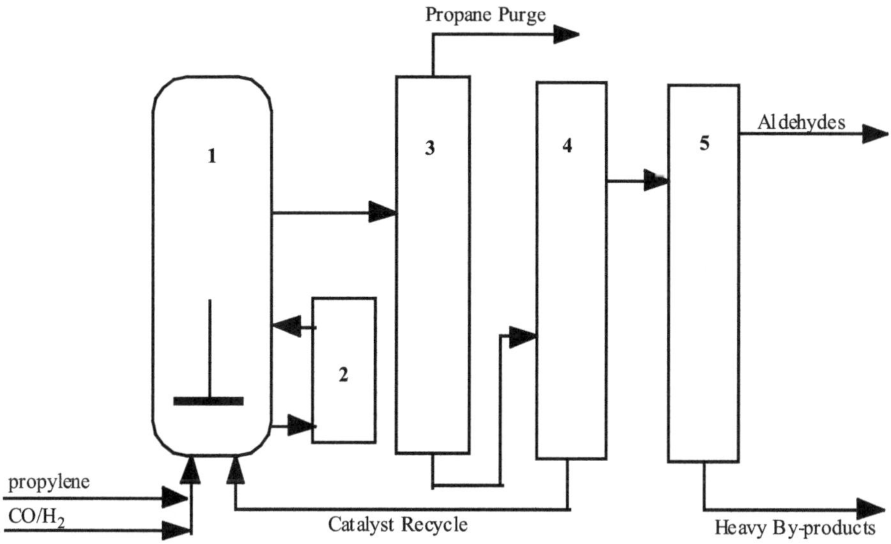

Figure 2.2. Block Flow Diagram for a Liquid Recycle Process

Propene and syn gas are fed to a reactor (1 in Figure 2.2) where the gases are intimately contacted with an organophosphorus-modified rhodium catalyst. The exothermic heat of reaction is controlled with heat exchanger (2). Effluent from the reactor passes to a column (3) where the solution is degassed. Propane in the cycle from hydrogenation of propene is vented along with some propene and syn gas. From the degassing column, the catalyst solution passes to column (4) where aldehyde products and condensation byproducts are separated from catalyst solution. The catalyst solution is recycled to the reactor, and the product mixture is transferred to column (5) where isolation of the butanal occurs.

Feed to tails ratio may be defined as the ratio between the liquid fed to column (4) and the liquid in the catalyst recycle. Higher feed/tails ratios contribute to higher conversion since with only catalyst and heavy solvent being recycled more of the reactor volume is available for product.

Whereas in Gas Recycle the product must be removed at the same temperature and pressure at which it is formed, in Liquid Recycle the separation of product (and byproducts) from catalyst is independent of the conditions under which the products were formed. This added degree of control brings a variety of benefits. Since large gas flows are no longer required in the reactor, the liquid expansion due to gassing is reduced and more catalyst can be contained in a specific reaction vessel. Reactor temperature and reactant concentrations can be tuned for optimum catalyst performance. The conditions in the separation system can likewise be tuned for optimum performance. In particular, more severe conditions will permit better control over the concentration of heavies in the catalyst solution.

The more concentrated the catalyst exiting the separation system (vaporizer) and being returned to the reactor, the higher the concentration of product in the effluent from the reactor. Higher product concentration means fewer passes of the catalyst through the vaporizer for a given production, and fewer passes means higher efficiencies in the conversion of raw materials to products since each time catalyst is removed from the reactor some unconverted reactants will be lost.

Another advantage of Liquid Recycle is that multiple reactors may be arranged in series with the effluent from one passing on to the next. The alkene concentration is less in the downstream reactors, but reaction conditions can be adjusted to optimize each reactor's performance. In back mixed reactors in continuous operation, the effluent from the reactor is the same as the catalyst solution throughout the reactor. By placing reactors in series, the first reactor can be optimized for high rates and later reactors for high conversion.

There are some downsides to Liquid Recycle operation. The first has been referred to as thermal degradation [11] or thermal shock, although this term suggests that only temperature is responsible, but remember that in chemistry the two key variables are temperature and concentration.[12] What one observes is that the catalyst may become less active or even less soluble when passed through a vaporizer or when exposed to carbon monoxide and hydrogen in the absence of alkene. The successful development of a homogeneous catalytic process requires the close cooperation of both chemists and engineers to manage the tradeoffs as product is separated from catalyst.

2.4.4 BIPHASIC SYSTEMS; WATER-ORGANIC

Considerable work has been conducted on a water-soluble catalyst using sulfonated phosphine-modified rhodium. Details of this chemistry will be described in Chapter 5. The general concept (Figure 2.3) is to make the catalyst water soluble, then after product formation, decant the product. In order for the water-soluble catalyst to be effective, the alkene must dissolve in the aqueous layer. This has been demonstrated on a commercial basis using propene. The low solubility of higher alkenes in the aqueous catalyst layer has proven problematic. The desirable characteristic of the ligand, water solubility, is needed in the separation step but is a disadvantage in the reaction step.

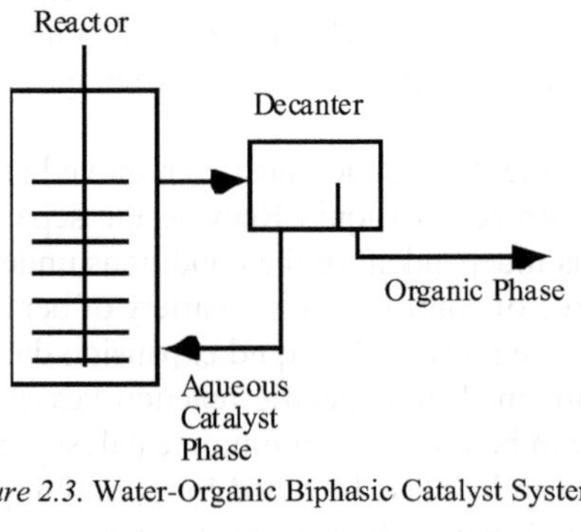

Figure 2.3. Water-Organic Biphasic Catalyst System

2.4.5 INDUCED PHASE SEPARATION

An approach that overcomes the disadvantage of having alkene and catalyst in separate phases in the reactor(s) is to use a phosphine ligand that is less highly sulfonated. One can prepare catalysts with monosulfonated phosphines which are organic soluble. During hydroformylation higher alkenes will be in the same phase as the catalyst and significantly higher rates will be obtained. To achieve separation, a small amount of water is added so that phase separation occurs.[13] After product separation, the catalyst is dried and then returned to the reactor (Figure 2.4).

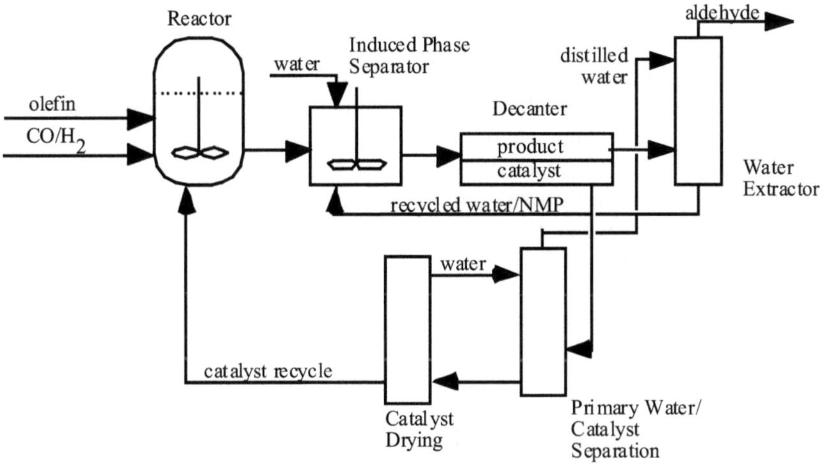

Figure 2.4. Induced Phase Separation Flow Diagram

In this process, catalyst solution leaving the reactor goes to a separator where the small amount of water is added to induce phase separation. The mixture passes to a decanter where the catalyst is separated from the product. The catalyst stream passes through two drying stages; the first stage produces distilled water that is fed to the water extractor, the second stage completes the drying of the catalyst which then is returned to the reactor. The product phase from the decanter is sent to the water extractor to remove the NMP used to facilitate solubilizing the catalyst.

Advantages of Induced Phase Separation are that very high molecular weight alkenes can be hydroformylated and the aldehyde product and byproducts can be separated without the catalyst suffering "thermal shock". Disadvantages include a more limited ligand selection and the removal of water that has a high heat of vaporization. In addition, this technology is, as is the water-soluble sulfonated catalyst, limited to the formation of nonpolar products.

2.4.6 NON-AQUEOUS PHASE SEPARATION

A major breakthrough in separation of products from catalyst, in particular heat sensitive products, came with the discovery of the NAPS or Non-Aqueous Phase Separation technology. NAPS provides the opportunity to separate less volatile and/or thermally labile products. It is amenable to the separation of both polar [14] and non-polar [15] products, and it offers the opportunity to use a very much wider array of ligands and separation solvents than prior-art phase separation processes. The phase distribution characteristics of the ligand can be tuned for the process. Two immiscible solvents are normally required to effect catalyst/product separation, but in some cases the product itself may have the appropriate polarity to behave as either the polar or non-polar solvent. For example, an aliphatic hydrocarbon such as hexane is a typical non-polar solvent while acetonitrile and methanol are typical polar solvents. The catalyst system is modified to have polarity opposite to the product. The ligand provides the basis for the desired catalyst separation selectivity.

2.4.6.1 NAPS Using a Non-Polar Catalyst

An alkene which will give a polar aldehyde product and syn gas are introduced into the reactor containing a non-polar ligand modified rhodium catalyst. Catalyst solution exiting the reactor enters a Flash stage where CO/H_2 are purged. The catalyst solution then enters an extractor where it is contacted with a polar solvent. The product aldehyde is captured in the polar solvent in the extractor, then concentrated in the Solvent Removal Column. Polar Solvent is recycled to the Extractor. The Non-Polar catalyst solution is recycled to the reactor (see Figure 2.5).

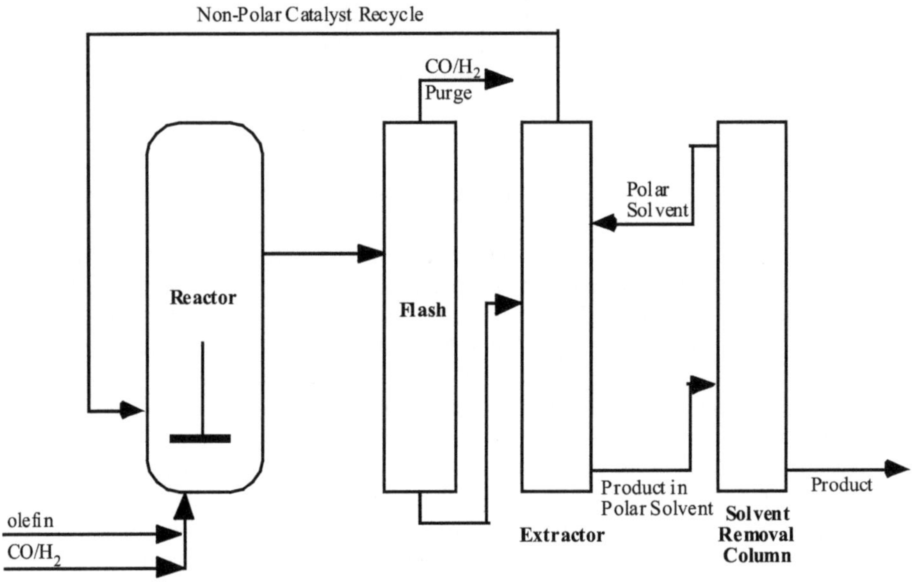

Figure 2.5. NAPS Flow Diagram for Polar Products

2.4.6.2 NAPS Using a Polar Catalyst

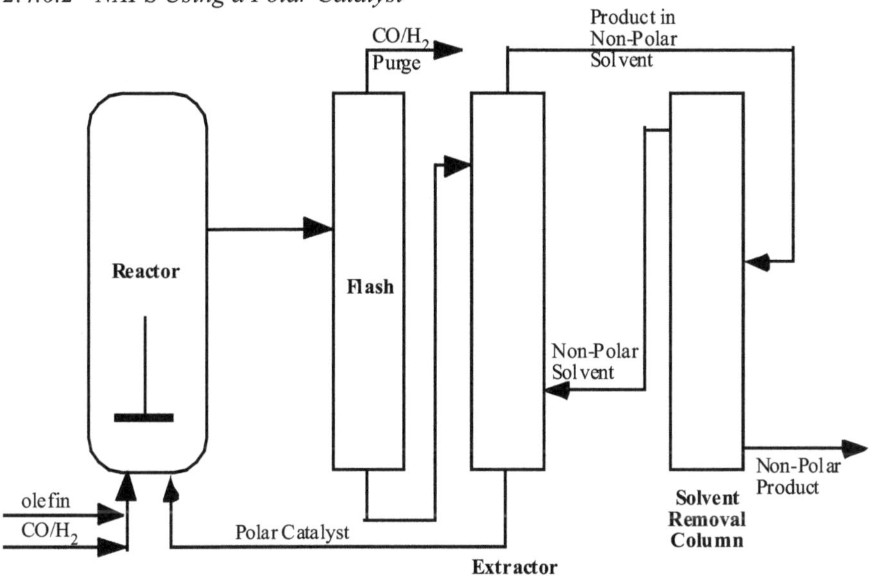

Figure 2.6. NAPS Flow Diagram for Non-Polar Products

In hydroformylating with a polar ligand modified rhodium catalyst to give a relatively non-polar aldehyde product, after the flash column, the catalyst solution is extracted with a non-polar solvent. Polar catalyst recycles from the extractor to the reactor. The non-polar solvent is removed and recycled to the extractor (see Figure 2.6).

2.4.6.3 Ligand Structure and Solubility Properties

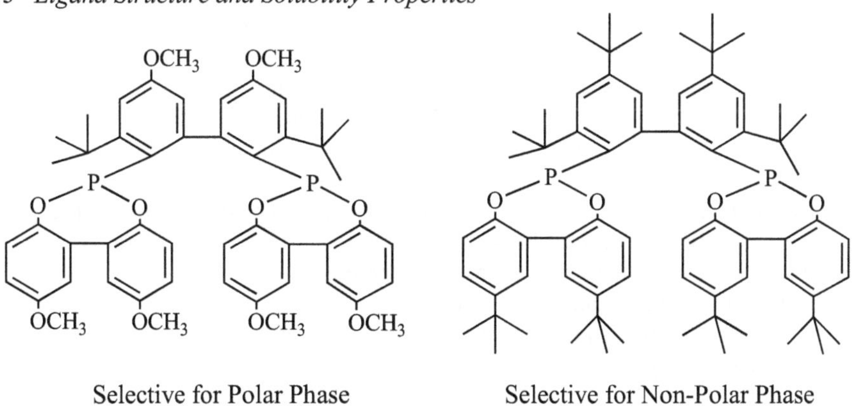

Figure 2.7. Ligands for different NAPS hydroformylation systems

Whereas ligand selection for the biphasic water-organic and the induced phase separation system are somewhat limited, there is very wide array of organophosphorus ligands to consider for Non-Aqueous Phase Separation. Two of these are shown in Figure 2.7. From the standpoint of separation efficiency, selection of appropriate ligand/solvent combinations may be evaluated using distribution coefficients to measure the partitioning of the product, ligand, catalyst and byproducts between potential polar and nonpolar solvents.[16]

2.5 Hypothetical processes - How might the Product be separated from the Catalyst?

In order to see how a product forming process might be coupled with a separation technology, we shall consider a matrix of processes and separations.

Hydroformylation processes include:

- Propene to butanal
- Octene to nonanal
- Allyl Alcohol to 4-hydroxybutanal
- Methoxyvinylnaphthylene to (2-(6-methoxy)naphthylpropanal, an intermediate in the formation of the anti-inflammatory drug, naproxen)

Separation technologies include:

- Gas Recycle
- Liquid Recycle
- Induced Phase Separation
- NAPS
- Water-Soluble Phosphine Catalyst
- Water Extraction

For homogeneous precious metal catalyzed processes, there are four major considerations. They are speed (or rate), selectivity, stability and separation. Of these four, speed is the most important since if you can obtain a high or very high reaction rate, you can use the flexibility that high rate provides by, for example, reducing reaction temperature and thereby slowing some byproduct forming reactions.

Selectivity refers to the fraction of raw material alkene that is converted to product aldehyde, but since hydroformylation typically gives both a linear and branched isomer, selectivity also refers to the relative amounts of each. The linear:branched (l:b) ratio is highly catalyst dependant. One must simultaneously consider whether the proposed catalyst will give the desired l:b selectivity and also whether the proposed catalyst is feasible for use with the catalyst/product separation technologies. For example, water extraction of a polar product, such as in the hydroformylation of allyl alcohol to give 4-hydroxybutanal, would not work well with a sodium salt of a sulfonated phosphine since both are water soluble.

Stability refers to the stability of the product, to the stability of the ligand and to the stability of the ligand-modified rhodium complex.

In separation technologies, as in medicine, the first consideration is to do no harm. Not only must one separate product from catalyst, one must also separate catalyst from product. In addition, one must separate heavy organic byproducts such as aldehyde dimers and trimers, and separate certain ligand decomposition products – in particular acidic ones arising from phosphite degradation. Finally, catalyst/product separation should be conducted under conditions such that the catalyst does not undergo undue deactivation or be converted to insoluble or much less soluble forms. Recall that in Liquid Recycle technology, the catalyst components may undergo a three to four-fold change in concentration. In addition, reagents including CO/H_2 and alkene that may have been providing some catalyst stabilization are at significantly lower levels in the separation system.

The different temperature and concentrations in the separation system may favor the formation of metal aggregates or clusters. Some may revert to a monomeric form in the reactor; others may show less or no catalytic activity. A consequence is that in addition to studying chemistry in the reactor, one must also study the chemistry of the separation system.

Separation technologies for different substrate types are compared in Table 2.1

TABLE 2.1. Matrix of Reactions and Separation Technologies

Separation process	Propene	Octene	Allyl Alcohol	2-vinyl naphthalene
Gas Recycle	OK	Impractically large gas flows	Insufficient volatility	Insufficient volatility
Liquid Recycle	OK	OK	Product thermally unstable	Product labiality
Induced Phase Separation	Not the best choice	OK	No, product soluble in catalyst phase	No suitable catalyst
NAPS	Possibly	OK	OK	Possibly
Water Soluble Phosphine	OK	Alkene not sufficiently soluble in aqueous catalyst	No, product soluble in catalyst phase	No. Catalyst is unsuitable
Water Extraction	Poor Choice	No	Yes	No

2.5.1 PROPENE HYDROFORMYLATION

From the standpoint of separation technologies, propene hydroformylation has the largest group of commercialized processes. The earliest commercialized process using a phosphine-modified rhodium catalyst was Gas Recycle. This was followed by Liquid Recycle and by a process using the catalyst in an aqueous phase. Propene can be hydroformylated and butanal separated using a variety of catalysts including those that give high l:b ratios and those that give low.

NAPS technology in which butanal is extracted with a non-aqueous solvent would probably also work technically, but it would be economically disadvantaged over processes in which butanal is separated by vaporization. In addition, since aldehyde byproduct formation can be controlled by vaporization of dimers and trimers and ligand decomposition products can be controlled by adjustments of reactor and separator conditions, neither of these problems would be uniquely solved using NAPS.

Water extraction would be a very poor choice for isolation of butanal, because. butanal solubility in water is relatively low. Considerable energy would be required to isolate butanal that is dissolved in the aqueous fraction. The solubility of aldehyde dimers and trimers is negligible in water, so that the buildup of these aldehyde condensation byproducts could limit catalyst lifetime.

Induced Phase Separation would work technically, but would be uneconomic relative to Liquid Recycle because of additional unit processes and increased energy requirements.

2.5.2 1-OCTENE HYDROFORMYLATION

The volatility of nonanal is sufficiently low to require impractically high levels of recycle gas. It would not be possible to remove the aldehyde dimers and trimers of nonanal by gas recycle. Practically speaking, pentanal is at the border of aldehydes that may be isolated this way. Removal of the dimers and trimers of pentanal would be problematic.

Another aspect of gas recycle that would need to be considered is the degree of catalyst solution expansion that would result from very high gas recycle flows. The catalyst solution might be blown out of the reactor, or if a very tall reactor were built, it would suffer from excessive capital costs for the large containment needed for the gas-expanded catalyst.

Liquid Recycle is practical for octene hydroformylation. 1-Octene is readily soluble in organic based catalyst solutions, and product aldehyde and its condensation products can be separated by vaporization.

Induced Phase Separation is also a good choice for octene hydroformylation. Octene can easily dissolve in the organic based catalyst solution, and with addition of small amounts of water, nonanal and its condensation products will readily separate from the sodium salt of a monosulfonated phosphine. To choose between Liquid Recycle and Induced Phase Separation would require a detailed technical and economic study that is outside the scope of this chapter.

NAPS is also a possibility for octene hydroformylation, but again a detailed technical and economic comparison would be required in order to chose among it, Liquid Recycle and Induced Phase Separation.

The use of a water-soluble phosphine based catalyst is not a preferred choice for octene hydroformylation. Although separation of nonanal and its condensation products from an aqueous catalyst should be facile, forming nonanal at a commercially viable rate could be challenging. In order to react, octene needs to be in the same phase as the catalyst, and octane has very low solubility in water.

Finally, water extraction of nonanal and its condensation products is totally impractical.

2.5.3 ALLYL ALCOHOL

$$HOH_2CHC{=}CH_2 \xrightarrow[\text{Rh catalyst}]{CO/H_2} \begin{array}{c} HOCH_2CH_2CH_2CHO \\ + \\ HOCH_2\underset{\displaystyle CH_3}{\underset{|}{C}}HCHO \end{array}$$

Equation 2.1. Allyl alcohol hydroformylation

The desired product in the hydroformylation of allyl alcohol is 4-hydroxybutanal. As with other alkenes, hydroformylation gives both a linear and a branched isomer (Equation 2.1).

Through dehydration (Equation 2.2) the branched isomer will yield methacrolein, an α,β-unsaturated carbonyl compound.

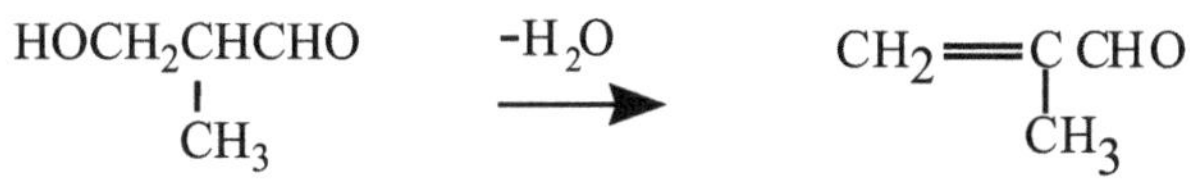

Equation 2.2. Hydroxyaldehyde dehydration to methacrolein

Dehydration is undesirable because α,β-unsaturated carbonyls are catalyst inhibitors. To make matters worse, phosphines can add to the α,β-unsaturated carbonyl (Equation 2.3) to give a product that is a dehydration catalyst, so the deactivation spiral continues.

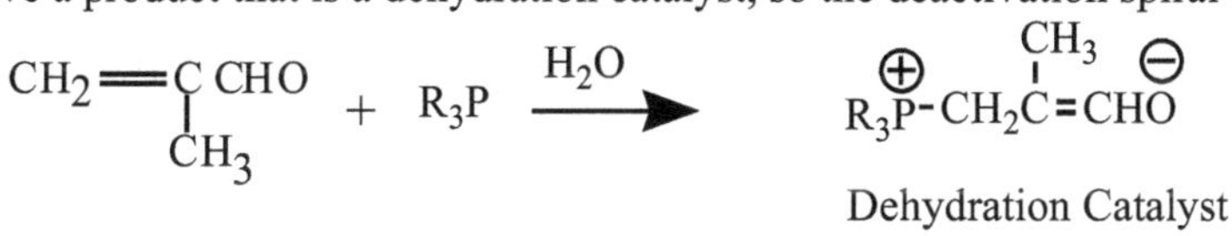

Equation 2.3. Initiation of the autocatalytic product decomposition cycle

Product separation by vaporization is not a good option because of the thermal sensitivity of the product. In addition, because of the autocatalytic nature of the methacrolein-phosphine adduct, it is imperative that its concentration be controlled.

Kuraray [17] appears to have solved this problem in a very clever way with chemistry that is not well understood. Their solution to the problem can be viewed as having two parts. As rhodium catalyst modifiers, they use both a stoichiometric amount of a bis-phosphine and excess triphenylphosphine. The second part is to use an aqueous extraction of the product. This provides at least two advantages. The first is that the products are not exposed to the type of high temperatures that are associated with vaporizers. The second, and this is speculation, is that the water also removes the phosphonium hydroxide.

2.5.4 METHOXYVINYLNAPHTHALENE

In this example, we will consider asymmetric hydroformylation to give an aldehyde intermediate with a high ee. Gas Recycle is out of the question because of the low volatility of the product. Vaporization in a Liquid Recycle process is theoretically possible, but impractical if we wish to maintain the high enantioselectivity of the product.

Up to this example we have given relatively little consideration to the performance of the catalyst. However, in order to obtain a product with high enantiomeric excess, the ligand used to modify rhodium must be selected with particular care. At a minimum it must contain an optically active center to have any hope of achieving enantiomeric excess [18] It must also show high selectivity towards the branched product, although for styrenes and vinyl naphthalenes this isomer is somewhat favoured on thermodynamic grounds.

Aqueous-phase catalysts are unsuitable since the vinyl naphthalene would have limited solubility in the phase containing the catalyst. Induced Phase Separation is also a questionable choice since at this time, suitable enantioselective ligands have not been identified.

NAPS can be run with a very wide variety of both polar and non-polar ligands that are enantioselective. One could consider the synthetic possibilities and solvents for the separation to select potential process/separation combinations. The mild separation conditions of NAPS are well suited to maintain enantiomeric excess. Water extraction of the hydroformylation product of methoxyvinylnaphthalene, (2-(6-methoxy)-naphthylpropanal) is not feasible.

2.5.5 SEPARATION TECHNOLOGY FOR LESS STABLE CATALYSTS

2.5.5.1 Mitsubishi TPPO/TPP Separation

Mitsubishi has patented a triphenylphosphine oxide-modified rhodium catalyst for the hydroformylation of higher alkenes with both alkyl branches and internal bonds.[19] Reaction conditions are 50-300 kg/cm^2 of CO/H_2 and 100-150 degrees C. The high CO/H_2 partial pressures provide stabilization for rhodium in the reactor, but rhodium stability in the vaporizer separation system is a different matter. Mitsubishi adds triphenylphosphine to stabilize rhodium in the vaporizer. After separation, triphenylphosphine is converted to its oxide before the catalyst is returned to the reactor.

2.5.5.2 Organic Polymer for Catalyst Stabilization

Rhodium precipitation in solubilized rhodium-phosphite complex catalyzed liquid recycle hydroformylation may be minimized or prevented by carrying out product recovery in the presence of an organic polymer containing polar functional groups such as amides, ketones, carbamates, ureas and carbonates.[20] Patent examples include the use of polyvinylpyrrolidone and vinylpyrrolidone-vinyl acetate copolymer with diorganophosphite-modified rhodium catalysts.

2.6 Real-World Complications

If one had only to separate the product from the catalyst, the situation would be vastly simpler than that observed in real life. Catalysts vary in their robustness. They can be poisoned, inhibited or form other less active species. Organophosphorus ligands can undergo a variety of degradation reactions.[21] Some degradation products are relatively inert in the process; others may significantly alter the reaction rate and/or selectivity. Conversion of raw materials to products is not perfect; alkene may be hydrogenated and dienes may polymerize. Products, particularly aldehydes, may undergo a variety of condensation reactions. Products, or subsequent byproducts, may react with the catalyst or ligand.

2.6.1 ORGANOPHOSPHORUS LIGAND DEGRADATIONS

2.6.1.1 *Oxidation*

Arylphosphines are converted with oxygen or peroxides to their corresponding phosphine oxides (Equation 2.4). Phosphine oxides are relatively inert and are of concern principally because they represent a loss of valuable reagents and secondly because as their concentration in the catalyst solution increases, the likelihood of precipitation increases. In a Liquid Recycle vaporizer, the concentration of low-volatility components may increase three to four-fold. Compounding this is that the process must be designed for planned and unplanned shutdowns when system heating may be absent. In addition, some plants may be exposed to extremely cold weather. The best approach for phosphine oxides is to restrict their formation by excluding oxygen and peroxides.

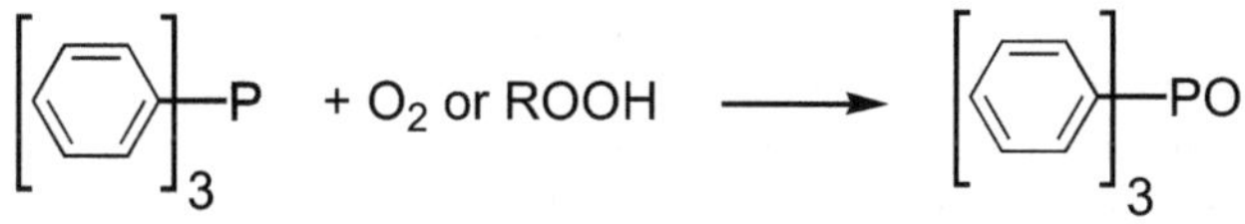

Equation 2.4. Oxidation of triphenylphosphine to its oxide with oxygen or a hydroperoxide

2.6.1.2 *Alkyldiarylphosphine Formation*

Arylphosphines in rhodium catalyzed hydroformylation reactions exchange an aryl group for an alkyl, principally linear alkyl, corresponding to the alkene being hydroformylated to give an alkyldiarylphosphine [22](see Equation 2.5).

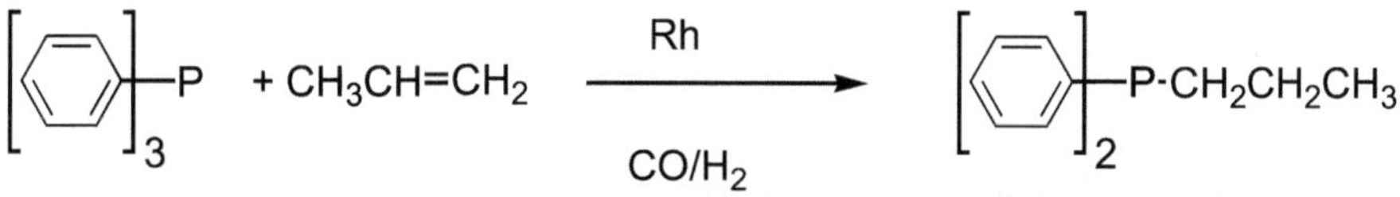

Equation 2.5. Formation of propyldiphenylphosphine under hydroformylation conditions

In the hydroformylation of propene using a triphenylphosphine-modified rhodium catalyst, the formation of propyldiphenylphosphine (PDPP) alters both rate and selectivity. It is both more basic and less sterically hindered than triphenylphosphine. Being more basic and less sterically hindered allows it to compete for coordination sites on rhodium even in the presence of a large excess of triphenylphosphine. In the hydroformylation of propene, PDPP is a minor annoyance. However, as discussed below, if using a monosulfonated ligand while hydroformylating a higher molecular weight alkene, significant separation complications may occur.

If similar chemistry were to occur while hydroformylating a C-12 alkene with TPPTS trisodium salt or the sodium salt of monosulfonated triphenylphosphine (TPPMS), the product could be a phosphine with both a long alkyl chain and an aromatic group containing a sodium sulfonate group. Such compounds could significantly alter the separating characteristics of the catalyst solution when water is added since the compounds resemble a detergent. Loss of the sulphonated aryl ring from TPPMS could also occur, leading to very low water solubility for the phosphine and compex.

2.6.1.3 Ligand Scrambling

Non-symmetrical arylphosphines may undergo scrambling of the aryl groups as illustrated in Equation 2.6.[23]

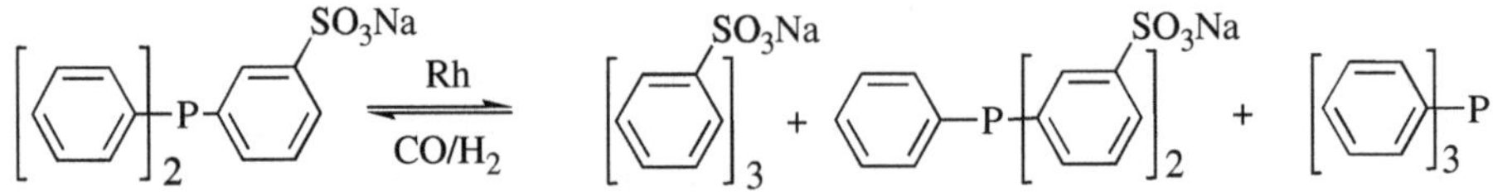

Equation 2.6. Scrambling of triphenylphosphine monosulfonated sodium salt under hydroformylation conditions

This chemistry can have very significant implications in the separation system. As described in the Section 2.4.4 on Induced Phase Separation, a monosulfonated triorganophosphine can be used to prepare an organic catalyst solution that favors dissolving and reacting of higher molecular weight alkenes. In the separation system small amounts of water are added to induce phase separation. [15] With the addition of water, the catalyst becomes water-soluble and can be separated from product by phase separation. Product and organic byproducts are in the organic phase; rhodium and ligand are in the polar phase.

If ligand scrambling occurs, two problems may develop. In the first, some disulfonated and trisulfonated phosphine may form. They and their rhodium complexes are much less soluble in the organic reaction solvent so precipitation may occur. In the separation system, the triphenylphosphine formed during ligand scrambling can give rhodium complexes that will be extracted into the organic product layer rather than the desired polar phase. Ligands can be designed which minimize or avoid scrambling [24].

2.6.1.4 Phosphine Reactions with Conjugated Systems

Phosphines may react with certain conjugated systems. This reaction may be conducted deliberately [25] to selectively remove an alkyldiarylphosphine in the presence of a triarylphosphine. Phosphines may react similarly with α,□-conjugated carbonyl reaction byproducts such as methacrolein or ethylpropylacrolein. A further concern is reaction with the conjugated system of a feedstream. This chemistry is undesirable when it consumes reagents. Additional harm arises when the reaction product promotes further side reactions, which consume the product.[26]

2.6.1.5 Phosphite Oxidation

When a phosphite is used as a catalyst modifier, it is susceptible to oxidation in the same manner as a phosphine. Unlike triphenylphosphine oxide, which is relatively innocuous except for precipitation when the solubility limit is reached, phosphite oxidation products may hydrolyze to give phosphoric acid. Since phosphites are esters, phosphoric acid can catalyst additional hydrolysis. Other than limiting formation of phosphite oxidation products, the best approach is to include some acidity control technology in the separation or reaction system.

2.6.1.6 Simple Phosphite Hydrolysis

Phosphites can undergo hydrolysis to phosphorus acid. Aldehyde condensation can give trace levels of water. The phosphorus acid in turn can catalyze further hydrolysis. Acidity control should be considered for any homogeneous catalytic process.

2.6.1.7 Poisoning Phosphite Formation

A phosphite degradation reaction that occurs during hydroformylation using an aryl-phosphite-modified rhodium catalyst involves replacement of one of the aryl groups with an alkyl group corresponding to the alkyl group of the hydroformylation product[.[27] This is illustrated in Equation 2.7

.

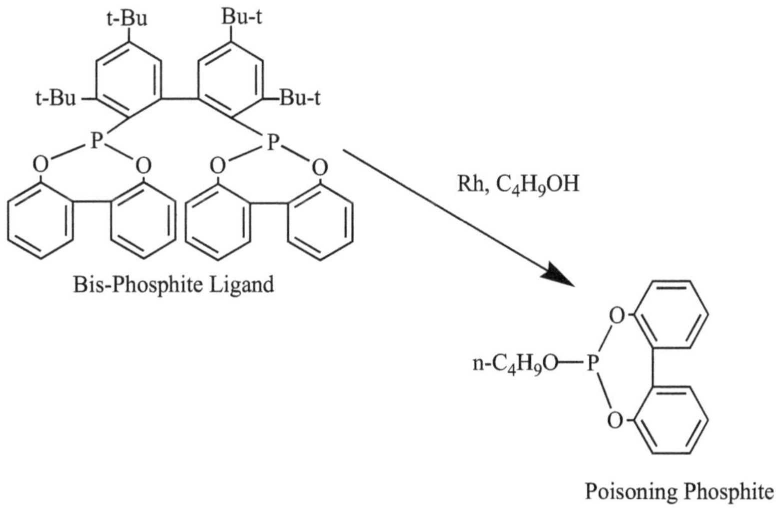

Equation 2.7. Formation of a poisoning phosphite during propene hydroformylation

Poisoning phosphites are particularly undesirable because their smaller steric bulk allows them to bind to the rhodium catalyst and inhibit hydroformylation.

2.6.1.8 Aldehyde Acid Formation

Degradation of poisoning phosphite [**27**] may lead to the formation of an aldehyde acid, as shown in Equation 2.8. The concentration of aldehyde acid and phosphorus or phosphoric acids should be monitored and controlled to minimize losses of the desired catalyst modifying ligand.

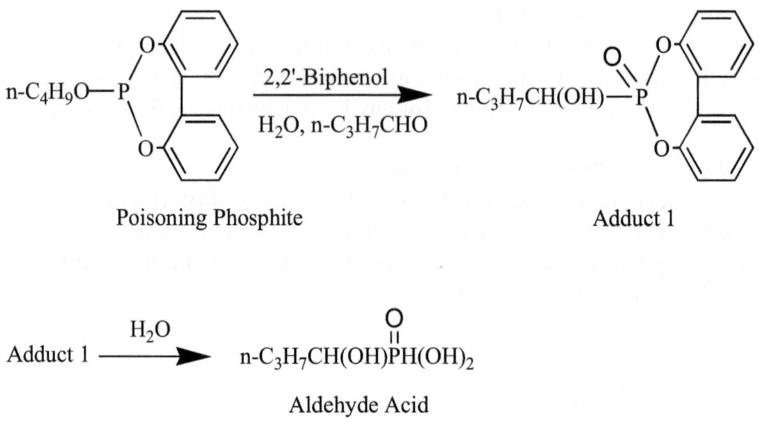

Equation 2.8. Formation of an aldehyde acid during hydoroformylation of propene using a phosphite modifier.

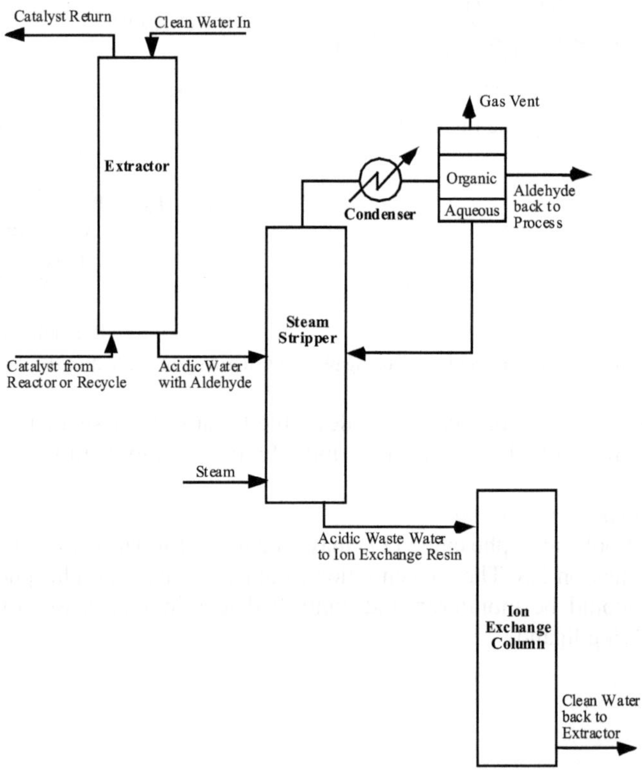

Figure 2.8. Acidity Control Using Water and an Ion Exchange Resin

2.6.1.9 Acidity Control

Acidity control is essential for the long-term stability of phosphite-modified catalysts. The acid may be extracted with water with subsequent recycle of water passage through an ion exchange resin (Figure 2.8)[28] or over a bed of, for example, calcium carbonate[29] to neutralize the acid. Alternately the aqueous fraction may simply be discarded.[30] Another variation is to include an acid-neutralizing agent in the aqueous fraction.[31]

In any case, the amount of water remaining in the catalyst solution recirculated to the reactor needs to be controlled to avoid undesirable levels of phosphite ligand hydrolysis.

2.6.2 SEPARATING BYPRODUCTS FROM REACTANTS OR PRODUCTS

In addition to separating product from catalyst, excess ligand and reaction solvent, one must also separate byproducts arising from the reactants or products. For example in hydroformylation, one must separate saturated hydrocarbon, isomerized alkene and aldehyde dimers and trimers.

2.6.2.1 Alkene Hydrogenation

Alkene hydrogenation as a side reaction, which other than the direct efficiency loss it represents, is not particularly challenging from the separations viewpoint. Lighter alkenes are usually separated in a flash stage prior to separation of the product from the catalyst. Higher molecular weight alkenes may be separated from the catalyst along with the product.

The presence of saturated hydrocarbon in the recovered alkene may limit its usefulness as a recycle stream for the process. This is particularly true for higher alkenes where there is less difference in volatility between the saturated and unsaturated compounds. The best approach is to minimize hydrogenation. The next option is to use a one-pass process, possibly using staged reactors, in which recovered alkene is not recycled. The higher the intrinsic activity of the catalyst, the easier it is to achieve this goal since the higher activity of the catalyst can somewhat compensate for the lower reaction driving force due to lower alkene concentrations in the later reaction stages.

2.6.2.2 Alkene Isomerization

Alkene isomerization has both positive and negative aspects. The positive aspect is where isomerization is needed prior to, for example, hydroformylation to give the desired product. The negative aspect of alkene isomerization is similar to that described in Section 2.6.2.1 on hydrogenation. The byproduct must be separated from both catalyst and product, and recycle opportunities may be limited. Not only is isomerization a direct efficiency loss, but when the isomerised alkene is purged, desired reactants will likely also be lost.

2.6.2.3 Aldehyde Dimerization and Trimerization

Aldehyde dimers and trimers are common byproducts produced during the hydroformylation of propene. Union Carbide addressed the problem in a creative way when it was discovered that the dimers and trimers could be used as the principal reaction solvent for hydroformylation.[32] Elimination of an extraneous solvent simplified the process. The Ester-diol Trimers may equilibrate, as shown in Equation 2.9 to give a mixture of diol, a dimer, and the diester of the diol, which is a tetramer of butanal.

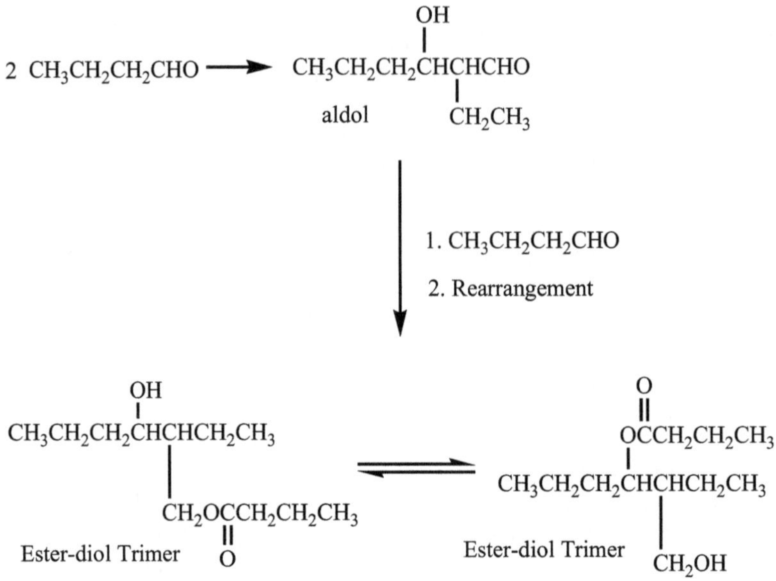

Equation 2.9. Trimer formation from butanal

2.6.2.4 Formation of Conjugated Carbonyls

Aldehyde dimer may undergo dehydration to give an α,□-unsaturated carbonyl. From butanal, the conjugated carbonyl is ethylpropylacrolein (Equation 2.10). The conjugated system of this material competes for coordination sites on the rhodium catalyst so that hydroformylation inhibition is observed.[**8**] The formation of 2-ethylhex-2-enal can be limited by minimizing the concentration of dimers. Dimers are removed along with the product in a liquid recycle separation system.

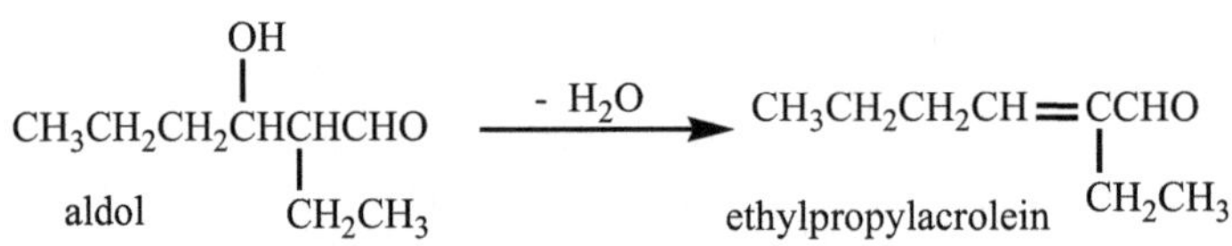

Equation 2.10. Formation of 2-ethylhex-2-enal *via* aldol condensation of butanal

2.6.3 INTRINSIC CATALYST DEACTIVATION

It was recognized during the development of propene hydroformylation that propene provided some stabilization for the catalyst. In the absence of the alkene, but in the presence of carbon monoxide and hydrogen, the catalyst can undergo what has been termed intrinsic deactivation.[33] Apparently after oxidative addition of triphenylphosphine to rhodium, diphenylphosphido bridged rhodium complexes are formed.

These complexes are not effective hydroformylation catalysts. A reaction regime in which this phenomenon is minimized was disclosed.[**33**]

A method to reduce degradation/deactivation of a phosphite modified rhodium hydroformylation catalyst in the separation system involves feeding a diene such as butadiene to the vaporizer to convert the phosphite-modified rhodium catalyst to a more stable form.[34] In the reactor, the diene is hydrogenated and catalyst activity is restored.

2.7 Further Separation Challenges

The separation challenge is often thought of as simply separating the product from the catalyst. In addition, one must separate the catalyst from the product. Precious metal losses with product are desirably in the low parts per billion (ppb). High value specialty products may economically permit significantly higher metal losses.

If a solvent is used as part of the catalyst solution, then it also must be separated from the product. Finally, the buildup of various byproducts from ligand degradation, from raw material side reactions and from subsequent reaction of the desired product must be addressed so that the catalyst solution remains fully functional to achieve an economic catalyst life.

2.7.1 RECOVERY OF METAL VALUES FROM A SPENT CATALYST

For the purposes of the following discussion, the *catalyst precursor* is the metal complex, purchased or prepared locally, that is charged to prepare the *catalyst solution*. The *active catalyst* is that which exists under reaction conditions and is involved in the catalytic cycle. *Deactivated catalyst* is that fraction of the metal which remains in the *catalyst solution* but which is not involved in the catalytic cycle.

When a catalyst has sufficiently deactivated to justify taking some action is determined by economics. Both Gas and Liquid Recycle hydroformylation plants may be operated to give essentially constant production rates as the catalyst deactivates. Hydroformylation is approximately first order in both rhodium and alkene concentration. As the rhodium catalyst deactivates, the alkene concentration may be allowed to increase to compensate for the declining catalyst activity. Action is taken when the alkene efficiency declines to the point where it approximates or exceeds the cost of catalyst replacement or reactivation.

A variety of techniques have been disclosed for both extending catalyst solution life and for catalyst activity recovery. For hydroformylation, the catalyst consists of rhodium and an organophosphorus ligand. In some circumstances, the value of the organophosphorus ligand in the catalyst solution may approach the value of the rhodium.

A consequence of the value of the ligand is that one of the simplest ways to restore catalyst activity is simply to add fresh catalyst precursor. Unfortunately, there are practical limits as the rhodium concentration increases. First one must consider metal complex solubility, particularly in the recycle catalyst solution in a liquid recycle system. Secondly, higher rhodium concentrations favor formation of various types of rhodium clusters.[**11**] As rhodium increments are added to a partially deactivated catalyst, catalyst lifetimes for the added increments are ever shorter. Catalyst management strategies are a sophisticated and highly confidential part of hydroformylation technology.

2.7.1.1 Catalyst Containment and Capture Technologies

Entrainment Separators. In any process in which the product is volatilized, including both Gas Recycle and Liquid Recycle, ppm or ppb levels of metal catalyst may be entrained in the vapors leaving the separation system. Entrainment separators (Figure 2.9) are often included to recover the metal. Vaporous product effluent from a gas recycle reactor may be sent to a separator where it is passed through a demisting pad to return some aldehyde and condensation product and particularly to prevent potential carry-over of catalyst.[**6**]

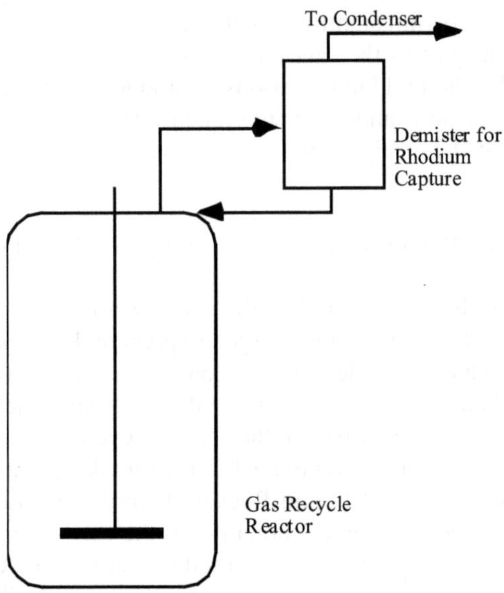

Figure 2.9. Entrainment separator for trace rhodium recovery

Selective Condensation of Vaporized Organophosphorus Ligand. Certain phosphorus ligands have sufficient volatility that portions may be volatilized when aldehyde and higher boiling aldehyde condensation byproducts are separated from the catalyst solution in, for example, a liquid recycle vaporizer. The phosphorus ligand may be condensed, recovered and returned to the catalyst solution [35] according to the procedure disclosed in US 5,110,990.

> "…the improvement comprising (a) selectively separating the phosphorus ligand and vaporized higher boiling aldehyde condensation by-products contained in said vaporized aldehyde product stream by thoroughly contacting said stream with a dispersed liquid having a lower boiling point than said higher boiling al-

dehyde condensation by-products so as to condense vaporized phosphorus ligand and vaporized higher boiling aldehyde condensation by-products contained in said volatilized aldehyde product stream, and (b) recovering the condensed phosphorus ligand and condensed higher boiling aldehyde condensation by-products so obtained from the volatilized aldehyde product stream, said dispersed liquid being employed in the form of droplets and in an amount such that the percent of phosphorus ligand so separated and recovered is at least about 1.2 times higher than the percent of higher boiling aldehyde by-products also so separated and recovered."

This process is illustrated in Figure 2.10, where butanal liquid is sprayed into vaporized butanal, heavies and a phosphine such as triphenylphosphine. Example No. 10 of Table 2 of US 5,110,990 illustrates how by using this technique, 99% of the triphenylphosphine in the stream may be recovered while allowing 97% of the contained aldehyde condensation by-products to remain in the vapor.

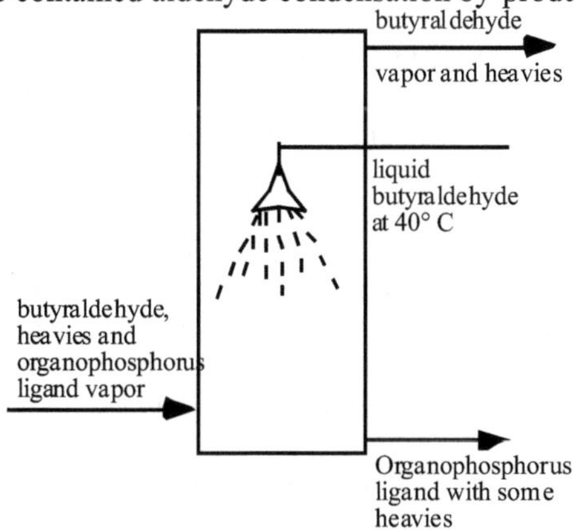

Figure 2.10. Selective recovery of phosphorus ligand

Trace Rhodium Recovery from Product or Byproduct Streams. As will be discussed later, there are what might be viewed as the ultimate rhodium recovery methods in which the organic matrix is burned, the rhodium recovered as an ash, then processed through a precious metal refinery before conversion into a catalyst precursor. Once rhodium is processed into an ash, there is significance expense associated with its conversion to a suitable catalyst precursor. Therefore, technologies which permit capture and reuse or reactivation and reuse are strongly preferred over more extreme procedures.

A technique that is suitable for capture of trace levels of precious metal is the use of an ion exchange resin on which is loaded a basic or acidic salt of a phosphine ligand.[36] A portion of the disclosure of US 5,114,473 reads:

"Ionic ligand-functionalized resins suitable for use in recovering rhodium from polar and non-polar liquid solution starting materials in accordance with the present invention can be prepared simply by contacting under ambient conditions an anion-exchange resin with an acidic or acidic salt derivative of an organophosphorus ligand or by contacting a cation-exchange resin with a basic or basic salt derivative of an organophosphorus ligand. For convenience, such derivatives will be referred to herein simply as ionic organophosphorus ligands. For example, the ionic organophosphorus ligand generally can be dissolved in water or a polar solvent such as methanol and the solution subsequently contacted with the ion-exchange resin. Sufficient contacting is provided simply by mixing an aqueous slurry of the resin and a solution of the ligand, or by passing the ligand solution in contact with a bed of the resin. The contacting need only be sufficient for the ion-exchange resin to take up the ionic ligand from the solution, e.g. to exchange the original ionic moiety of the resin for the ionic ligand. Suitable conditions for carrying out this contacting can be identified using routine experimentation. Such contacting need only be sufficient to effect ion-exchange or reaction between the active sites of the resin and the ionic organophosphorus ligand. The resin then is separated from the aqueous solution and is appropriately washed and then dried as needed before use."

Advantages of this technique include the use of ionic rather than covalent bonds to incorporate phosphine into the resin. In addition, one can elute rhodium by passing through a solution of ionic phosphine. This simultaneously displaces the rhodium and regenerates the resin.

Recovery and/or Reactivation of Spent Catalysts. Technologies are available for dealing with spent catalyst. There are procedures for separating active catalyst from deactivated catalyst, and there are procedures for reactivating deactivated catalyst. If neither of these is suitable, then the catalyst solution can be concentrated and returned to a precious metal refiner such as Johnson Matthey in England for recovery.

Extraction of Active Catalyst Using TPPMS and Conditioning Agent.

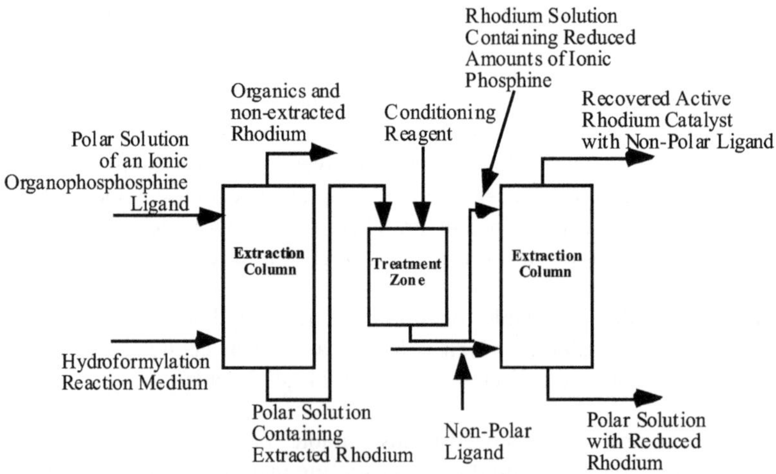

Figure 2.11. Active Catalyst Recovery

As outlined in Figure 2.11, a polar solution of an ionic organophosphine ligand such as triphenylphosphine monosulfonated sodium salt in an aqueous solution will extract active rhodium from a hydroformylation reaction solution. The aqueous phase containing the rhodium may then be treated with what is described [37] as a conditioning reagent, such as for example, maleic acid. The phosphine ligand adds to maleic acid to give an ylid. Rhodium is isolated from the aqueous layer by extraction with an organic solvent containing, for example, triphenylphosphine.

Wiped-Film Evaporator/O_2 Reactivation of Catalyst. In this technology [38], spent organophosphine-modified rhodium catalyst is first concentrated in a wiped-film evaporator where most of the organophosphine is removed. The rhodium concentrate is contacted with air to break down phosphido bridges in rhodium clusters. This air treated concentrate may then be used as a catalyst precursor. This procedure is suitable in circumstances where most of the aldehyde dimers, trimers and tetramers are sufficiently volatile to be removed in a wiped-film evaporator.

Capture of Active Catalyst Using Solid Acidic Support with H_2 Elution. The limit on the practical life of a catalyst solution may be determined by several factors including the presence of poisons or inhibitors, the buildup of less soluble materials such as the oxidation products of organophosphorus ligands, or an increase in the concentration of heavy aldehyde condensation products in the catalyst solution. In the latter case, there may be substantial amounts of active catalyst, but it is in a solvent that is unsuitable. Alternately, active rhodium catalyst may have been carried over with product. Technology has been disclosed [39] that permits the isolation of an active metal hydride catalyst. Steps include:

(a) removing a stream from a reactor, said stream comprising the homogeneous metal hydride catalyst;
(b) contacting the stream with a solid acidic absorbent under process conditions which allow at least some of the metal to become bound to the adsorbent;
(c) subjecting the metal bound to the absorbent, under process conditions which allow desorption of the metal, to a fluid stripping medium comprising hydrogen and solvent; and
(d) recovering the active metal hydride catalyst.

Apparently an ion-exchange resin will allow the absorption of the metal hydride species onto its surface by protonation and subsequent elimination of hydrogen (Equation 2.11). This hydrogen elimination is a reversible reaction. The metal species remains as a labile species that can be desorbed by hydrogen in a fluid-stripping medium.

$$HM(X)_n + -SO_3H \longrightarrow -SO_3M(X)_n + H_2$$

Equation 2.11. Absorption of a metal hydride onto an ion exchange resin

In a preferred aspect of the invention, the catalyst solution is passed through an oxidizer where it is contacted with air to break down clusters before being brought into contact with the absorbent. For a rhodium catalyst having triphenylphosphine as a ligand, air will break down the clusters by oxidation of phosphido bridges.

The separation of rhodium hydride complex from a stream comprising catalyst and high molecular weight aldehyde condensation products using a carboxylic acid functionalized ion exchange resin in illustrated schematically in Figure 2.12.

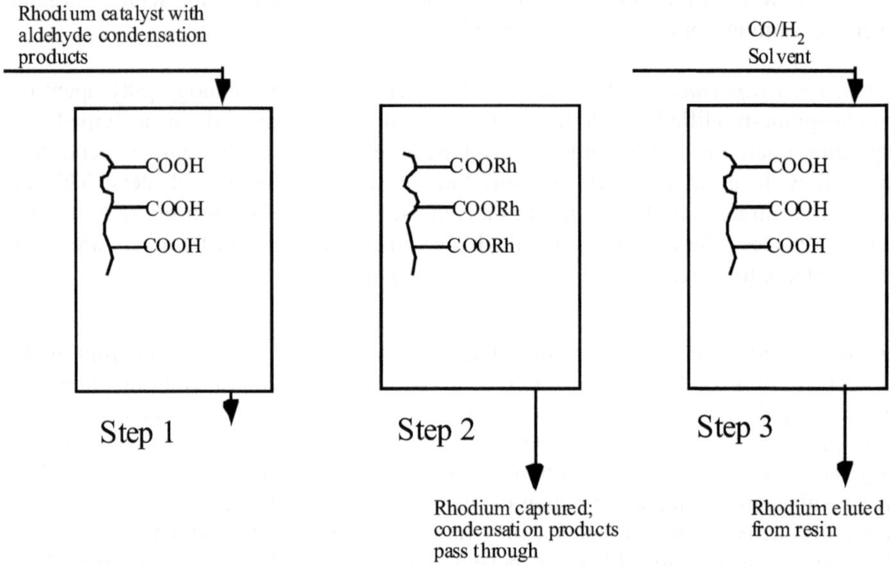

Figure 2.12. Recovery of Active Rhodium from a Spent Catalyst Solution

Catalyst Reactivation Using Propargyl Acetate. The Wiped-Film Evaporator/O_2 reactivation procedure and the Capture of Active Catalyst Using Solid Acidic Support with H_2 Elution procedure (see above) both involve the separation of uncomplexed phosphine from rhodium complex. Since the value of the uncomplexed phosphine is significant, technology that does not require separation of phosphine during catalyst reactivation is desirable.

Catalyst reactivation without the need for concentration of the catalyst solution and/or separation of the organophosphine ligand has been disclosed [40] using a variety of compounds including alkynes having the formulae shown in Figure 2.13.

$$\mathrm{R{-}C{\equiv}CCH_2X} \qquad \mathrm{HC{\equiv}CCH_2O\overset{\overset{\displaystyle O}{\|}}{C}R}$$

Figure 2.13. Alkynes used for catalyst reactivation

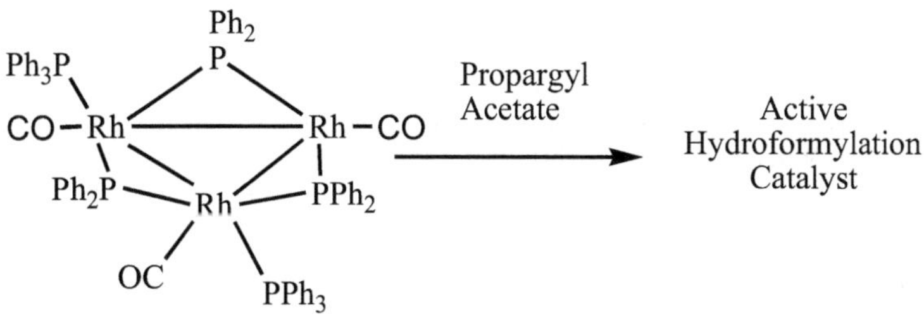

[Structure for illustrative purposes only.]

Equation 2.12. Reactivation of a phosphido bridged deactivated catalyst

Such compounds may be contacted with partially deactivated hydroformylation catalyst under non-hydroformylation conditions to effect disruption of the phosphido bridges of a rhodium cluster (Equation 2.12). After the treatment period, the catalyst solution is again suitable for hydroformylation.

Preparation of a Rhodium Hydride Catalyst Precursor from Spent Catalyst. If, for whatever reason, none of the reactivation procedures above are suitable, one is faced with the option of returning the catalyst concentrate to a vendor for conversion into a catalyst precursor. If the recovery involves reduction of the catalyst concentrate to a rhodium ash, significant expense is involved. Procedures that avoid rhodium ashing may be more economic.

In one such procedure a rhodium complex concentrate prepared from a 400 ppm rhodium containing hydroformylation catalyst solution, for which catalytic activity had declined to about 30 percent of its initial value, was concentrated in a wiped-film evaporator to about 27,700 ppm rhodium. This concentrate was oxygenated with tertbutylhydroperoxide. After isolation and treatment with triphenylphosphine, a 70% yield of $[HRh(CO)(PPh_3)_3]$ was obtained.[41]

2.8 Concluding Remarks

Consider the effect of every impurity or potential impurity in every feedstock, in the solvents in the ligand and in the catalyst. What harm, if any, will befall the product or catalyst if these impurities enter the process, and in particular, if they accumulate? Is the product stable in both the reaction and separation system? If byproducts form, what affect will they have on the process and catalyst life?

After considering the direct and byproduct chemistry, the next most important thing to do is to pilot the process *with all recycle loops closed.* Remember the principal that the only things in chemistry that matter are temperature and concentration. By closing the recycle loops, the concentrations of process components build to equilibrium levels, and process parameters can be adjusted sooner.

2.9 References

[1] New Syntheses with Carbon Monoxide, ed. J. Falbe, Springer-Verlag, New York 1980, p21ff.
[2] Applied Homogeneous Catalysis with Organometallic Compounds, ed. Boy Cornils and Wolfgang A. Herrmann, Wiley-VCH Verlag GmbH, 2000, p 69ff
[3] US 3,239,566, March 8, 1966, assigned to Shell Oil Company; L. H Slaugh and R. D. Mullineaux, *J. Organometal. Chem.*, 1968, **13**, 469-77.
[4] *Chemical Engineering*, Dec. 5, 1977, p110-115.
[5] R. L. Pruett and J. L. Smith, US 3,527,809, Sept. 8, 1970, assigned to Union Carbide Corporation
[6] E. A. V. Brewester and R. L. Pruett, US 4,247,486, Jan. 27, 1981, assigned to Union Carbide Corporation
[7] *Symp. Paper – Inst Chem Eng, Northwest Branch*, 1980, **190(3)** 243-6
[8] D. R. Bryant, US 4,143,075, March 6, 1979, assigned to Union Carbide Corporation
[9] Indications, Winter 1982/83, 20-28
[10] Applied Homogeneous Catalysis with Organometallic Compounds, ed. Boy Cornils and Wolfgang A. Herrmann, Wiley-VCH Verlag GmbH, 2000, p 79
[11] P. E Garrou, *Chem. Rev.* 1985, **85**, 3, p175
[12] D. R. Bryant and E. Billig, US 4,277,627, July 7, 1981, assigned to Union Carbide Corporation
[13] A. G. Abatjoglou, D. R. Bryant and R. R. Peterson, US 5,180,854, Jan. 19, 1993, assigned to Union Carbide Corporation
[14] J. N. Argyropoulos, D. R. Bryant, D. L. Morrison and K. E. Stockman, US 5,952,530, Sept. 14, 1999, assigned to Union Carbide Corporation
[15] J. N. Argyropoulos, D. R. Bryant, D. L. Morrison, K. E. Stockman and A. G. Abatjoglou, US 5,932,772, Aug 3, 1999, assigned to Union Carbide Corporation
[16] J. S. Kanel, J. N. Argyropoulos, A. G. Phillips, B. M. Roesch, J. R. Briggs, M. M. Lee, J. M. Maher and D. R. Bryant, US 6,294,700, Sept. 25, 2001; J. S. Kanel, J. N., Argyropoulos, A. G. Phillips, B. M. Roesch, J. R. Briggs, M. M. Lee, J. M. Maher and D. R. Bryant US 6,303,829, Oct. 16, 2001;;J. N. Argyropoulos, J. S. Kanel, M. L. Tulchinsky, D. J. Miller, D. L. Morrison, P. Foley and D. R. Bryant, US 6,303,830, Oct. 16, 2001; J. N. Argyropoulos, J. S. Kanel, M. L. Tulchinsky, D. J. Miller, D. L. Morrison, P. Foley and D. R. Bryant; US 6,307,108, Oct 23, 2001; J. S. Kanel, D. R. Bryant, B. M. Boesch and A. G. Phillilps, US 6,307,109 Oct. 23, 2001; J. N. Argyropoulos, J. S. Kanel, M. L. Tulchinsky, D. J. Miller, D. L. Morrison, P. Foley, D. R. Bryant, A. G. Phillips, B. M. Roesch, J. R. Briggs, M. M. Lee and J. M. Maher, US 6,307,110, Oct. 23, 2001; J. N. Argyropoulos, J. S. Kanel, M. L. Tulchinsky, D. J. Miller, D. L. Morrison, P. Foley, D. R. Bryant, A. G. Phillips, B. M. Roesch, J. R. Briggs, M. M. Lee and J. M. Maher,US 6,310,260, Oct. 30, 2001, all assigned to Union Carbide Corporation
[17] M. Matsumoto, M. Tamura,. Cent. Res. Lab., Kuraray Co. Ltd., Okayama, Japan. Journal of Molecular Catalysis (1982), 16(2), 195-207.
[18] J. E. Babin and G. T. Whiteker, US 5,360,938, Nov. 1, 1994, assigned to Union Carbide Corporation
[19] K. Tano, K. Sato and T. Okoshi, US 4,528,403, July 9, 1985, assigned to Mitsubishi Chemical Industries Ltd. ; C. Miyazawa, H. Mikami, A. Tsuboi and K. Hamano, US 4,822,917, April 18, 1989, assigned to Mitsubishi Chemical Industries, Ltd. ; T. Onada, Chemtech 1993, **23**(9), 34
[20] US 4,774,361, Sep. 27, 1988, assigned to Union Carbide Corporation
[21] T Sakakura, et al., *J. Organometallic Chem.,* 1984 **267** 171-177.
[22] A. G. Abatjoglou, D. R Bryant, *Organometallics* 1984, **3**, 923-926
[23] A. G Abatjoglou, D. R. Bryant, *Organometallics* 1984, **3**, 932-934; P. E Garrou, Chemical Reviews, 1985, **85**, 3, 171-183.
[24] R. E. Murray, US 4,668,823, May 26, 1987, assigned to Union Carbide Corporation
[25] D. R. Bryant and R. A. Galley, US 4,283,304, Aug. 11, 1981, assigned to Union Carbide Corporation
[26] A. G. Abatjoglou and D. R. Bryant, *Arabian J. Sci. Eng.* 1985 **10** 427.
[27] J. E. Babin, J. M. Maher and E. Billig, US 5,364,950, Nov. 15, 1994, assigned to Union Carbide Corporation
[28] D. R. Bryant, J. C. Nicholson and E. Billig, US 5,763,677, June 9, 1998; D. R. Bryant, J. C. Nicholson and E. Billig US 5,763,680, June 9, 1998 both assigned to Union Carbide Corporation
[29] D. R. Bryant and J. C. Nicholson, US 5,763,671, June 9, 1998; D. R. Bryant and J. C. Nicholson, US 5,789,625, Aug. 4, 1998 both assigned to Union Carbide Corporation
[30] D. R. Bryant and J. C. Nicholson, US 5,786,517, July 28, 1998 assigned to Union Carbide Corporation

[31] D. R. Bryant, T. W. Leung, E. Billig, T. C. Eisenschmid and J. C. Nicholson, US 5,741,942, April 21, 1998; D. R. Bryant, T. W. Leung, E. Billig, T. C. Eisenschmid, J. C. Nicholson, J. R. Briggs, D. L. Packett and J. M. Maher, US 5,874,640, Feb. 23, 1999 assigned to Union Carbide Corporation

[32] R. L. Pruett and J. A. Smith, US 4,148,830, April 10, 1979 assigned to Union Carbide Corporation

[33] D. R. Bryant and E. Billig, US 4,277,627, July 7, 1981 assigned to Union Carbide Corporation

[34] E. Billig, D. R. Bryant, C. A. Beasley, D. L. Morrison, M. D. Warholic and K. E. Stockman, US 6,090,987, July 18, 2000, assigned to Union Carbide Corporation

[35] M. A. Blessing, G. J. Dembowski and G. K. Finnell, US 5,110,990, May 5, 1992, assigned to Union Carbide Corporation

[36] A. G. Abatjoglou and D. R. Bryant, US 5,114,473, May 19, 1992, assigned to Union Carbide Corporation

[37] D. J. Miller and D. R. Bryant, US 4,935,550, June 19, 1990, assigned to Union Carbide Corporation

[38] D. R. Bryant and R. A. Galley, US 4,297,239, Oct. 27, 1981, assigned to Union Carbide Corporation

[39] S. P. Crabtree and R. Wild, WO 02/096555 A1, Publication Date 05. 12. 2002, assigned to Davy Process Technology Limited

[40] D. J. Miller, D. R. Bryant, E. Billig and B. L. Shaw, US 4,861,918, Aug. 29, 1989, assigned to Union Carbide Corporation

[41] J. D. Jamerson, E. Billig and D. R. Bryant, US 4,446,074, May 1, 1984, assigned to Union Carbide Corporation, Example 5.

CHAPTER 3 SUPPORTED CATALYSTS

Immobilisation of Tailor-made Homogeneous Catalysts

J.N.H. REEK[1], P.W.N.M. VAN LEEUWEN[1], A.G.J. VAN DER HAM[2] and A.B. DE HAAN[2]
[1]*University of Amsterdam, van't Hoff Institute for Molecular Sciences, Nieuwe Achtergracht 166, 1018 WV, Amsterdam, The Netherlands.*
[2]*Separation Technology Group, Faculty of Science and Technology, University of Twente PO Box 217, 7500 AE Enschede, The Netherlands*

3.1 Introduction

The enormous increase in knowledge of organometalic chemistry acquired over the past decades has had a great impact on catalysis, and numerous homogeneous transition metal catalysts for commercially interesting conversions are nowadays available. One of the factors hampering the commercialisation of the newly discovered catalytic conversions is the introduction of an additional separation step required to remove the homogeneous catalyst from the product, a problem that is not an issue for heterogeneous catalysts. The importance of the separation problem has been acknowledged for more than three decades [1] and, ever since, intensive research has been devoted to trying to solve this problem. Several elegant strategies to overcome this drawback have been explored including two-phase catalysis using various solvent combinations and the use of supercritical fluids. The most natural solution to combine the advantages of homogeneous and heterogeneous catalysis is the immobilisation of homogeneous catalysts to obtain tailor-made catalysts that can be applied in fixed bed or slurry type reactors. Indeed, this approach is probably the most widely studied as is evident from the numerous papers, reviews and books on this topic [2]. Remarkably, there are many patents on immobilised catalysts, but so far there is only one commercial process based an anionic rhodium complex $[RhI_2(CO)_2]^-$ that is bound to an ion exchange resin and is used for the carbonylation of methanol [3, 9a]. This indicates that there is a common belief that the immobilisation of catalyst on a support is a suitable solution for the separation problem, but there are several difficulties which prevent the step to real commercial success: (a) metal leaching from the support, (b) reduced activity and selectivity upon catalyst immobilisation, (c) limited mechanical strength and swelling properties of the support, and (d) non-innocence of the support. Considering the amount of work published in the literature it is impossible to give a complete overview of this area. We will therefore limit this chapter to important contributions that are illustrative of the field and we will attempt to give a clear view of the current state of the art including a view of the remaining challenges. Since hydroformylation is the model reaction used through-out this book, we will strong focus strongly on supported catalysts developed for this reaction, and we will only briefly discuss a few other reactions.

D. J. Cole-Hamilton and R. P. Tooze (eds.), Catalyst Separation, Recovery and Recycling, 39–72.

3.2 Short Historical Overview

A literature search on "supported catalysts" revealed a few early examples (1923) of nickel and palladium catalysts supported on various substances such as $BaSO_4$ and kieselguhr [4]. These are the first examples of supported catalysts in which the metal is directly attached to the support. Much later, after the development of transition metal complexes as catalysts, methodologies have been explored to immobilise ligands on supports in order to obtain supported transition metal complexes. The first publications and patents involve the immobilisation of cationic platinum complexes on sulfonated polystyrene (Merrifield) via ionic interactions [1]. This work was basically inspired by Merrifield, who had just reported the preparation of polymer supported enzymes [5]. The work rapidly expanded into an active area of research, and at the end of the seventies several reviews had already appeared on this topic [2]. The use of structured, insoluble supports in the form of zeolites and supported liquid phase catalysis as a strategy to immobilise catalysts were also reported in the late sixties [6]. In the early seventies the first supported chiral ligands appeared in the literature [7]. The preparation of immobilised catalysts via sol-gel procedures were introduced much later [8]. Several recent reviews on catalyst immobilisation have appeared [9], discussing various important aspects in more detail, including molecular accessibility [10], immobilised chiral catalyst for asymmetric C-C bond formation [11] and interphase catalysis [12].

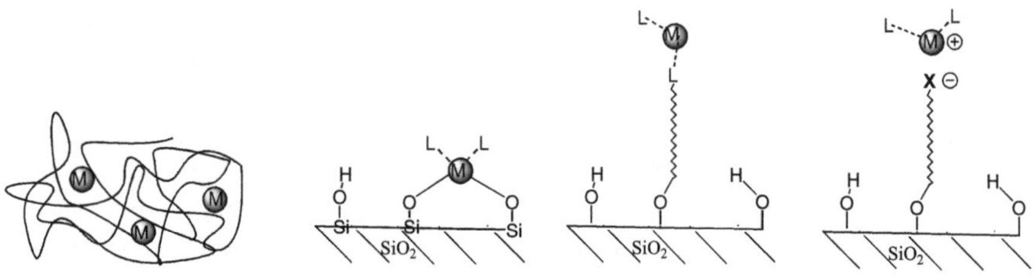

Figure 3.1. Several approaches that lead to supported catalysts: catalyst entrapment, anchoring via an interaction between the metal and the support, anchoring a ligand to a support, and anchoring via ion pairing (or other noncovalent interactions).

There are many ways to make polymer supported catalysts (Figure 3.1). Catalysts can be entrapped, retained by the polymer via ionic interactions or covalently attached to polymeric material. In the latter case the catalyst can be grafted onto a pre-derivatised support or can be co-polymerised. Alternatively, the catalyst can be attached to a support via hydrogen bonds, or it can be dissolved in a thin layer of water adhered to the support (SAPC, supported aqueous phase catalysis). The various ways of immobilisation each have their advantages. One can imagine that entrapped catalysts are easy to prepare since known catalysts can be used without further modification and they likely retain their properties, but catalyst leaching is a severe problem. In contrast, covalently attached ligands can eliminate ligand leaching and metal leaching if the metal-ligand interaction is sufficiently strong, but now a new ligand has to be prepared that enables anchoring to the support. Catalyst anchoring via ionic interactions can work well if ionic catalysts are used, but the properties of the catalyst might change especially if the catalyst performance depends on the counter ion. In addition, catalyst leaching can be expected if salts are produced during the reaction. For hydrogen bonded catalysts one needs to introduce a binding motive, which indeed involves the synthesis of new ligands, and also here competition with substrates, by-products or solvents presents limitations. The type of support ranges from the classical organic supports such as polystyrene (possibly cross-linked with disubstituted reagents or functionalised with polyethylene glycol spacers (TentaGel)) to inorganic supports such as alumina and silica. The choice of support is of crucial importance as all have advantages and disadvantages. For example, the mechanical strength and thermal stability of polystyrene are rather limited, which is a problem that nowadays can be managed through various technical solutions. In addition, these types of organic polymers have swelling properties that depend on the solvent used, and the swelling in turn influences the accessibility of the active sites. So there are many variables in the area of catalyst immobilisation, and there is no general solution that works best for all catalytic conversions. We will discuss various strategies applied to the rhodium catalysed hydroformylation of alkenes and, where required, we will add examples of other reactions. We will focus on catalyst performance, deactivation and metal/catalyst leaching, since these are crucial parameters when it comes to commercial applications.

3.3 Polystyrene Supported Catalysts

The use of polystyrene as support has the main advantage that it is relatively cheap and commercially available in many variations. For industrial applications polystyrene supported catalysts have some serious disadvantages. For fixed bed reactors the swelling properties of the organic support will lead to blocking of the inter-bead channels reducing the flow through of the reactor bed. The low mechanical strength of the polymer support will result in high resistance in fixed catalyst beds, and to attrition of particles in fluidised beds. These problems become even worse if the (local) temperature is high leading to degradation of the polymer. That polystyrene can be used on a large scale is proven by the fact that ion-exchange materials have found wide application. For laboratory use, polystyrene is an excellent support since most reactions are carried out batch-wise and at relatively small scales. A recent review on the use of polystyrene supported catalysts lists all the reactions that have been explored so far [13], which includes hydroformylation, hydrogenation, hydrosilylation, oligomerisation, Heck arylation, Suzuki coupling, Hartwig-Buchwald amination, allylic substitution, metathesis, cycloaddition, and aldol and Mannich type reactions.

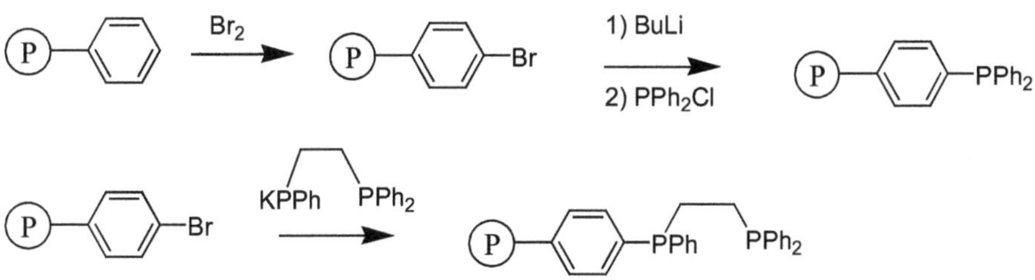

Scheme 3.1. Direct functionalisation of polystyrene with phosphine ligands.

Early work reported by Grubbs showed that polystyrene supported phosphine ligands can be prepared by direct functionalisation of polystyrene via bromination and subsequent lithiation (scheme1) [14]. Rhodium complexes of this material were shown to be active in hydrogenation. The activity strongly depended on the catalyst loading and the degree of cross-linking. Pitman et al. reported an unusually high selectivity (linear branched ratio, l:b, up to 16) in the rhodium catalysed hydroformylation of 1-

pentene upon using polystyrene supported phospine ligands [15]. The explanation is straight-forward; the local concentration of phosphine and rhodium within the resin is much higher then in the homogeneous phase under similar conditions because the resin beads only occupy a small portion of the reaction volume. Indeed if the degree of phosphination of the resin is low, the catalyst properties were close to those of the homogeneous analogue. Hartley et al. studied the rhodium catalysed hydroformylation of 1-hexene using phosphinated polypropylene in detail and they reported similar effects [16]. At relatively low temperature (65 °C) the selectivity for the linear aldehyde is 3.5 times higher (l:b=16) than that of the corresponding homogeneous catalyst. At higher temperatures (85 °C) the selectivity dropped to l:b=6. With respect to catalyst leaching they report that the use of [Rh(acac)CO_2] as precursor is much better then [RhH(PPh_3)$_3$CO], because of the absence of free triphenylphosphine. Still rhodium leaching is between 5% and less then 1%, which is far too high for bulk processes. Clearly a stronger metal ligand interaction is required. This can be achieved by using bidentate ligands. Pitman et al. prepared one of the first styrene supported bidentate ligands via a simple procedure (Scheme 3.1) [17]. Both the homogeneous and the supported dppe ligands gave rise to low selectivities in hydroformylation of 1-pentene, which we now can attribute to the small bite angle of the bidentate ligand[18]. Interestingly, the supported catalyst could be recycled 20 times with only a little loss in activity. In addition, the leaching of rhodium from the supported catalyst was investigated by extensive extraction of the support with toluene under a relatively low pressure of 1 bar of CO. No change of rhodium content was observed using atomic absorption spectroscopy, suggesting that metal-ligand complexation is indeed strong with this chelating ligand and metal leaching will be at the ppm level or lower. For bulk chemical processes leaching is only acceptable at the ppb (1 part in 10^9) level.

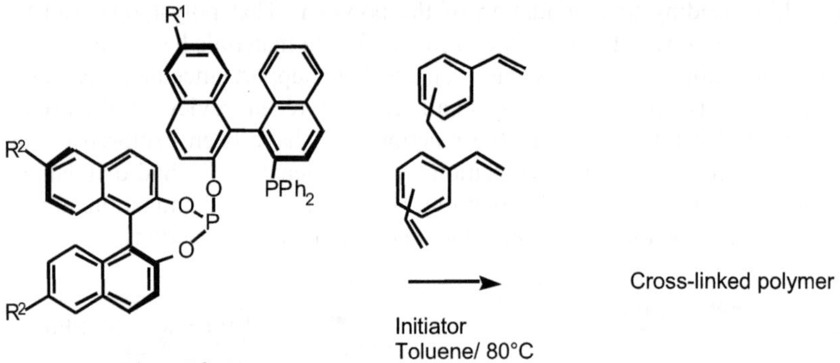

1a R^1=H, R^2=H
1b R^1=vinyl, R^2=H
1c R^1=H, R^2=vinyl
1d R^1=vinyl, R^2=vinyl

Figure 3.2. Chiral phosphine-phosphite ligands on a highly cross-linked polystyrene polymer

For small scale applications asymmetric hydroformylation is more important, and immobilisation of expensive ligands can be worthwhile. An illustrative example of a chiral hydroformylation catalyst immobilised on a highly cross-linked polystyrene support was reported by Nozaki and Takaya [19]. Several BINAPHOS ligands carrying vinyl groups have been prepared and co-polymerised with ethylstyrene and divinylbenzene (Figure 3.2). By using different ligands (**1b-1d**) the degree of freedom of the ligand in the polymer matrix was varied. The influence of the degree of cross-linking and the polymer/catalyst preparation method on the catalyst performance was investigated (Table 3.1). Generally, the selectivity (90 % ee, l:b=0.2) obtained with the polymer-supported catalysts is similar to those obtained using the homogeneous system (run 1, Table 3.1).

TABLE 3.1. Asymmetric hydroformylation of styrene using polystyrene supported rhodium catalysts based on (R,S)-BINAPHOS)[a]

Run	Catalyst	l:b	ee (%)
1	Rh(acac)(**1a**)	0.12	92
2	(PS-**1b**)-Rh(acac)	0.19	89
3[b]	(PS-**1b**)-Rh(acac)	0.25	81
4[c]	(PS-**1b**)-Rh(acac)	0.20	89
5[d]	(PS-**1b**)-Rh(acac)	0.23	84
6	(PS-**1c**)-Rh(acac)	0.12	89
7	(PS-**1d**)-Rh(acac)	0.14	68
8	(PS)-[Rh(acac)(**1b**)]	0.18	90
9	(PS)-[Rh(acac)(**1c**)]	0.11	97
10	(PS)-[Rh(acac)(**1d**)]	0.15	85
11[d]	(PS)-[Rh(acac)(**1b**)]	0.18	90

[a] Conditions: 6.20 mmol styrene in benzene, 0.0031 mmol catalyst, 20 bar CO/H_2 (1:1), 60 °C, 12 hours, conversions to aldehydes >99%. [b] Hexane as solvent. [c] Lower cross-linking degree. [d] Higher cross-linking degree.

Two methods of catalyst-preparation were tested: the co-polymerisation of the ligand and subsequent formation of the rhodium acetylacetonate complex (runs 2-7) and co-polymerisation of the rhodium acac complex of the ligand (runs 8-11). This generally gave the same results, except for ligand **1d** (cf. run 7 and 10). Ligand **1d** is cross-linked with the matrix at three positions, which give rise to constrained conformations. Apparently, this leads to lower enantioselectivity in the hydroformylation reaction. In contrast, the other ligands have more freedom and can therefore form the proper catalyst conformation after the polymerisation. The degree of cross-linking has no significant influence on the selectivity of the reaction (runs 2, 4, 5, and 11). The reaction performed in hexane instead of benzene yielded slightly lower ee. The recycling experiments clearly show the drawback of these types of supports. The polymers were partly crushed by the stirring bar and approximately 50% of the initially charged rhodium was removed after the first run.

In subsequent work the same supported catalysts were used in different reactor set-ups [20] (Figure 3.3). A vapour-phase reactor in which the supported catalyst was mounted on a bed was used for the hydroformylation of volatile alkenes such as *cis*-2-butene and trifluoropropene. The initial activities and selectivity's were similar to those of the homogeneous solutions, i.e. a TOF of 114 and 90% ee in the hydroformylation of trifluoropropene was reported. No rhodium was detected in the product phase, which means less then 0.8% of the loaded rhodium had leached. The results were, however, very sensitive to the conditions applied and, especially at longer reaction times, the catalyst decomposed. In a second approach the polymer supported complex was packed in a stainless steal column and installed in a continuous flow set-up.

Volatile substrates were again reacted in the gas-phase, and selectivities equal to those from the homogeneous phase reactions were obtained. For less volatile substrates supercritical CO_2 was used as mobile phase and the substrates were added via a stepwise injection. For styrene a small drop in selectivity was observed compared to the reaction carried out in benzene (85% vs 92% ee). The most important point is that in these reactors the mechanical breakdown of the support was not observed.

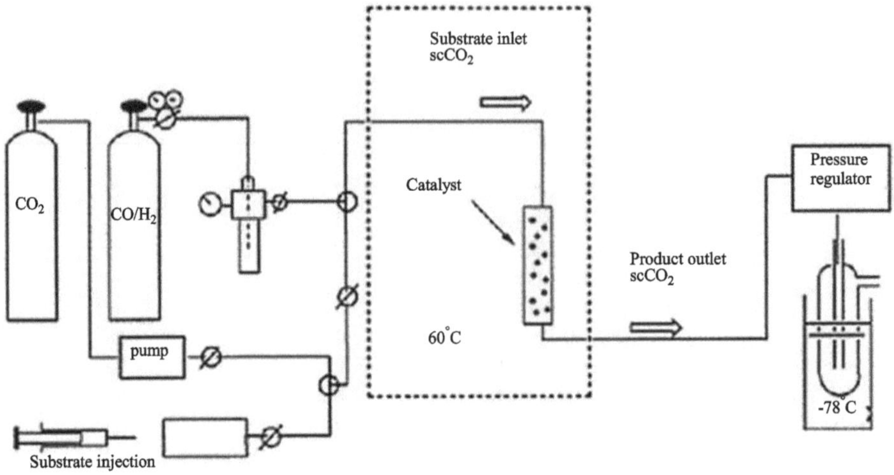

Figure 3.3. Asymmetric hydroformylation of non-volatile substrates using a stepwise injection into continuous flow reactor with $scCO_2$ as mobile phase

3.4 Silica Supported Catalyst

The covalent anchoring of catalysts onto inorganic silica surfaces was already being studied in the late seventies [21]. The most common strategy involves the functionalisation of ligands with alkoxysilane groups, which can subsequently be used to graft the ligand to the surface of silica type materials via a simple condensation step. Nowadays many building blocks required for this synthetic route are commercially available. In the first examples in which hydroformylation catalysts were anchored to an inorganic support a slightly different procedure was followed. Chloromethyl silicone polymers were converted to the iodo-analogues and subsequently reacted to form the diphenylphosphinemethyl polysiloxanes [22]. After treatment with $[RhCl(CO)(PPh_3)_2]$ these materials were used as catalysts in the hydroformylation of 1-hexene (68 bar CO/H_2 (1:1), 100 °C, benzene). Under these conditions metal dissociation from the support was a major problem. The leached metal formed very active rhodium carbonyl species in solution that was responsible for the low selectivity.

An alternative strategy to obtain silica immobilised catalysts, pioneered by Panster [23], is via the polycondensation or co-condensation of ligand functionalised alkoxysilanes. This co-condensation, later also referred to as the sol-gel process [24], appeared to be a very mild technique to immobilise catalysts and is also used for enzyme immobilisation. Several novel functional polymeric materials have been reported that enable transition metal complexation. 3-Chloropropyltrialkoxysilanes were converted into functionalised propyltrialkoxysilanes such as diphenylphosphine propyltrialkoxysilane. These compounds can be used to prepare surface modified inorganic materials. Two different routes towards these functional polymers can be envisioned (Figure 3.4). One can first prepare the metal complex and then proceed with the co-condensation reaction (route I), or one can prepare the metal complex after the polymerisation step (route II). The former route results in far better catalysts due to the higher structural homogeneity. In addition, more stable catalysts are obtained, showing less metal leaching.

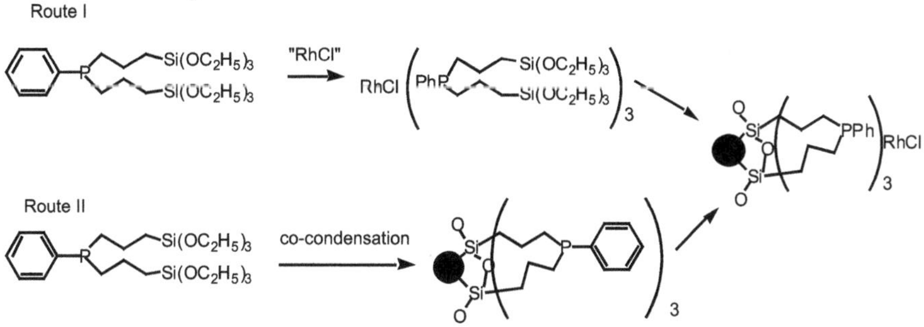

Figure 3.4. Two different routes to prepare sol-gel immobilised transition metal complexes

Panster studied the hydroformylation of 1-octene using a polysiloxane-anchored rhodium phosphine-amine based catalyst in a trickle-bed reactor [25]. The selectivity for the linear product was low (l:b =1.5), but the catalytic performance was retained over a period of 1000 h, and almost no rhodium leaching was detected. Initial leaching of rhodium was in the range of 2-3 ppm, which declined to 0.3 ppm after the start-up phase. Contrary to the homogeneous phase reaction, the presence of an excess of phosphine ligands did not result in a higher selectivity for the linear product, which was attributed to the rigid structure of the polymeric material. Using this type of material the ligands can be considered as site-isolated [26], i.e. they are fixed at such a distance that they are unable to coordinate to the same metal.

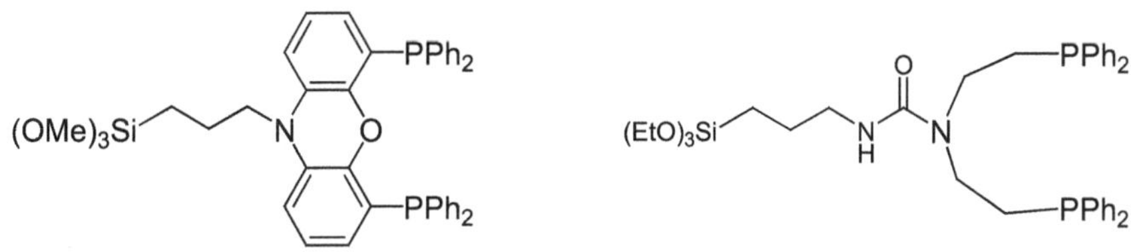

2 **3**

The immobilisation of **2** and **3** on silica has been studied in detail. Both ligands were immobilised using a sol-gel procedure [27] (I, Scheme 3.2) and **2** was also grafted onto commercially available silica in four different ways [28]. In the first three approaches, **2** was covalently tethered to silica to obtain silica(**2**) by refluxing a suspension of **2** and silica in toluene. The subsequent complexation of the rhodium precursor was performed under three different conditions. In the first approach the rhodium precursor [Rh(acac)(CO)$_2$] and silica(**2**) were simply mixed together and stirred in a THF suspension (material II). In the second, silica(**2**) was first reacted with dimethoxydimethylsilane in order to modify the acidic silanols on the silica surface. To this end a suspension of silica(**2**) and dimethoxydimethylsilane was refluxed in toluene and subsequently stirred in a solution of [Rh(acac)(CO)$_2$] in THF at room temperature (material III). In the third approach, [Rh(acac)(CO)$_2$] was added to a pre-stirred mixture of silica(**2**) and triethylamine in THF (IV). The fourth method involved the synthesis of the diphosphine rhodium complex Rh(**2**)(acac)(CO) prior to the immobilisation on predried silica (V). Interestingly, in the examples in which the Rh(acac)(CO)**2** complex was used for the immobilisation, cationic rhodium complexes were formed during the process. These were converted to neutral hydride species under proper hydroformylation conditions.

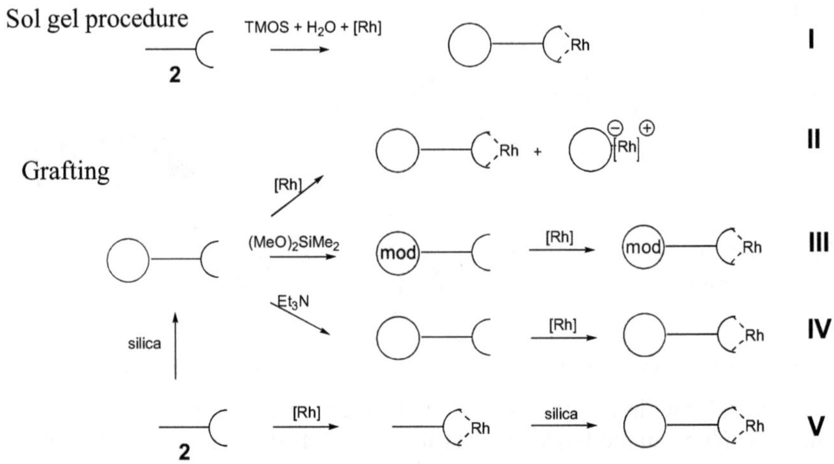

Scheme 3.2. Schematic representation of different routes to prepare silica immobilised [Rh(**2**) (CO)]$^+$.

The selectivity for the linear aldehyde of sol-gel immobilised **2** was found to be as high as 93%, which is similar to that of the homogeneous catalysed reaction (Table 3.2, entries 1 and 6). Sol-gel immobilised **3** gave a much lower selectivity (70%) for the linear aldehyde, and in the absence of ligand the selectivity to the linear aldehyde is only 26%. This proves that catalysts based on these ligands retain the selectivity upon immobilisation, which means that the selectivity is dominated by the size of the P-Rh-P bite-angle of ligand [18].

TABLE 3.2: Hydroformylation of 1-octene using silica immobilised [Rh(**2**)CO]$^+$ [a]

Entry	Catalyst	Time (h)	Conv. (%)	TOF[b] (h^{-1})	l:b Ratio	l-aldehyde (%)	b-aldehyde (%)	l-alcohol (%)	2-octene, octane (%)
1	sol-gel-**2** (**I**)	2	20	18.3	65	94.6	1.5	3.6	0.2
2	**II**	0.5	97	n.d.	2	40.1	22.0	0	37.9
3	**III**	23	24	8.0	19	85.5	4.5	0	9.9
4	**IV**	22	18	8.7	37	96.2	2.6	1.0	0.2
5	**V**	22	37	13.2	37	90.7	2.6	5.1	1.6
6	hom. **2**	2	19	283	32	93.3	2.9	0	3.7
7	sol-gel-**3**	2	72	119	2	70.0	28.9	0.1	1.0

[a] Ligand to rhodium ratio is 10, catalysis performed in 13 ml of toluene using 1ml of 1-octene as the substrate at 80 °C and 50 bar CO/H_2. Samples were analysed by means of GC and GC-MS.[b] Average turnover frequencies were calculated as (mol product)(mol catalyst)$^{-1}$$h^{-1}$

The method of catalyst immobilisation appeared to affect its performance in catalysis. Catalyst obtained by method II showed a low selectivity in the hydroformylation of 1-octene (l:b aldehyde ratio was even lower than 2) at a very high rate and high yields of isomerised alkenes (Table 3.2, entry 2), whereas procedure IV resulted in a catalyst that was highly selective for the linear aldehyde (with a l:b ratio of 37) (entry 5). In accordance with examples from literature it is likely that procedure II gave rise to the ionic bonding of ligand-free rhodium cations on the slightly acidic silica surface [29], i.e. a direct connection of the rhodium to the silica material. If the rhodium phosphine complex is prepared prior to anchoring (V) no ligand-free rhodium is attached to the silica. The effect of the acidic silanols on the catalyst preparation can be eliminated via chemical modification of the silica surface (III) or upon addition of a neutralising base (IV). Pre-modification of the silica using dimethoxydimethylsilane largely improved the catalyst selectivity (l:b aldehyde ratio of 19) at the cost of some activity. The addition of triethylamine (IV) also resulted in a very selective hydroformylation catalyst with a l:b aldehyde ratio of 37.

It is interesting to note that using the sol-gel procedure (I) the pre-formation of the rhodium diphosphine complex suppressed the formation of ligand free rhodium-cations on the silica surface. This approach gave rise to a well-defined, very selective hydroformylation catalyst. All immobilised catalysts were 10 to 40 times slower than the homogeneous catalyst under the same conditions, the sol-gel procedure yielding the fastest catalyst of this series.

TABLE 3.3. Hydroformylation of 1-octene using sol-gel and silica immobilised rhodium catalyst **2** in subsequent catalytic runs.[a]

Entry [cycle]	Meth of prep	Time (h)	Conversion (%)	TOF[b] (h^{-1})	l:b Ratio	1-aldehyde (%)	b-aldehyde (%)	1-alcohol (%)	Side-products (%)
1 [1]	I	2	20	18	65	94.6	1.5	3.6	0.2
2 [4]	I	2	12	12	62	89.6	1.6	8.8	0
3 [1]	I	18	38	23	22	61.0	4.1	29.6	5.3
4 [2]	I	18	30	17	25	77.9	3.7	15.1	3.4
5 [1]	V	22	37	13	37	90.7	2.6	5.1	1.6
6 [3]	V	72	61	8	27	79.9	3.5	13.2	3.4
7 [1]	III	23	24	8	19	85.5	4.5	0	9.9
8[4]	III	72	44	5	16	84.8	5.4	0	9.8
9 [1]	I[c]	24	69	35[d]	32	92.8	3.0	2.5	1.7
10 [4]	I[c]	24	67	35[d]	35	94.5	2.7	1.3	2.2
11	I[e]	2.5	7	4	33	92.0	2.8	0	5.2

[a] Ligand to rhodium ratio is 10, catalysis performed in 13 ml of toluene using 1 ml of 1-octene as the substrate at 80 °C and 50 bar CO/H_2. Samples were analysed by means of GC and GC-MS analysis. [b] Average turnover frequencies were calculated as (mol product)(mol catalyst)$^{-1}$h^{-1}. [c] 1 ml of 1-propanol added to the catalyst mixture [d] Initial turnover frequency. [e] 1 ml of triethylamine added to the catalyst mixture.

Sol-gel and silica immobilised Rh(**2**) were tested in successive hydroformylation reactions of 1-octene to investigate the effect of the support on the recyclability of the catalyst (Table 3.3). For both systems the selectivity was found to be similar to that of the homogeneous analogue. Upon recycling, the high regioselectivity for the linear aldehyde was maintained (l:b ratios ranging from 22 to 65) while only a small amount of alkene isomers was formed as side products. A small decrease in hydroformylation activity upon recycling along with an increasing formation of 1-nonanol (4-13 %) was observed, especially at longer reaction times (Table 3.3), a phenomenon not obtained with the analogous homogeneous catalysts. This strongly suggests that the silica support plays a role in this secondary reaction. It is likely that the acidic silica increasingly suppresses the formation of [HRh(**2**)$(CO)_2$] from $[Rh(\mathbf{2})CO]^+$. The cationic rhodium complex is responsible for hydrogenation activity converting the aldehyde into the corresponding alcohol, and is not active in hydroformylation [30].

When the silica material with the capped (acidic) silanols was used (III) the influence of the silica on the selectivity after recycling was largely suppressed. No hydrogenation of the aldehyde was observed as a secondary reaction in any of the successive catalytic runs (Table 3.3, entry 7-8). The modification of the surface silanols with alkylsilanes did not yield an optimal catalyst system. The overall selectivity for the linear aldehyde was slightly lower compared with the other catalysts; more isomerisation and branched aldehyde was obtained.

A more subtle method to reduce the influence of the silica comprises the addition of 1-propanol, which leads to an efficient suppression of hydrogenation activity. A high overall selectivity for the linear aldehyde was obtained in subsequent batch-wise runs (Table 3.3, entries 9-10). Interestingly, in other examples alcohols or amines were added to promote the hydrogenation of aldehydes via a transfer hydrogenation reaction [31], showing that this co-solvent can have different functions. In the present example alcohol prevents the formation of the cationic hydrogenation catalyst from the neutral hydroformylation catalyst by deactivating the acidic silanols on the silica surface. The addition of triethylamine as a base also enabled the suppression of the hydrogenation activity (Table 3.3, entry 11).

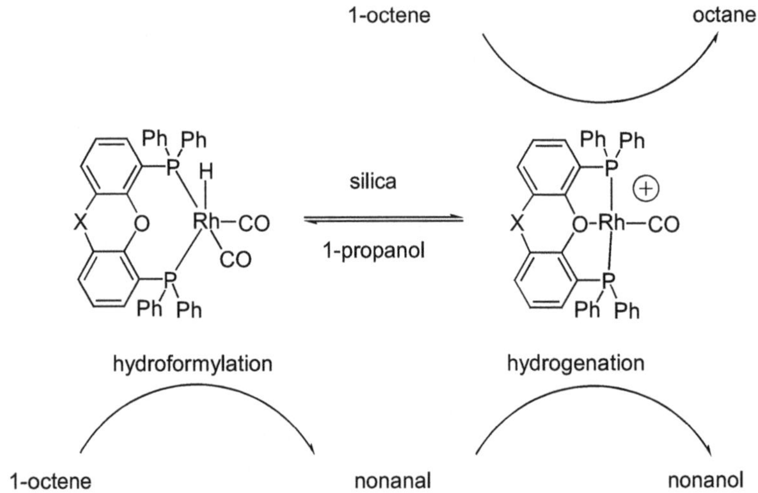

Scheme 3.3. Schematic representation of the interconversion of the catalyst systems $[Rh(\mathbf{2})(CO)]^+$ and $[HRh(\mathbf{2})(CO)_2]$ and the products generated from each catalyst species.

It is evident that the silica support influences the catalytic performance and it is important to understand the details of the processes involved. For the sol-gel material it was shown by ^{31}P NMR spectroscopy that the immobilised cationic complex completely transforms to the neutral rhodium-hydride species under a CO/H_2 atmosphere (Scheme 3.3). On dried silica, however, this conversion might not be complete since the dried support is more acidic [32]. It is therefore very likely that the neutral and cationic rhodium complexes co-exist on the silica support. ^{31}P NMR measurements on homogeneous rhodium complexes have shown that a simple protonation indeed converts the neutral rhodium hydride species into the cationic complex.

The existence of two different rhodium species co-existing on the silica support can be used as an advantage by controlling their relative amount. Under standard hydroformylation conditions, the cationic species and the neutral hydride complex are both present in significant amounts. Hence hydroformylation and hydrogenation will both proceed under a CO/H_2 atmosphere. Indeed a clean one-pot reaction of 1-octene to 1-nonanol was performed, using the supported catalyst for a hydroformylation-hydrogenation cascade reaction. 98 % of the 1-octene was converted in the hydroformylation reaction and 97 % of the linear nonanal was subsequently hydrogenated to 1-nonanol resulting in an overall selectivity of 90 % for the linear alcohol. Importantly, no heavy-end side-products were observed in this reaction owing to the mild conditions applied. In the first few hours of the cascade process, mainly the hydroformylation of 1-octene to the aldehyde was observed. The hydrogenation towards the corresponding alcohol started at higher aldehyde concentrations. When approximately 90 % of the 1-octene was consumed (after 60 hours) the hydroformylation activity had decreased significantly, which is in line with the first order rate dependency in substrate concentration. In contrast, the hydrogenation of the aldehyde product proceeded after 60 hours. As a result, the aldehyde concentration decreased again after approximately 40 hours, which in turn caused a decrease of the hydrogenation rate.

TABLE 3.4. Results from switching between hydrogenation, hydroformylation and hydroformylation-hydrogenation sequence reactions using sol-gel immobilised $[Rh(\mathbf{2})CO]^{+}$ [a]

Entry[b]	Time (h)	Conversion Octene (%)	Conversion Aldehyde (%)	l:b	l-aldehyde (%)	b-aldehyde (%)	l-alcohol (%)	Octane[c] (%)
1	172	97	75	23	18.5	3.6	66.7	11.2
2	24	100	100	-	0	0	100	100
3	68	60	16	18	65.2	4.5	13.7	16.6
4	2	98	10	-	0	0	9.7	98
5	96	96	0	18	90.7	5.1	0	4.3

[a] Ligand to rhodium ratio is 10, catalysis performed at 80 °C, in 13 ml of toluene using 1 ml of 1-octene (plus 1 ml of 1-nonanal entries 2 and 4). Samples were analysed by means of GC and GC-MS analysis. [b] Entries 1, 3, and 5; 50 bar syn gas. Entries 2 and 4; 50 bar H_2. [c] Numbers include isomers of 1-octene since these are not separable from octane on GC.

The ratio of the two co-existing catalyst species could be controlled by modification of the conditions and the catalyst was recycled by a simple filtration procedure, enabling the use of the supported catalyst for different subsequent conversions. This was demonstrated by a series of catalyst experiments using 1-octene as the substrate and polysilicate immobilised Rh(**2**). A first reaction under standard conditions, thus with the catalyst as a mixture of the neutral and cationic species, resulted in a product mixture that consisted for 66.7 % of 1-nonanol and 18.5 % of 1-nonanal (Table 3.4, entry 1). After this reaction, the system was recycled and transformed into a hydrogenation catalyst just by washing it with toluene and subsequently adding a mixture of 1-octene and 1-nonanal in toluene to the catalyst mixture. After a reaction time of 24 h under an H_2-atmosphere a complete hydrogenation of both substrates was observed to octane and 1-nonanol, respectively (Table 3.4, entry 2). This modulation is completely reversible; in the third catalyst cycle the system was applied again as a mixed catalyst by changing the atmosphere from H_2 to CO/H_2 and using 1-octene as the substrate. This resulted in a switch of the catalyst from the hydrogenation mode back to the hydroformylation-hydrogenation sequence mode. The regioselectivity for the linear aldehyde and alcohol (overall l:b ratio of 18) was largely restored (Table 3.4, entry 3). This provides evidence that the catalyst is stable during the switching procedure, since even a small amount of decomposed rhodium catalyst would result in a dramatic drop in selectivity. In the fourth cycle, the hydrogenation catalyst 5 was restored (Table 3.4, entry 4). This time, the hydrogenation reaction of a 1:1 mixture of 1-octene and 1-nonanal was stopped after 2 h and a chemoselectivity for the alkene reduction over aldehyde reduction was observed (96 % alkene and 10 % aldehyde was hydrogenated). From these four catalytic runs it can be concluded that the immobilised catalyst system is switched easily and repeatedly between the hydrogenation mode and the hydroformylation-hydrogenation cascade mode. In the fifth run the catalyst was switched to a pure hydroformylation catalyst (Table 3.4, entry 5) by using an atmosphere of CO/H_2 and the addition of 1 ml 1-propanol, along with a fresh batch of 1-octene in toluene. After 96 hours, 1-octene was almost completely converted to the linear aldehyde with a complete suppression of its hydrogenation to the alcohol! These results nicely demonstrate the importance of detailed knowledge of the effect of immobilisation on the catalyst. In this example the immobilised catalyst Rh(**2**) can not only be recycled, but it can also be switched reversibly between two different states leading to different products. In addition, the catalyst is very stable and metal leaching is below the detection limit. An important advantage of the supported catalyst is that solubility issues of the catalyst in the reaction medium do not play a role. As such, a hydroformylation experiment was performed in pure 1-octene and under these conditions the activity was similar as that of the homogeneously catalyzed reaction (TOF=287 mol/mol/h) [27].

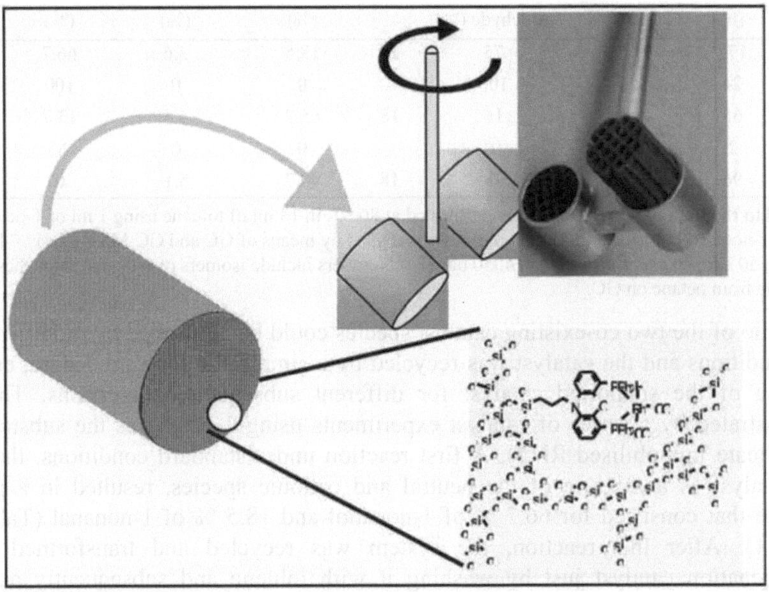

Figure 3.5. Schematic representation of the ROTACAT concept: a homogeneous complex, immobilised on monoliths, and the implementation of these monoliths in the blades of a mechanical stirrer. Inset: a picture of the stirrer with a silicon-carbide monolith

In recent years monolithic reactors have found wide application in gas-solid processes such as automotive emission control [33] and industrial off-gas purification [34]. In addition, these systems are being considered for gas-liquid-solid processes such as hydrogenations and oxidations [35]. Recently, Moulijn et al. reported on a novel reactor configuration in which blocks of monoliths are arranged in a stirrer-like fashion [36]. Using this setup a reasonably large surface area of the immobilised catalyst is retained, but many problems of handling a suspension of finely divided supported catalyst species are eliminated. In an attempt to increase the applicability of immobilised catalyst Rh(**2**) a new concept (ROTACAT: an immobilised homogeneous rotating catalyst) was developed that unites the above-described processes (Figure 3.5) [37]. Rh(**2**) was covalently anchored to monoliths and these monoliths were implemented in two hollow tubes that were constructed as blades of a mechanical stirrer. Two types of monoliths (Si and SiC) were explored that differ in mechanical strength and surface area. The Si monoliths have a higher surface area (120 m^2/g vs 0.5 m^2/g) whereas the SiC monoliths are mechanically far more stable. The SiC monoliths required a special treatment since the surface area of this type of monolith is too low (less than 0.5 m^2/g according to BET surface area measurements). Only very little of ligand **2** could be loaded on a SiC monolith of 3.5 g. To increase the surface area the SiC monoliths were first wash-coated with a porous silica top-layer of approximately 15 weight %. The silicon top layer, causing an increase in surface area to 17 m^2/g (determined by BET surface area measurements), enabled a sufficient ligand loading (0.02-0.04 mmol/g monolith). The Si monoliths could be charged with a sufficient amount of ligand without pre-treatment (~0.05 mmol g^{-1} monolith). The mechanical stirrer of a 200 ml autoclave, consisting of two parallel cylindrical tubes, was charged with two monoliths. A sufficient amount of **2** was immobilised on the monoliths by agitation in a toluene solution of **2** at 80 °C under an inert atmosphere. The immobilised ligand was subsequently functionalised with the rhodium precursor by stirring a toluene solution of [Rh(acac)$(CO)_2$] at 80 °C under 50 bar of CO/H_2 (1:1) for 1 hour in an autoclave, equipped with the ligand functionalised rotor.

In the hydroformylation of 1-octene, using Rh(**2**) immobilised on the SiC monoliths, the overall selectivity for 1-nonanal was 89 ± 3 %. As was found previously for the homogeneous system and the silica-immobilised system, the l:b aldehyde ratio was very high (ranging from 20 to 46). Remarkably, the selectivity for the linear product using Rh(**2**) immobilised on the Si monoliths was lower (72 %). The regioselectivity for the linear aldehyde was rather high (l:b ratio of 14) but alkene isomerisation was observed to a large extent (18 %). Furthermore, significant hydrogenation of 1-nonanal to the alcohol was observed (4.8 %). Similar to sol-gel immobilised catalyst the hydrogenation could be completely suppressed by the addition of a small amount of 1-propanol during catalysis, suggesting that the lower catalyst selectivity of the Si- monolith immobilised catalyst is a result of the higher acidity of this material.

The SiC ROTACAT was studied in more detail. The initial rate of hydroformylation under standard conditions was found to be low (between 1 and 2 mol mol^{-1} h^{-1}). Compared to the sol-gel immobilised analogue this is about 15 times lower. In an octene/toluene 1:1 mixture the rate of the reaction increased to 15 mol mol^{-1} h^{-1} as a result of the higher concentration of octene. Under optimised conditions the reaction rate reached 97 mol mol^{-1} h^{-1} when the catalysis was performed in neat 1-octene. The SiC ROTACAT could be reused in ten consecutive catalytic cycles showing no systematic deterioration in hydroformylation performance. To demonstrate the advantage of this multipurpose reactor set-up, the same catalyst was used for hydroformylation reactions in the gas phase (propene), in aqueous phase (propene) and again in toluene. In all cases clear colourless product solutions were drawn from the autoclave that did not contain rhodium or phosphine, as indicated by means of Atomic Emission Spectroscopy. The catalyst showed a remarkable long-term stability. It was found that this catalyst system (2 mg of Rh!) was reusable for over six months without decrease in catalytic performance. This makes ROTACAT among the most stable hydroformylation systems reported in the literature. Importantly, the bond-strength of the chelating diphosphine to the rhodium is sufficiently high and the rhodium-diphosphine complex is solidly anchored on the monolithic rotor. The system enables a straightforward recycling procedure (compared to slurry reactions) in which required washings do not damage the catalyst.

One strategy to reduce the negative effects of catalyst immobilisation on the activity is the use of supercritical fluids such as $scCO_2$. $scCO_2$ has solvation properties comparable to higher alkanes, and at the same time the diffusion is much higher because of the low density [38]. Importantly, dihydrogen and carbon monoxide dissolve very well in $scCO_2$. At Nottingham, a set-up for continuous processes with $scCO_2$ as mobile phase and supported catalyst as stationary phase has been built and successfully applied to hydrogenations, Friedel-Crafts alkylations and etherification reactions [39]. The reactor set-up has also been used to study the properties of immobilised catalyst Rh(**2**) in the hydroformylation of 1-octene [40]. For this purpose the rhodium complex was immobilised on silica with particle size 200-500 μm. Typically 1g of silica with a 0.4% Rh loading was loaded into a 5 ml supercritical flow reactor.

TABLE 3.5. Results from the hydroformylation of 1-octene using silica-immobilised Rh(**2**) using a continuous flow reactor with $scCO_2$ as the mobile phase [a]

Entry	TOF[b]	Linear aldehyde[c] (%)	Branched aldehyde[c] (%)	Alkene isomers[c] (%)	Linear alcohol[c] (%)	l:b ratio	1-Octene conversion (%)
1[d]	39	96.1	2.4	1.5	0	40	3.6
2	87	92.9	3.0	3.8	0.3	32	9.4
3[e]	112	94.4	2.4	2.5	0.7	40	10.1
4[f]	117	92.6	3.8	2.5	1.0	24	10.3
5[g]	44	90.7	4.4	3.7	1.3	21	14.3
6[h]	160	93.5	2.8	2.9	0.8	33	4.1
7[i]	93	96.0	1.9	1.1	0.9	50	14.3
8[j]	96	91.3	4.1	4.3	0.3	23	4.6

[a] Ligand : Rh ratio is 10:1 and the catalysis was performed at 80°C, 120 bar CO_2 at 0.65 l/min flow rate (at 20°C, 1 atm), 50 bar overpressure syngas and an 1-octene flow rate of 0.05 ml/min (substrate:syngas = 1:10) unless otherwise stated. Catalyst: see text. Values shown are average numbers over a period of 3-6 hours. [b] Average turnover frequencies were calculated as (mol aldehyde/mol catalyst)$^{-1}$/h^{-1}. [c] Determined by means of GC analysis using decane as an internal standard. [d] Reaction temperature is 70°C. [e] Syngas overpressure is 25 bar. [f] 0.3 l/min CO_2 flow rate (at 20°C, 1 atm). [g] 180 bar CO_2. [h] Reaction temperature is 90°C. [i] 1-Octene flow rate of 0.03 ml/min. [j] 1-Octene flow rate of 0.1 ml/min.

The catalyst system converted 1-octene selectively to linear nonanal. The average l:b aldehyde ratio was 40:1. 1-Octene conversions up to 14% per pass were obtained (Table 3.5) and only a few % of octene isomers and a trace amount of alcohol were observed as by-products. At 70 °C and a 1-octene flow rate of 0.05 ml/min the rate of hydroformylation was moderate (39 mol mol^{-1} h^{-1}, Table 3.5, entry 1). The rate increased to 87 mol mol^{-1} h^{-1} with the catalyst bed at 80 °C (entry 2) and improved further to 112 mol mol^{-1} h^{-1} on decreasing the syngas pressure from 50 to 25 bar (entry 3). The latter effect is consistent with the negative order in CO pressure that is commonly observed in hydroformylation reactions. A TOF of 117 mol mol^{-1} h^{-1} was observed on increasing the residence time of the substrate in the reactor by decreasing the CO_2 flow rate (entry 4). An increase of the CO_2 pressure resulted in a decrease in TOF to 44 mol mol^{-1} h^{-1} (entry 5). This suggests that a higher pressure, which results in a higher density of CO_2, changes the transport properties in the reactor. The highest TOF (160 mol mol^{-1} h^{-1}) was observed at 90 °C (entry 6). On decreasing the 1-octene flow rate from 0.1 to 0.03 ml/min an increase in the l:b ratio from 23 to 50 (entries 2, 7 and 8) was observed.

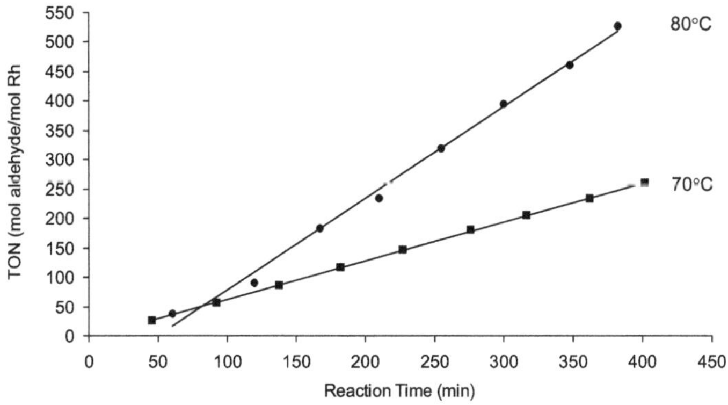

Figure 3.6. Turn over number (TON) displayed as function of time for the hydroformylation of 1-octene using a set-up for continuous processes with $scCO_2$ as mobile phase and supported catalyst **2**

It is remarkable that the hydroformylation rate is at least four times higher than the batch reaction with the use of the same immobilised catalyst in toluene and only half the rate of the homogeneous analogue. The high rate in $scCO_2$ is probably caused by enhanced mass-transport properties and the lower viscosity of the solvent medium. The catalyst appeared to be very robust, as its performance is constant over at least 30 hours. In Figure 3.6, the turnover number (TON), is plotted against reaction time. The TON increased linearly with time at both 70 and 80 °C. Moreover, the catalyst was used for six non-consecutive days with no observable decrease in either activity or selectivity. Furthermore, no rhodium leaching was found (detection limit of used technique (ICP-AES) is 0.2% of the total amount of rhodium of the catalyst). This again demonstrates that the rhodium-diphosphine bond in this catalyst also remains stable under hydroformylation conditions if $scCO_2$ is used as the solvent.

The process is potentially interesting in the manufacture of chemicals and this approach has several advantages compared to conventional homogeneously catalysed reactions. Firstly, $scCO_2$ is a clean, environmentally benign medium that can be easily separated from the organic phase [38]. Secondly, the application of an immobilised homogeneous catalyst in the flow reactor provides a direct and quantitative separation of the products from the catalyst and avoids any solubility limitations of homogeneous catalysts. Furthermore, the catalyst is very robust and leaching levels are very low. The use of high pressures adds to the cost of the process, but, in addition to the above benefits, it should be mentioned that the space-time yields of the immobilised system are one order of magnitude higher than those of selective homogeneous catalyst systems.

3.5 Catalysis in Interphases

The term "chemistry in interphases" was first introduced in the field of reverse-phase chromatography [41]. In 1995 Lindner et al. transferred the concept to the area of transition metal catalysis [42] and in a recent review the concept is explained in detail [43]. The interphase is defined as a region within a system in which the stationary and a mobile component penetrate on a molecular level without the formation of a homogeneous mixture. In these regions the reactive centre on the stationary phase becomes mobile, thereby simulating the properties of a solution. Therefore, this should lead to a system that is conceptually in between homogeneous and heterogeneous. In practice, this means that the stationary phase consists of an inert rigid matrix functionalised with a flexible spacer with a reactive centre at the end. The mobile phase can be chosen in a more classical way: a solvent or a gas containing the reactants. Of course one can also decide to use $scCO_2$ as the mobile phase.

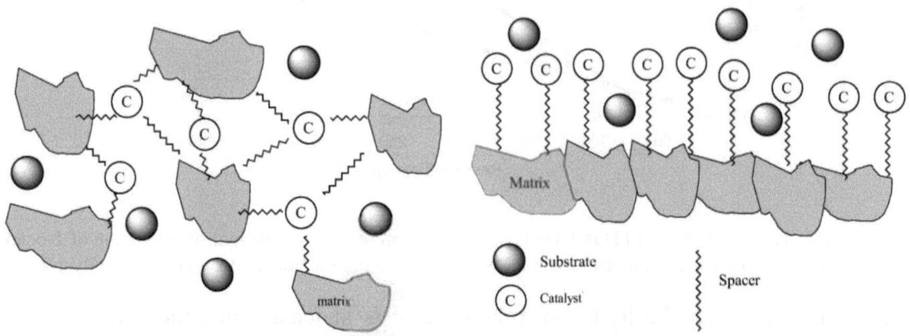

Figure 3.7. The concept of chemistry in interphases; visualisation of two different ways to obtain such a material, sol-gel procedure using different reagents (left) and silica grafted materials with sufficiently long spacers (right). Solvent occupies the space between the spacers

Organic-inorganic hybrid polymers are suitable as matrices (Figure 3.7 left), but also suitable are rigid porous materials with sufficiently long organic spacers between the active centre and the support (Figure 3.7 right). It is noteworthy that these types of materials were already studied before the introduction of the concept of chemistry in interphases. An example is shown in Figure 3.8, where phosphite functionalised polystyrene was grafted on silica [44]. This material was used in a continuous flow reactor and the rhodium catalysed hydroformylation of styrene was carried out for a period of ten days (syngas pressure 30 bar, 100 °C, benzene). In this period no metal leaching (<0.05%) nor catalyst deactivation was observed. In toluene the catalyst was about 8 times more active, but now a significant amount of metal leaching occurred. The metal leaching was explained by incomplete coordination of the rhodium due to contraction of the polymer in this solvent.

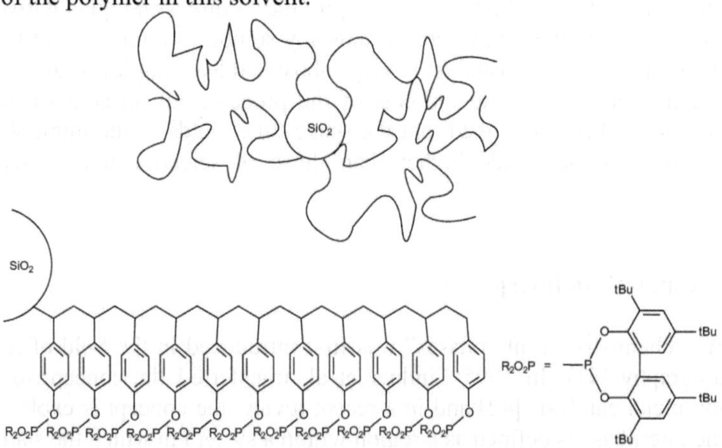

Figure 3.8. An early example of a hybrid support applied in the rhodium catalysed hydroformylation operated in a continuous flow reactor. Polystyrene containing phosphite ligands were grafted on inorganic silica, such that the catalyst will behave as a homogeneous catalyst when using a compatible solvent

Lindner et al. carried out detailed studies on the use of the sol-gel procedure to obtain new materials as suitable matrices [45]. A co-condensation of $Si(OR)_4$, $Si(OR)_3$-spacer-$Si(OR)_3$ and $Si(OR)_3$-spacer-ligand resulted in new hybrid materials, the properties of which strongly depended upon the spacer and the ratio of the different components used. Most of these materials were characterised with IR, EXAFS, and EDX spectroscopy, scanning electron microscopy and solid state NMR spectroscopy. Solid state ^{31}P NMR was used to quantify the mobility of the interphases; the changes in line-widths reflect the flexibility of the material. It is noteworthy that these matrices lost the characteristic properties of silica and showed significant swelling depending on the solvent used. In one of the contributions Linder et al. used hemi-labile ligands such as **4**, and their ruthenium complexes for the sol-gel process and the materials obtained were characterised in detail using solid state NMR spectroscopy [46]. These materials proved to be efficient catalysts for the ruthenium catalysed hydrogenation and some interesting features were observed. The catalytic behaviour was strongly dependent on the solvent used. A significant increase in activity was obtained in polar solvents. There was only a very small effect of the hydrogen pressure and a strong effect of substrate concentration (if bulky phenyl ethyne was used as a substrate). This suggests that the diffusion of these large molecules is involved in the rate limiting step. Importantly, the supported catalyst showed higher activity then the homogeneous analogue. This was attributed to the decrease in formation of inactive rhodium dimer due to the spatially isolation of the metal centres on the stationary phase.

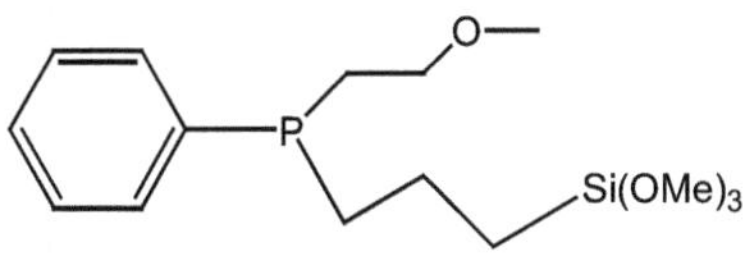

4

Lindner et al. also prepared catalysts for the hydroformylation of olefins using a sol-gel-procedure to obtain rhodium complexes **5a-c** (Figure 3.9). The bifunctional co-condensation agent $(MeO)_3Si(CH_2)_6Si(OMe)_3$ and a carbonylhydrido-(trisphosphine)rhodium complex with T-silyl-functionalised diphenylphosphine ligands containing three- or six-membered hydrocarbon spacers were employed as starting materials. The polymer-bound complexes **5a-c** were good catalysts for the hydroformylation of 1-hexene and they have been studied ina wide variety of solvents (toluene, THF, acetone, ethanol, methanol and water). Polar organic solvents such as methanol afforded the highest activities (TOF=340 mol(sub) mol(cat)$^{-1}$ h^{-1}) with low selectivities (l:b=3.3), whereas the selectivity toward the formation of 1-heptanal was optimal in solvents of medium polarity such as ethanol (l:b=5.5) or acetone at the expense of the activity (TOF (ethanol) =234 mol(sub) mol(cat)$^{-1}$ h^{-1}). In water the activity was rather low (TOF=20 mol(sub) mol(cat)$^{-1}$ h^{-1}). The high activity in solvents of medium polarity was explained by the optimal swelling of the polymers in these solvents, and the state of the reactive centres was suggested to be similar to those in solution. In line with this, wide-line separation (WISE) NMR spectroscopic investigations revealed a substantial decrease in line-width for the swollen polymer. In addition to hydroformylation a high isomerisation activity was observed (typical 20-30%).

These materials were also prepared via a sol-gel process in the presence of free phosphine ligands leading to materials with non-coordinated ligands. These supported catalysts gave rise to higher selectivities for the linear aldehyde in the hydroformylation of 1-hexene; a l:b ratio of 14:1 with an average turnover frequency of 164 mol(sub) mol(cat)$^{-1}$ h^{-1} was obtained with these materials. In addition the isomerisation side reaction was suppressed to a large extent (5%). The longer spacers between the catalyst and the solid support in material **5c** were introduced for higher mobility and thus improved performance. However, l:b ratio (6) and turnover frequencies (118) obtained with **5c** were lower than those observed with **5b**. This was attributed to the lower actual concentration of phosphine functions in **5c** (mmoles of phosphine per unit volume of swollen catalyst bead at constant P/Rh) in the swollen polymer. Olefins higher than 1-hexene were also hydroformylated with **5c** giving similar results.

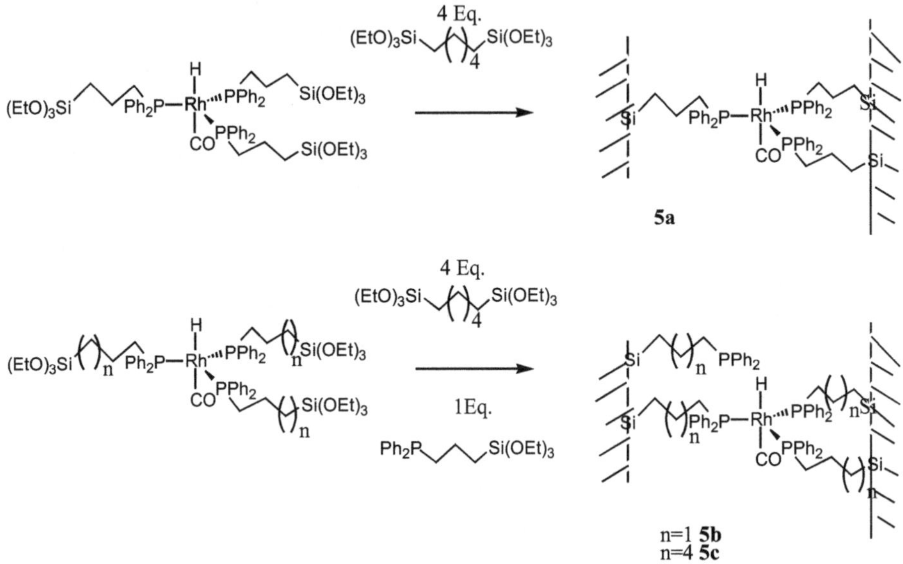

Figure 3.9. Sol-gel processed rhodium complexes using various procedures. These catalysts gave different results in the hydroformylation

In a similar study with modified phosphine ligands several hybrid polymers were prepared [47]. A novel triphenylphosphine analogue was prepared by the introduction of a spacer at the para-position of one of the phenyl rings. Subsequent reaction with a rhodium precursor yielded the catalytically active T-silyl-functionalised rhodium(I) compound $ClRh(CO)[PPh_2C_6H_4NHCONH(CH_2)_3Si(OEt)_3]_2$. The modified complex was co-condensed with $(MeO)_2MeSi(CH_2)_6SiMe(OMe)_2$, $(MeO)_2MeSi(CH_2)_3(1,4\text{-}C_6H_4)(CH_2)_3SiMe(OMe)_2$, $(MeO)_3Si(CH_2)_6Si(OMe)_3$, and $(MeO)_3Si(CH_2)_3(1,4\text{-}C_6H_4)(CH_2)_3Si(OMe)_3$. Structural and mobility investigations of these novel stationary phases were carried out by multinuclear solid-state NMR spectroscopy (^{13}C, ^{29}Si, ^{31}P). Complexes were successfully employed in the hydroformylation of 1-hexene with good turnover numbers. If the reaction was carried out at 60 bar pressure, mobility became less important, and highly cross-linked supports gave better results. This shows that the concept of "chemistry in interphases" works under limited conditions only. No leaching of the transition metal complexes during the reactions was observed by atomic absorption spectroscopy, indicating that the metal-ligand interaction was sufficient for catalyst recycling.

The use of dendrimers as supports to anchor transition metal catalysts has attracted considerable attention over the past decades [48] (see also Chapter 4 of this book). Several groups studied the use of dendrimers immobilised on insoluble supports [49], and this type of material meet the requirements for catalysis in interphases. Alper reported the use of diphosphine functionalised polyamidoamine (PAMAM) dendrimers that were anchored onto a silica support [50] as polydentate ligands in alkene hydroformylation. The dendrimer was actually synthesised on the support, which simplified the synthetic procedures but complicates detailed analysis of the system. The Rh-PPh_2-PAMAM-SiO_2 catalyst was shown to be an active catalyst for the hydroformylation reaction and it was, as expected, more selective towards the branched product. The difference in chain-length (n = 0, 2, 4) (see Figure 3.10) had mainly an impact on the recyclability of the catalyst and it was suggested that this is due to steric congestion of the more compact dendrimers. Indeed, colored product solutions were obtained indicating that slow metal leaching occurred. Additional Soxhlet extraction experiments showed that rhodium leached under catalytic conditions only; both CO and substrate promote metal leaching from the support. Unfortunately, the properties of the supported catalysts have not been compared with the analogous homogeneous compounds.

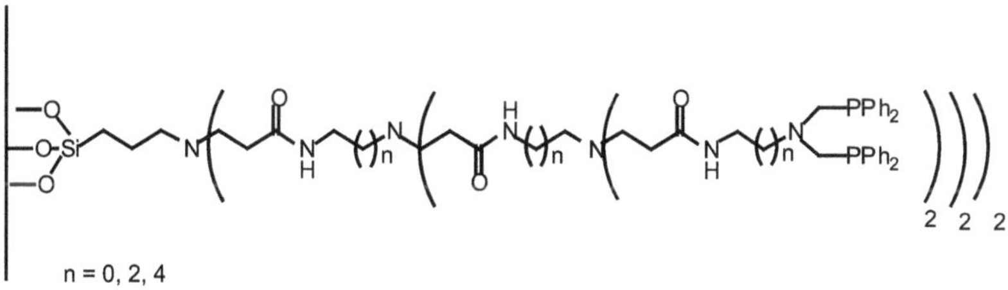

Figure 3.10. Phosphine functionalised PAMAM dendrimers anchored on a silica support

Arya et al. used solid phase synthesis to prepare immobilised dendritic catalysts with the rhodium centre in a shielded environment to mimic nature's approach of protecting active sites in a macromolecular environment (e.g. catalytic sites inside enzymes) [51]. Two generations PS immobilised rhodium-complexed dendrimers, **6** and the more shielded **7**, were synthesised.The PS resin immobilised rhodium-complexed dendrimers were used in the hydroformylation of styrene, p-methoxystyrene, vinyl acetate and vinyl benzoate using a total pressure of 70 bar 1:1 CO/H_2 at 45 °C in CH_2Cl_2.

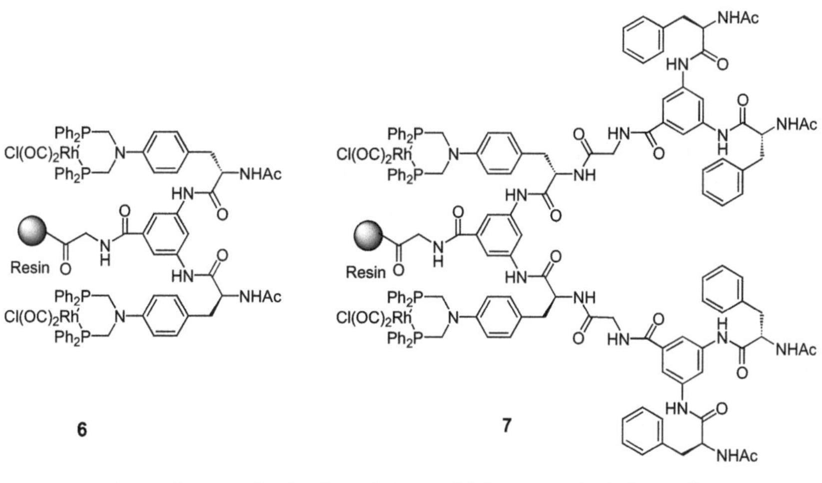

Using styrene as substrate both **6** and **7** could be recycled four times, apparently without loss in activity or selectivity (>99% conversion after 20 hours; b:l ratio 18 : 1). When p-methoxystyrene was used as a substrate the first three cycles did not suffer

from any deactivation (>99% conversion after 20 hours, b:l ratio | 17:1). The fourth cycle did not reach full conversion after 20 hours for **6** (56% conversion, b:l ratio = 30:1) nor for **7**, although this more shielded dendritic catalyst did reach a higher conversion (85% conversion, b:l ratio = 30:1). In the hydroformylation of vinyl acetate catalysts **6** and **7** gave similar results. Similar selectivities (b:l | 17:1) and slightly declining activities per cycle were observed (99% conversion for the 2nd cycle → 85 % conversion for the 5th cycle). Using vinyl benzoate as substrate the immobilised dendrimers **6** and **7** could be recycled with constant activity and selectivity for three cycles (>99% conversion after 20 hours, b:l | 25:1). In the fourth and fifth cycle deactivation of both catalysts was observed. A dramatic drop in activity was found for **6** (4th cycle, conversion 43%; 5th cycle, conversion 20%). A smaller drop in activity was observed for **7**, which was speculated to be due to the larger dendritic environment (4th cycle, conversion 91%; 5th cycle, conversion 83%). Determination of the turn-over numbers of both systems should give a conclusive support for this dendritic effect.

The studies on catalysis in interphases have taught us several lessons. The most important points are that the catalyst must be connected to the insoluble support via a sufficiently long spacer enabling good mixing with the solution that contains the reactants. The spacer should consist of solvophylic material and should not adsorb to the surface. The second point is that the properties of the insoluble support can be fine-tuned by using organic network modifiers. This enables better penetration of the solvents, making the catalyst more accessible to the substrates. These properties can be measured by NMR techniques, but only a limited number of studies have been reported in which these detailed studies have been combined with catalysis. Therefore, it is very hard to judge if examples in the literature meet the requirements for the classification "catalysis in interphase", which makes the border between immobilised catalysts and these systems vague. An important question still to be answered is what the properties of the matrices are with respect to thermal and physical stability. How these materials behave in fixed bed reactors is still a question to be answered.

3.6 Ordered Mesoporous Support

Most examples discussed so far made use of amorphous inorganic supports or sol-gel processed hybrid polymers. Highly disperse materials have recently become accessible via standard processes and, as a result, materials with various controlled particle size, pore diameter are now available. Micelle-templated synthesis of inorganic materials leads to mesoporous materials such as MCM-41, MCM-48, MSU, and these have been extensively used as solid supports for catalysis [52]. Modifications of the polarity of the material can increase the reactivity of the embedded centre, or can decrease its susceptibility to deactivation. In rare cases, enhanced stereo- or even enantioselectivities have resulted from complex immobilisation, via mechanisms related to the "confinement concept" [53]. Accurate control of pore shapes and dimensions is needed to control precisely the environment of the active site, and it is clear that for such aims zeolites and ordered mesoporous molecular sieves offer much better perspectives than relatively disordered, amorphous materials such as silica or alumina. In the case of ordered mesoporous materials, there are several illustrations of the effect of pore diameter on the reactions of embedded complexes [52].

Mesoporous supports have been studied in ligand-free rhodium-catalysed hydroformylation aiming at the stabilisation of specific catalytic species to induce stereochemical control of the regioselectivity. Among the procedures for preparation of Rh-zeolites, dilute aqueous exchange at room temperature with $RhCl_3$ or $[Rh(NH_3)_5Cl]Cl_2$ is most popular [54]. In large-pore zeolites such as Y and mordenite, Rh can be uniformly distributed across the zeolites, while medium pore-size zeolites (ZSM-34 and ZSM-11) show rhodium enrichment at the surface due to hydrolysis. In small pore zeolites (e.g., erionite, zeolite A, ZK-5) surface hydrolysis dominates, and most rhodium is found on the external surface. Depending on the pretreatments of these zeolites, e.g., calcination, autoreduction, reduction at various temperatures, under H_2 or CO, various species can be formed, e.g., lattice-coordinated Rh^{3+}, RhO_2, or Rh_2O_3 particles, metallic Rh^0 clusters, or carbonyl complexes such as $Rh^I(CO)_2$ or $[Rh_6(CO)_{16}]$. These zeolites have been applied for gas and liquid-phase reactions [55]. As an example, gas-phase hydroformylation of 1-hexene was performed with a zeolite CaA synthesised in the presence of Rh. Enhanced selectivity for the linear aldehyde was observed in comparison with an exchanged zeolite, but loss of volatile Rh-carbonyls resulted in a gradually decreasing activity. Liquid phase 1-hexene hydroformylation was investigated with $[Rh(NH_3)_5Cl]^{2+}$ exchanged NaY zeolite after carbonylation treatment. This material contains $[Rh_6(CO)_{16}]$ clusters associated with the zeolite crystals. This catalyst was characterised by a high hydroformylation selectivity and a lack of regioselectivity. This is very similar to the behavior of ligand free rhodium carbonyl complexes, which were indeed present in the reaction mixture. While for Rh^{III}-exchanged NaY and CaA the observed activity is caused by eluted rhodium, experiments with poisons and additives suggested that in materials with NaX zeolites exchanged with $[Rh(NH_3)_5(H_2O)Cl]Cl_2$ the activity was from homogeneous as well as heterogeneous sources. At high reaction temperatures, leaching from the zeolite decreased, possibly because of the formation of larger Rh clusters that are not easily converted into soluble species [56]. Unfortunately, under such conditions, the chemoselectivity is largely in favor of substrate isomerisation, probably catalysed by residual zeolite acidity. We can tentatively conclude that rhodium carbonyls entrapped in zeolites are not an ideal choice for immobilisation since the neutral character of the active species results in severe leaching due to weak binding to the zeolite.

In a recent contribution of Chaudhari et al. several mesoporous and microporous materials were used to encapsulate $[HRh(CO)(PPh_3)_3]$ in the channels of the materials [57]. They functionalised MCM-41and MCM-48 at the inner wall with aminopropyl groups and at the outer wall with diphenylsilane moieties (Figure 3.11). As a reference support MCM-48 with aminopropyl groups at the outside of the material was prepared. The materials obtained were characterised in detail using CP-MAS NMR, FT-IR, TEM, XPS and powder XRD. The catalytic performance was tested in the hydroformylation of styrene, 1-hexene, 1-octene and 1-dodecene. Generally the activity obtained was high, whereas the selectivity was moderate to low. The effect of the location of the catalyst was high: the activity of the MCM-48 material with the catalyst at the outer wall was higher (reported TOF=583 mol(styrene) $mol(cat)^{-1}$ h^{-1}, at full conversion, 100 °C) then that of the encapsulated material (TOF=325 mol(styrene) $mol(cat)^{-1}$ h^{-1}) and the supported catalysts showed similar activity (l:b=0.41). Importantly, the encapsulated catalyst showed significantly less leaching. Typically, the total amount of rhodium leaching after six consecutive cycles was <0.05% for the encapsulated system, whereas the material with the catalyst at the outer wall showed around 6% leaching. In addition, the selectivity and the activity were retained indicating that the encapsulated catalyst was also stable.

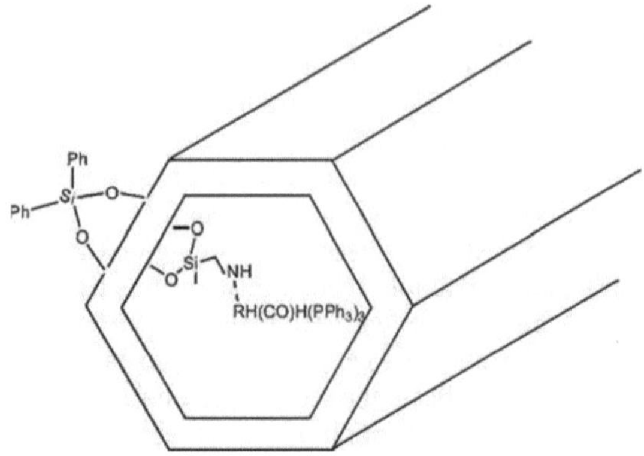

Figure 3.11. Schematic representation of MCM materials functionalised at the outside with diphenylsilane and with the rhodium catalyst encapsulated at the inside of the pores

TABLE 3.6. Comparison of the activity and selectivity in the hydroformylation of linear alkenes using homogeneous $[HRh(CO)(PPh_3)_3]$ and $[HRh(CO)(PPh_3)_3]$ encapsulated in MCM-48 [a]

Catalyst	Substrate	Substrate concentr. [b]	selectivity aldehyde (%)	Aver. TOF[c]	l:b ratio	1-Octene conv. (%)
$[HRh(CO)(PPh_3)_3]$	1-hexene	3.20	98.8	2467	2.6	98.8
Rh-MCM-48	1-hexene	0.64	99.5	338	2.3	100
$[HRh(CO)(PPh_3)_3]$	1-octene	2.55	98.2	1391	2.4	98.8
Rh-MCM-48	1-octene	0.51	99.4	270	1.7	100
$[HRh(CO)(PPh_3)_3]$	1-decene	2.11	92.2	961	1.4	98.2
Rh-MCM-48	1-decene	0.42	99.2	109	1.5	99
$[HRh(CO)(PPh_3)_3]$	1-dodecene	1.80	89.1	630	0.8	97.4
Rh-MCM-48	1-dodecene	0.36	98.6	88	1.0	99

[a] Reaction conditions: catalysts, 0.96 g/L $[HRh(CO)(PPh_3)_3]$ or 8 g/L Rh-MCM-48 (containing 0.69 wt/wt % Rh); P_{CO}, P_{H2} = 2.04 MPa; T=100°C; Toluene. [b] mol/L [c]Average turnover frequencies were calculated as (mol alkene/mol catalyst)$^{-1}$/h^{-1}.

The activity of the encapsulated catalyst was also compared to that of the homogeneous analogue (Table 3.6). Although it is hard to compare the difference in activity because the reactions were performed at different substrate concentrations and the TOF was determined at too high conversions, it seems that the activity of the MCM-48 encapsulated catalyst is lower then the homogeneous analogue. It is noteworthy that the selectivity of the encapsulated catalyst is different from that of its homogeneous counterpart. For 1-hexene the l:b is lower whereas that for 1-dodecene is higher, which was attributed by the authors to the geometric constraints of the substrates accessing the complex inside the support.

3.7 Non-covalently Supported Catalysts

One of the only examples of a commercial process using immobilised homogeneous catalysts comprises an anionic rhodium complex $[RhI_2(CO)_2]^-$ that is bound via ionic interactions to an ion exchange resin [3] and is used for the carbonylation of methanol.

The metal leached from the support is re-adsorbed to a guard bed of ion exchange material. This shows that the concept of noncovalent anchoring is very attractive, especially if the homogeneous catalyst needs no further modification.

Augustine et al. reported a similar approach in which cationic rhodium complexes were immobilised on support via ionic interactions [58]. Heteropoly acids such as phosphotungstic acid (PTA) appeared to be very suitable materials for this purpose, but alumina was also demonstrated to work well. Upon mixing a cationic rhodium complex such as [Rh(dipamp)(COD)]BF_4 with the PTA material, a supported chiral rhodium catalyst was obtained that was active in the asymmetric hydrogenation of methyl 2-acetamidoacrylate. Interestingly, the activity and ee were higher (after the first run) than that of the homogeneous parent complex. The degree to which the nature of the HPA influences the performance of the reaction suggests that a direct interaction between the HPA and the metal is responsible for the increased performance of the catalyst upon immobilisation [59]. The catalyst was recycled 15 times without loss in performance. Analysis of the product mixtures revealed that no rhodium leached from the support (detection limit 1 ppm). The major advantage of this approach is its generality and several pre-formed rhodium catalysts were immobilised in a similar manner. The strategy is limited to ionic catalysts and it can therefore not be used to immobilise neutral hydroformylation catalysts.

Bianchini reported a fascinating strategy, denoted as supported hydrogen-bonded catalysts (SHB-catalysts), in which the immobilisation procedure involved the grafting of tailor-made catalysts equipped with a sulfonate tail on high surface area silica (Figure 3.12) [60]. The interaction between the catalyst and the support is based on hydrogen bonds between the silanols of the surface and the oxygens of the sulfonate group of the ligand. The existence of these interactions was supported by IR, EXAFS and CP MAS ^{31}P NMR studies.

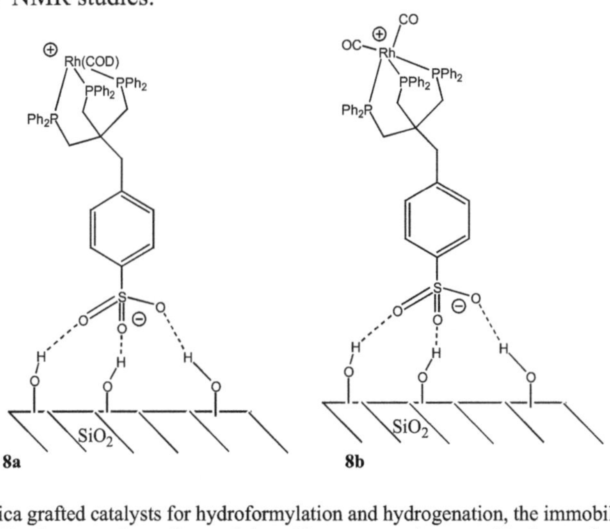

8a **8b**

Figure 3.12 .Silica grafted catalysts for hydroformylation and hydrogenation, the immobilisation is based on hydrogen bonding between the surface and the sulfonate group (supported hydrogen-bonded (SHB) catalysts)

The grafting procedure is straightforward, which is a major advantage of this method. The silica was pretreated to remove water from the support, after which it was added to a dichloromethane solution containing the catalyst. After 6 hours of mixing the supported catalyst was isolated by filtration. The remaining solvent was colourless while the supported had turned yellow. Catalyst **8a** was a good catalyst for the hydrogenation of styrene, 100 % conversion was reached after 3 hours using 1 mol % of catalyst, 120 °C and 30 bar of molecular hydrogen. Catalyst **8a** was also active in the hydroformylation of 1-hexene. Using 1 mol% of catalyst at 120 °C and 30 bar of syngas the substrate was completely converted in three hours. The branched aldehyde was the main product (l:b=0.8, 41 % branched) and also some 2-ethylpentanal (15 %) and hydrogenated product (4 %) were formed. No rhodium was detected in the product phase by atomic absorption analysis, and the recycled catalyst was reused for both hydrogenation and hydroformylation giving identical results as the first runs. The results obtained with the supported catalysts **8** were different from those obtained in the homogeneous phase and in two-phase processes. In THF these catalysts were less chemoselective, hydrogenation being the side reaction, and showed a lower activity. The transformation of the mononuclear rhodium complex into binuclear species was proposed to be responsible for the lower activity in the homogeneous phase, and the better performance of the SHB catalysts was ascribed to the site isolation of the active sites preventing the formation of such dimers. Attempts to use these catalysts in solid-gas hydroformylation reactions failed, whereas solid-gas hydrogenation of ethene was successful. One of the additional advantages of this approach is that the catalyst can be removed from the support by washing with methanol, which enables detailed analysis of the complex after the reaction.

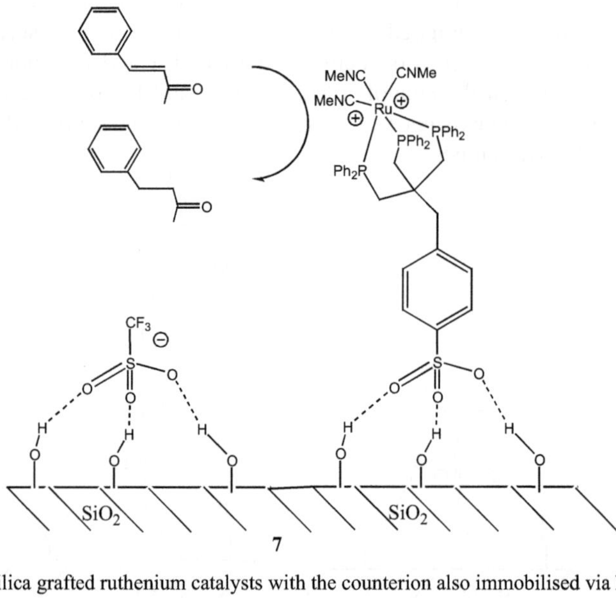

7

Figure 3.13. Silica grafted ruthenium catalysts with the counterion also immobilised via hydrogen bonds

In subsequent work the approach was extended to dicationic ruthenium complexes that were active in hydrogenation [61]. The complex is now no longer zwitterionic and the counterion is also bound to the support via hydrogen bonds (Figure 3.13) and therefore the metal complex should be more strongly bound to the support. The hydrogenation of trans-4-phenyl-3-buten-2-one (benzylideneacetone) was more chemoselective with the supported catalyst than the homogeneous parent catalyst, at the expense of activity (factor 14). The supported catalyst yielded mainly benzylacetone as the product, whereas the homogeneous catalyst gave 100 % of the saturated alcohol. The work has been extended to asymmetric hydrogenations, but in at least half of the examples the enantioselectivity was lower than that of the homogeneous analogue [62].

Figure 3.13 already suggests that the use of sulfonated ligands is not required for this approach. Indeed Rege et al. showed that chiral cationic rhodium complexes can be immobilised on silica by using triflate as counterion that forms hydrogen bonds with the support [63]. In fact, this strategy is a hybrid between the two previously described approaches. By stirring a dichloromethane solution containing [(R,R)-Me-(DuPHOS)Rh(COD)]OTf in the presence of MCM-41 the catalyst was immobilised. Powder X-ray diffraction revealed that the MCM-41 structure remained unchanged, whereas elemental analysis showed that the material contained 1 % (wt) rhodium, consistent with the theoretical amount. The BET surface area decreased from 953 m^2g^{-1} to 854 m^2g^{-1}. Interestingly, the mesopore volume decreased indicating that the catalyst was in the pores of the material. The performance of the immobilised catalyst equalled that of the homogeneous analogue and in some examples the selectivity and activity was even better. The catalyst was recycled with retention of catalyst performance. No active catalyst leached from the support, and the catalyst was shown to be stable for 16 h under hydrogen-free conditions.

3.8 Supported Aqueous Phase Catalysis

Davis and Hanson developed a new concept of immobilising homogeneous catalysts, denoted as supported aqueous phase catalysts (SAPC) [64]. They reasoned that in aqueous biphasic catalysis the reaction mainly takes place at the interface. In order to increase this interface they used a high-surface-area hydrophilic support (Figure 3.14). These materials have a thin film of water adhered to the surface, in which the water-soluble catalyst is dissolved. The substrate and the product are in an organic solvent such as toluene, and reaction takes place at the water-organic interface. The supported catalyst has a very large interfacial area, which results in very efficient catalysis for organic substrates. Furthermore, the catalyst stays completely on the support.

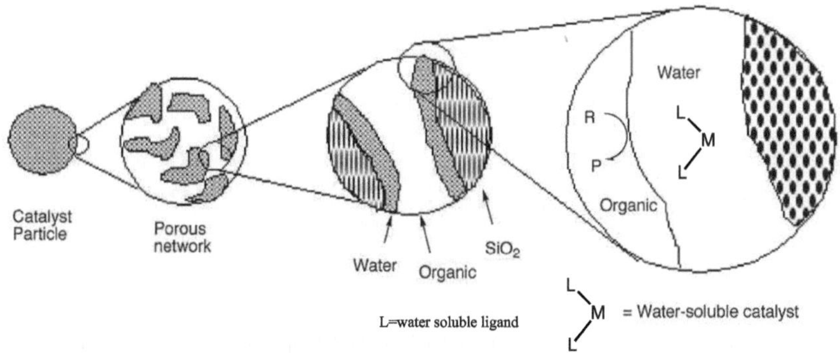

Figure 3.14. Schematic representation of the concept of supported aqueous phase catalysis

The important issues are the generality of the concept, the robustness of the catalyst system and the influence of the thickness of the water layer [65]. This water layer has an enormous impact on the catalytic activity (Figure 3.15). It was observed that if the layer is too thin the activity of the catalyst is much lower due to a decrease of the catalyst mobility. The catalyst is bound to the silica resulting in a heterogeneous system. This was verified by ^{31}P NMR spin relaxation measurements, which show that the spin relaxation declines with increasing water content [66]. It was observed in the hydroformylation of 1-heptene using a [HRh(CO)(TPPTS)$_3$-SAPC] that the TOF increases by a factor as high as 100 when going from 2.9 to 9 wt. % water (Table 3.7). If higher water contents were adhered to the support the water layer became too thick and the substrate had to diffuse into the water layer, or the catalyst had to diffuse to the interface. The result is a decrease in catalyst-product contact time leading to lower activities. This sensitivity towards water is a drawback of this otherwise attractive concept, since it requires an additional parameter to be controlled.

TABLE 3.7. Hydroformylation performance of SAPC compared with homogeneous and biphasic systems

Catalyst system	Substrate	T (°C)	TOF (mol mol^{-1} h^{-1})
SAPC (2.9 wt.% H_2O)[a]	1-heptene	75	0.75
SAPC (9 wt.% H_2O)[a]	1-heptene	75	72
Homogeneous[b]	1-heptene	75	288
SAPC[c]	Mixture[h]	100	432[e]/432[f]/396[g]
Homogeneous[d]	Mixture[h]	100	1656[e]/1800[f]/1800[g]
Biphasic[c]	Mixture[h]	125	17[e]/5[f]/1[g]

[a] [HRhCO(TPPTS)$_3$], P=7 bar (H_2/CO=1:1) [16d]. [b] [HRhCO(PPh$_3$)$_3$] in toluene, P=7 bar (H_2/CO=1:1) [16b]. [c] [HRhCO(TPPTS)$_3$], P=51 bar (H_2/CO=1:1) [18]. [d] [HRhCO(PPh$_3$)$_3$] in hexane, P= 51 bar (H_2/CO=1:1) [18]. [h] The substrate was a 1:1:1 mixture of 1-hexene, 1-octene and 1-decene. [e] Heptanals. [f] Nonanals. [g] Undecanals.

Horváth performed experiments using substrates with different solubilities in water and showed that, under optimal conditions, this solubility did not influence the activity [67]. These experiments clearly support the fact that the reaction takes place at the organic-water interphase. Furthermore, he performed a hydroformylation reaction in a continuous system and even under reaction conditions no leaching of rhodium complex was detected. Water obviously leaches if the SAPC is used in a continuous flow system, which in a practical application should be compensated for by using water-saturated organic solvents.

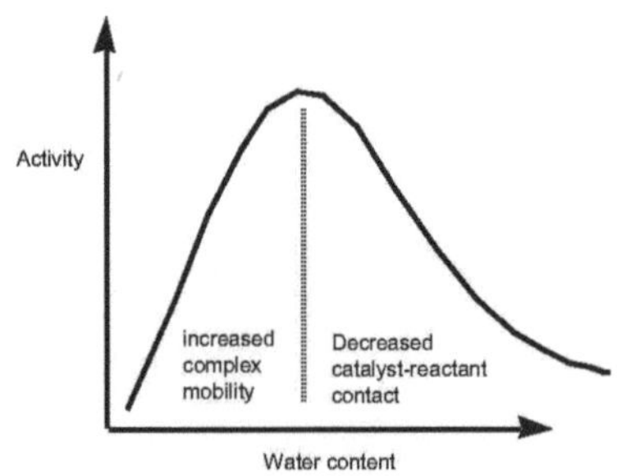

Figure 3.15. The influence of the water contents of the hydrophilic support on the relative catalytic activity

A water-soluble chelating diphosphine ligand (**9**) based on the xanthene backbone was also studied as supported aqueous phase catalysts. It was shown that this ligand performed well as SAPC since it is much more selective than other SAPC systems reported in literature [68]. Recycling experiments showed that these catalysts retained their activity and selectivity for at least ten consecutive runs, whereas under similar conditions the TPPTS based catalyst showed a reduced performance in the fourth run.

This shows that the strong chelating effect of the bidentate ligand efficiently retains the metal attached to the support.

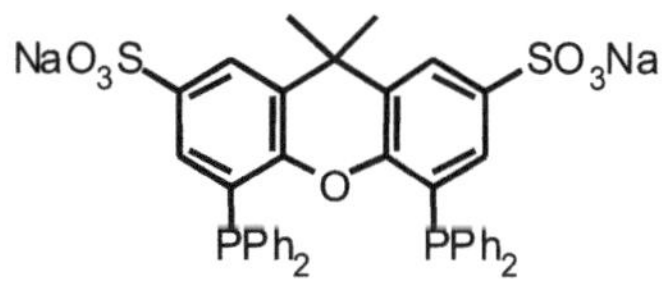

9

Mortreux and co-workers compared the activity of the SAPC catalysts with that of the homogeneous analogue in the hydroformylation of methyl acrylate [69]. They observed an activity for the SAPC that was strongly dependent on the amount of water present in the system. More remarkably, the optimised activity of the SAPC was higher than that of the homogeneous systems. This effect was ascribed to the polar interactions between the substrate and the silica support. This effect was not observed for nonpolar substrates such as propene, which supported the hypothesis.

In order to show the versatility of the method Davis extended the concept to other hydrophilic liquids such as ethylene glycol and glycerol [70]. The reactions then take place at the hydrophilic-hydrophobic liquid interface. In this specific example the supported-phase concept was used for asymmetric reduction using a ruthenium catalyst.

3.9 Process Design [71]

A process for the hydroformylation of 1-octene to nonanal was designed for an immobilised homogeneous catalyst. The production capacity was fixed at 100 kton of nonanal. Kinetic data reported for the rhodium catalyst complex of N-(3-trimethoxysilane-n-propyl)-4,5-bis(diphenylphosphino)-phenoxazine immobilised on silica, (**2**) was used as a starting point. Other process specifications are given in Table 3.8.

TABLE 3.8. Process and reactor specifications for hydroformylation of 1-octene

Operating pressure (bar)	50
Operating temperature (°C)	80
Syngas / 1- octene ratio	10:1
Catalyst: Rh complex with **2** immobilised on silica	0.1 wt%
1-octene conversion per pass	70%
Selectivity toward n-nonanal	93%
by-products: iso-nonanal / 1-nonanol	4% / 3%

A systematic process design procedure going through the levels of black box, conceptual design, and index flow sheet design, resulted finally in the process flow sheet (PFD) shown in Figure 3.16, which was simulated in Aspen Plus. In this process pure 1-octene (#1) is mixed with a recycle stream of 1-octene (#31), pressurized to 50 bar in pump P-101 and heated to 80 °C in E-102 before it enters reactor R-101. Pure syngas (CO: H_2= 1:1) is mixed with a recycle stream of syngas (#21), pressurised in compressor C-101 to 50 bar, mixed with syngas recycle stream (#14) and heated in E-

101 to 80 °C before entering the reactor R-101. In reactor R-101 the hydroformylation reaction of 1-octene with syngas to form nonanal takes place (for details see Table 3.8). The reactor consists of a packed bed in which large amounts of syngas is contacted with the liquid 1-octene. It is assumed that the operationing conditions are such that the reaction is kinetically controlled. The proposed first order kinetics in 1-octene favour the plug flow character for the reactor. Since the reaction is also highly exothermic the reactor needs to be effectively cooled to keep the reaction temperature at 80 °C. Combined this leads to a co-current trickle flow reactor column with in-situ cooling. The reaction product stream (#10) is cooled in E-103 to 40 °C and is fed to a gas-liquid separator V-103. The gas stream (#12) containing mainly CO, H_2 and a minor amount of 1-octene is recycled after purging 0.01% (#13) and recompressing in C-102. The liquid stream (#15) is expanded to 3 bar, heated in E-104 to 60 °C before separating the gas from the liquid in V-104. The gas stream (#19) is recycled after purging a small part of the stream. The liquid product stream (#22) is depressurised to 1 bar, heated in E-106 to 100 °C and separated in distillation columns T-101 and T-102. Column T-101 removes the compounds like 1-octene and minor amounts of CO and H_2 which are lighter than nonanal. These lights (#25) are cooled in E-106 to 25 °C, the 1-octene condenses and is separated from the gas stream in gas-liquid separator V-105. The gas stream (#27) is purged. The liquid stream (#29) with 98%+ 1-octene is recycled after purging 0.01% and is mixed with fresh 1-octene. Finally, Column T-102 separates the product, the nonanal mixture (#33, 95%+ nonanal, 4% 2-Me octanal) from the rest which is mainly 1-nonanol. At this stage nonanal and 2-Me octanal are not separated but sold as a mixture. The bottom stream of column T-102 contains nonanol, heavies and Rh leached from the reactor. This stream should be treated to recover the lost rhodium. Details of the different process units are shown in Table 3.9.

TABLE 3.9. Specifications of different units in the 1-octene hydroformylation process as shown in Figure 3.16.

Heat exchangers	**E-101**	**E-102**	**E-103**	**E-104**	**E-105**	**E-106**
Duty (kW)	809	458	-1280	215	455	-669
U (W/m²,C)	165	600	625	600	600	600
A (m²)	108	12	93	6	21	59
Heat exchangers	**E-107**	**E-108**	**E-109**	**E-110**		
Duty (kW)	-230	1590	-2190	3290		
U (W/m²,C)	600	600	600	600		
A (m²)	4	27	36	55		
Flash Drum	**V-103**	**V-104**	**V-105**			
orientation	Vert.	Horz.	Horz.			
Diameter (m)	1.0	0.3	0.25			
Height (m)	4.6	1.2	1.2			
Compressor	**C-101**	**C-102**	**C-103**	**Pump**		**P-101**
P at inlet (bar)	3	45			P at inlet (bar)	1
P at outlet (bar)	50	50			P outlet (bar)	50
Flow (m3/hr at P_{inlet})	1640	1430			Flow (m3/hr)	21
Power (kW)	630	270			Power (kW)	54
Efficiency	72%	72%			Efficiency	52 %
Distillation columns	**T-101**	**T-102**	**Reactor**			**R-101**
Reflux ratio R	0.5	2	Type: Gas Liquid Solid Reactor			
Number of trays	13	46	Pressure (bar)			50
Condensor duty (MW)	-0.23	-2.2	Temperature (C)			80
Reboiler duty (MW)	1.6	3.3	Catalyst volume (m³)			70*
Diameter (m)	0.83	1.42	Amount of Rh (kg)			70
Height (m)	7.5	24	Duty (MW)			-3.6

* The catalyst volume can be significantly reduced by increasing the Rh concentration on the catalyst. In case of 0.4 wt % Rh the catalyst volume will be 18 m³.

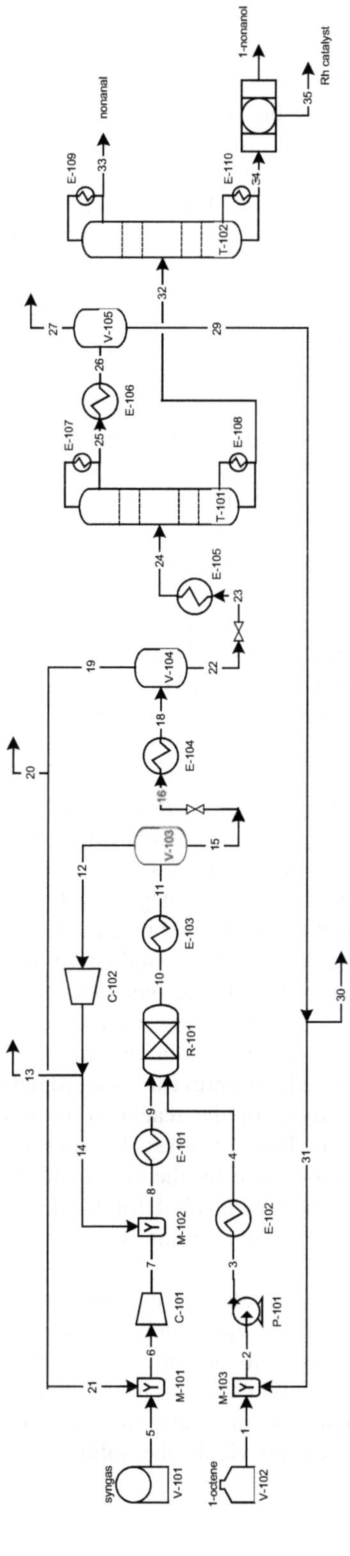

Figure 3.16. Process flow sheet of hydroformylation process of 1-octene to 1-nonanal

For the supported catalyst it is expected that the ligand does not leach since it is chemically bonded to the carrier. In contrast, the rhodium metal bound to the ligand is subject to leaching due to the reversible nature of the complex formation. The amount will depend on the equilibrium between rhodium dissolved in the organic phase and that bound to the ligand. When an equilibrium concentration of 10 ppb Rh is attained, the yearly loss of Rh for a 100 kton production plant will be about 1 kg Rh per year. Compared to the reactor contents of rhodium (see Table 3.9, 70 kg Rh) this would result in a loss of 1.5% of the inventory per year, which would be acceptable.

3.10 Concluding Remarks

The immobilisation of a catalyst on a support is a suitable solution for the separation problem and solutions exist nowadays for most of the problems that prevent the step to commercialisation. Metal leaching from the support has been the major problem for a long time. From the examples given in this chapter it is clear that bidentate ligands are required to retain the metal on the support, and also the use of an excess ligand compared to metal does suppress metal leaching. Most experiments were run as batch reactions, and these are not necessarily representative of continuous processes. Leaching for several cases was below the detection limit, which is roughly 0.1-1 ppm of rhodium in the liquid product. As indicated, for an industrial application for low-value products, this number should perhaps be two orders of magnitude lower and more accurate measurements are required to show that metal leaching is even as low as 1-10 ppb. For high-value products ppm levels of metal losses are economically acceptable, but health, safety and environmental issues should also be taken into consideration. With respect to catalyst anchoring one can conclude that catalysts for hydroformylation cannot be immobilised via ionic interactions between the metal and the support since the active species comprises a neutral species. For cationic (rhodium) catalysts such as those for hydrogenation, this is a proper strategy which is very easy to apply.

With respect to the selectivity of immobilised catalysts one can conclude that the selectivity is generally dominated by the ligand coordination, which remains identical to the homogeneous analogue if chelating bidentate ligands are used. This is clearly demonstrated by the Xantphos type ligands and the BINAPHOS ligands. Upon using monodentate ligands, there is a risk of obtaining monophosphine coordinated species since site-isolation of the ligands can prevent biscoordination. The acidity of the silica can change the chemoselectivity of the reactions as was observed for immobilised Xantphos-rhodium hydroformylation catalyst. Detailed analysis showed that cationic rhodium complexes were responsible for the effect and the formation of these species was suppressed by the addition of alcohol (or triethyl amine) to the liquid phase. Immobilisation of catalysts generally results in a lower activity. This effect can be minimised in several ways.

The lesson learned from the "catalysis in interphase" approach is that the spacer between the insoluble support and the actual catalyst should be sufficiently long and soluble in the solvent of interest to obtain active catalysts. The use of supercritical fluids can also be very beneficial for the activity. Upon using Xantphos immobilised on silica in scCO2 for example, the rates are only half of those of the homogeneous catalyst. Expressed as space-time yields the solid catalysts are almost an order of magnitude faster, admittedly neither of them has been optimised. Thus, there is clearly a benefit in the supercritical system, which has to be weighed against the extra costs of the high pressures involved. The $scCO_2$ runs, however, were continuous runs and the losses over longer periods were below 0.2 %, which was again the detection limit and the actual result may be better than this. Another interesting feature that has so far received little attention is the fast heat transfer in $scCO_2$, in addition to the fast mass transfer. The hydroformylation reaction is highly exothermic and when the rates start approaching those of the homogeneous systems this certainly becomes a limiting factor for heterogeneous hydroformylation catalysts. For lower alkenes the starting alkenes could form the supercritical phase as well, thus reducing the extra costs involved in the high pressures.

For several silica-supported catalysts in condensed phase, including the SAPC system, the rates are disappointing. This can be assigned to slow mass transfer, and perhaps to incomplete rhodium hydride formation as we have discussed and observed. The sol-gel catalyst is relatively fast and is sometimes only a few times slower than the homogeneous one. Since only limited ways of preparation were tested, there is probably more scope for sol-gel catalysts. Space-time yields are promising at the present state of affairs.

The limited mechanical strength and the swelling properties of the support are not necessarily a problem. Silica and alumina have proven their value as supports in various reactor set-ups. The example given by Nozaki clearly shows that polystyrene can also be used in well designed reactors. Considering the recent progress in the field we are confident that the problems related to immobilised catalysts no longer present limitations that prevent commercialisation of processes using these materials.

3.11 References

[1] (a) W.O. Haag, and D. D. Whitehurst, *German Patent* 1969, 1.800.371 and 1.800.379 and 1.800.380. (b) W.O. Haag, and D. D. Whitehurst, *Belgian Patent* 1969, 721.686.

[2] (a) F.R. Hartley, P.N. Vezey, *Adv. Organomet. Chem.*, 1977, **15** 189. (b) F.R. Hartley, 'Supported Metal Complexes. A New Generation of Catalysts,' Reidel, Dordrecht, 1985. (c) Y. Iwasawa, 'Tailored Metal Catalysts (Ed. R. Ugo, B.R. James,)', Reidel, Dordrecht, 1986. (d) W. Keim, B. Driessen-Hölscher, in 'Handbook of heterogeneous catalysis,' vol 1 G. Ertl, H. Knözinger, J. Weitkamp (Ed.), Wiley-VCH, Weinhein, 1997, p. 231. (e) R. H. Grubbs, *CHEMTECH*, 1977, 512. (f) C. C. Leznoff, *Chem. Soc. Rev.* 1974, **3**, 65.

[3] H. Sugiyama, F. Uemura, N. Yoneda, T. Minami, T. Maejima, K. Hamato, *Japan patent*, 235250 1997.

[4] (a) R. Thomas, *J. Soc. of Chem. Ind.*, 1923, 42. (b) Th. Sabalitschka, W. Moses, *Ber.*, 1927, **60B** 786.

[5] R. B. Merrifield, *Science,* 1965, **150**, 178.

[6] (a) N. Takahashi, I. Okura, T. Keij, *J. Am. Chem. Soc.* 1975, **97,** 7489 (b) G.J.K. Acres, G.C. Bond, B.J. Cooper, J.A. Dawson *J. Catal.* 1966, **6**, 139. (c) P.R. Rony *Chem. Eng. Sci.* 1968, **23**, 1021. (d) K.K. Robinson, F.E. Paulik, A, Hershman, J.F. Roth *J. Catal.* 1969, **15**, 245. (d) L.A. Gerritsen, J.M. Herman, J.J.F. Scholten *J. Mol Catal.* 1980, **9**, 241. (e) Th. G. Spek, J.J.F. Scholten *J. Mol. Catal.* 1977, **3**, 81. (f) J. Hjortkjear, M.S. Scurrell, P. Simonsen *J. Mol. Catal.* 1979, **6**, 405.

[7] E. Bayer, M. Mutter, *Nature,* 1972, **237**, 512.

[8] (a) U. Deschler, P. Kleinschmit, P. Panster, *Angew. Chem. Int. Ed. Engl.* 1986, **25**, 236. (b) S. Wieland, P. Panster *Cat. Org. React.* 1994, **62**, 383.

[9] (a) D. J. Cole-Hamilton, *Science,* 2003, **299**, 1702. (b) Thematic issue *Chem. Rev.* 2002, **102**, vol 10 (c) Y.R. de Miguel, E. Brulé, R. G. Margue, J. *Chem. Soc., Perkin Trans. 1,* 2001, 3085. (d) S. V. Ley, I. R. Baxendale, R. N. Bream, P. S. Jackson, A. G. Leach, D. A. Longbottom, M. Nesi, J. S. Scott, R. I. Storer and S. J. Taylor, *J. Chem. Soc., Perkin Trans. 1*, 2000, 3815. (e) R. T. Baker, W. Tumas, *Science,* 1999, **284**, 1477. (f) S. Kobayashi, R. Akiyama, *Chem Commun,* 2003, 449. (g) *J. Mol Catal A.* 2001, **177**, thematic issue.

[10] (a) B. Corain, M. Zecca, Jeřabek, *J. Mol Catal. A.* 2001, **177**, 3. (b) U. Kunz, C. Altwicker, U. Limbeck, U. Hoffmann, *J. Mol Catal. A.* 2001, **177**, 21.

[11] (a) S. Bräse, F. lauterwasser, R. E. Ziegert, *Adv. Synth. Catal.* 2003, **345**, 869. (b) Q-H. Fan, Y.-M. Li, A. S. C. Chan, *Chem. Rev.* 2002, **102**, 3385.

[12] Z.-L. Lu, E. Lindner, H. A. Mayer, *Chem. Rev.* 2002, **102**, 3543.

[13] (a) C. A. Mcnamara, M. J. Dixon, M. Bradley, *Chem. Rev.* 2002, **102**, 3275. (b) N. E. Leadbeater, M. Marco, *Chem. Rev.* 2002, **102**, 3217.

[14] (a) R. H. Grubbs, L. C. Kroll, *J. Am. Chem. Soc.* 1971, **93,** 3062. (b) R. H. Grubbs, S.-C. H. Su, *J. Organom. Chem.* 1976, **122**, 151 (c) R. H. Grubbs, E. W. Sweet, *J. Mol Catal.* 1977, **3**, 259.

[15] C. U. Pittman Jr., R. M. Hanes, *J. Am. Chem. Soc.* 1975, **98,** 5402.

[16] (a) F. R. Hartley, S. G. Murray, P. N. Nicholson, *J. Mol Catal.* 1982, **16**, 363. (b)

[17] C. U. Pittman Jr., A. Hirao, *J. Org. Chem.* 1978, **43,** 640.

[18] P.W.N.M. van Leeuwen, P.C.J. Kamer, J.N.H. Reek, and P. Dierkes, *Chem. Rev.* 2000, **100**, 2741.

[19] (a) K. Nozaki, Y. Itoi, F. Shibahara, E. Shirakawa, T. Ohta, H. Takaya, T. Hiyama, *J. Am. Chem. Soc.* 1998, **120**, 4051. (b) K. Nozaki, F. Shibahara, Y. Itoi, E. Shirakawa, T. Ohta, H. Takaya, T. Hiyama, *Bull. Chem. Soc. Jpn.* 1999, **72**, 1911.

[20] F. Shibahara, K. Nozaki, T. Hiyama, *J. Am. Chem. Soc.* 2003, **125**, 8555.

[21] (a) K. G. Allum, R. D. Hancock, I. V. Howell, S. Mckenzie, R. C. Pitkethly, P. J. Robinson, *J. Organomet. Chem.* 1975, **87**, 203. (b)) K. G. Allum, R. D. Hancock, I. V. Howell, T.E. Lester, S. Mckenzie, R. C. Pitkethly, P. J. Robinson, *J. Catal.* 1976, **43**, 331. (c) J. P. Collman, J. A. Belmont, J. I. Brauman, *J. Am. Chem. Soc.* 1983, **105**, 7288.

[22] M.O. Farrell, C.H. van Dyke, L.J. Boucher, S.J. Metlin, *J. Organomet. Chem.* 1979, **172**, 367.

[23] U. Deschler, P. Kleinschmit, P. Panster, *Angew. Chem. Int. Ed. Engl.* 1986, **25**, 236.

[24] (a) J. Blum, D. Avnir, H. Schumann, H. *CHEMTECH* 1999, **32**. (b) U. Schubert, *New J. Chem.* 1994, **18**, 1049. (c) J. Blum, A. Rosenfeld, N. Polak, O. Israelson, H. Schumann, D. Avnir, *J. Mol. Catal. A,* 1996, **107**, 217. (d) E. Lindner, M. Kemmler, H.A. Mayer, P. Weger, *J. Am. Chem. Soc.* 1994, **116**, 348. (e) C.M. Ingersoll, F.V. Bright *CHEMTECH* 1997, 26.

[25] S. Wieland, P. Panster *Cat. Org. React.* 1994, **62**, 383.

[26] R. H. Grubbs, C. Gibbons, L. C. Kroll, W. D. Bonds, Jr., C. H. Brubaker, Jr., *J. Am. Chem. Soc.* 1973, **95**, 2373.

[27] A.J. Sandee, L.A. van der Veen, J.N.H. Reek, P.C.J. Kamer, M. Lutz, A.L. Spek, P.W.N.M. van Leeuwen, *Angew. Chem. Int. Ed. Engl.* 1999, **38** 3231.

[28] A.J. Sandee, J.N.H. Reek, P.C.J. Kamer, P.W.N.M. van Leeuwen, *J. Am. Chem. Soc.* 2001, **123**, 8468.

[29] L. Bemi, J.A. Davies, C.A. Fyfe, R.E. Wasylishen, *J. Am. Chem. Soc.* 1982, **104**, 438. (b) J.-Z. Yao, Y.-Y. Chen, B.-S. Tian, J. *Organomet. Chem.* 1997, **534**, 51.

[30] P.W.N.M. van Leeuwen, C. Claver (Ed.) Rhodium Catalyzed Hydroformylation, Kluwer Acedemic Publishers, Dordrecht, 2000.

[31] K. Kaneda, T. Imanaka, S. Teranishi, *Chem. Lett.* 1983, 1465. (b) J.-A. Anderson, A.W.S. Currie, *Chem. Commun.* 1996, 1543.

[32] The presence of triethylamine also reduced the reaction rate.

[33] R.J. Farrauto, R.M. Heck, *Kinetics and Catalysis* 1998, **39**, 594.

[34] A. Cybulski, J.A. Moulijn, Structured catalysts and reactors (1998) New York: Marcel Dekker.

[35] H.A. Smits, A. Stankiewicz, W.Ch. Glasz, T.H.A. Fogl, J.A. Moulijn, *Chem. Eng. Sci.* 1996, **51**, 3019.

[36] R.K. Edvinsson, M.J.J. Houterman, T. Vergunst, E. Grolman, J.A. Moulijn, *AlChe Journal* 1998, **44**, 2459.

[37] A.J. Sandee, R.S. Ubale, M. Makkee, J.N.H. Reek, P.C.J. Kamer, J.A. Moulijn, P.W.N.M. van Leeuwen, *Adv. Synth. Catal.* 2001, **343**, 201.

[38] (a) R. Noyori, *Chem. Rev.* 1999, **99**, 475. (b) J.A. Darr and M. Poliakoff, *Chem. Rev.* 1999, **99**, 45. (c) B.M. Bhanage, Y. Ikushima and M. Shirai, *Chem. Commun.* 1999, 1277. (d) A. Baiker, *Chem. Rev.* 1999, **99**, 453.

[39] (a) M. G. Hitzler and M. Poliakoff, *Chem. Commun.* 1997, 1667. (b) M.G. Hitzler, F.R. Smail, S.K. Ross and M. Poliakoff, *Org. Proc. Res. Dev. 2* 1998, 137. (c) M.G. Hitzler, F.R. Smail, S.K. Ross and M. Poliakoff, *Chem. Commun.* 1998, 359. (d) W.K. Gray, F.R. Smail, M.G. Hitzler, S.K. Ross and M. Poliakoff, *J. Am. Chem. Soc.* 1999, **121**, 10711.

[40] N.J. Meehan, A.J. Sandee, N.J.H. Reek, P.C.J. Kamer, P.W.N.M. van Leeuwen, M. Poliakoff, *Chem. Commun.* 2000, 1497.

[41] (a) J. A. Marqusee, K. A. Dill, *J. Chem. Phys.* 1986, **85**, 434. (b) J. G. Dorsey, K. A. Dill, *Chem. Rev.* 1989, **89**, 331.]

[42] E. Lindner, M. Kemmler, T. Schneller, H. A. Mayer, *Inorg. Chem.* 1995, **34**, 5489.

[43] E. Lindner, T. Schneller, F. Auer, H. Mayer, *Angew. Chem. Int. Ed.* 1999, **38**, 2154.

[44] (a) T. Jongsma, M. Fossen, G. Challa, P.W.N.M. van Leeuwen, *J. Mol. Catal.* 1993, **83**, 17. (b) T. Jongsma, H. van Aert, M. Fossen, G. Challa, P.W.N.M. van Leeuwen, *J. Mol. Catal.* 1993, **83**, 37.

[45] (a) E. Lindner, A. Bader, H. A. Mayer, *Inorg. Chem.* 1991, **30**, 3783. (b) E. Lindner, A. Jager, T. Schneller, H. A. Mayer, *Chem. Mater.* 1997, **9**, 81. (c) E. Lindner, F. Auer, A. Baumann, P. Wegner, H. A. Mayer, H. Bertagnolli, U. Reinohl, T. S. Ertel, A. Weber, *J. Mol. Catal.* 2000, **157**, 97. (d) E. Lindner, S. Brugger, S. Steinbrecher, E. Plies, M. Seiler, H. Bertagnolli, P. Wegner, H. A. Mayer, *Inorg. Chim Acta* 2002, **327**, 54.

[46] E. Lindner, M. Kemmler, H. A. Mayer, P. Wegner, *J. Am. Chem. Soc.* 1994, **116**, 348.

[47] E. Lindner, T. Salesch, S. Brugger, F. Hoehn, P. Wegner, H. A. Mayer, *J. Organomet. Chem.* 2002, **157**, 97.

[48 (a) Oosterom, G. E.; Reek, J. N. H.; Kamer, P. C. J.; van Leeuwen, P. W. N. M. *Angew. Chem. Int. Ed.* 2000, **40**, 1828. (b) R. Kreiter, A. W. Kleij, R. J. M. Klein Gebbink, G. van Koten, in "Dendrimers IV: Metal Coordination, Self-Assembly, Catalysis"; F. Vögtle, C. A. Schalley, Ed.; Springer-Verlag, Berlin, 2001, Vol. 217, p. 163. (c) D. Astruc, F. Chardac, *Chem. Rev.* 2001, **101**, 2991. (d) R. M. Crooks, M. Zhao, L. Sun, V. Chechik, L. K. Yueng, *Acc. Chem. Res.* 2001, **34**, 181. (e) L. J. Twyman, A. S. H. King, I. K. Martin, *Chem. Soc. Rev.* 2002, **31**, 69. (f) R. van Heerbeek, P. C. J. Kamer, P. W. N. M. van Leeuwen, J. N. H. Reek, *Chem. Rev.* 2002, **102**, 3717.

[49] A.S.H. King, L.J. Twyman, *J. Chem. Soc., Perkin Trans. 1,* 2002, 2209.

[50] (a) S. C. Bourque, F. Maltais, W. -J. Xiao, O. Tardif, H. Alper, P. Arya, L. E. Manzer, *J .Am. Chem. Soc.* 1999, **121**, 3035. (b) S. C. Bourque, H. Alper, *J. Am. Chem. Soc.* 2000, **122**, 956. (c) P. Arya, N. V. Rao, J. Singkhonrat, H. Alper, S. C. Bourque, L. E. Manzer, *J. Org. Chem.* 2000, **65**, 1881.

[51] (a) P. Arya, G. Panda, N. V. Rao, H. Alper, S. C. Bourque, L. E. Manzer, *J. Am. Chem. Soc.* 2001, **123**, 2889. (b) S.-M. Lu, H. Alper *J. Am. Chem. Soc.* 2003, **125**, 13126.

[52] (a) D.E. De Vos, M. Dams, B. F. Sels, P. A. Jacobs, *Chem. Rev.* 2002, **102**, 3615, and references cited therein. (b) J. M. Thomas, *Angew. Chem. Int ed.*1999, **38**, 3589.

[53] (a) J.M. Thomas, T. Maschmeyer, B. F. G. Johnson, D. S. Shepard, *J. Mol. Catal.* 1999, **141**, 139. (b) B.F.G. Johnson, S.A. Raynor, D.S. Shephard, T. Maschmeyer, J.M. Thomas, G. Sankar, S. Bromley, R. Oldroyd, L. Gladden, M.D. Mantle *Chem. Commun.* 1999, 1167.

[54] J. F. Goellner, B. C. Gates, G. N. Vayssilov, N. Rősch, *J. Am. Chem. Soc.* 2000, **122**, 8056.

[55] (a) W. A. Weber, B. C. Gates, *J. Phys. Chem. B* 1997, **101**, 10423. (b) L. F. Rao, A. Fukuoka, N. Kosugi, H. Kuroda, M. Ichikawa, *J. Phys. Chem.* 1990, **94**, 5317. (c) L. F. Rao, M. Pruski, T. S. King, *J. Phys. Chem. B* 1997, **101**, 5717 (d) E. J. Rode, M. E. Davis, B. E. Hanson, *J. Catal.* 1985, **96**, 563. (e) E. J. Rode, M. E. Davis, B. E. Hanson, *J. Catal.* 1985, **96**, 574. (f) R. J. Davis, J. A. Rossin, M. E. Davis, *J. Catal.* 1986, **98**, 477.

[56] M. E. Davis, ,P. M. Butler, J. A. Rossin, B. E. Hanson, *J. Mol. Catal.* 1985, **31**, 385.

[57] K. Mukhopadhyay, A. B. Mandale, R. V. Ghaudhari, *Chem. Mater.* 2003, **15**, 1766.

[58] (a) S. K. tanielyan, R. L. Augustine, U.S. Patent 1998, 08/994,025, PCT WO-9828074. (b) R. L. Augustine, S. K. Tanielyan, S. Anderson, H. Yang, *Chem. Commun,* 1999, 1257.

[59] R. L. Augustine, S. K. Tanielyan, N. Mahata, Y. Gao, A. Zsigmond, H. Yang, *Appl. Catal. A.* 2003, **256**, 69.

[60] C. Bianchini, D.G. Burnaby, J. Evans, P. Frediani, A. Meli, W. Oberhauser, R. Psaro, L. Sordalli, F. Vizza, *J. Am. Chem. Soc.* 1999, **121**, 5961.

[61] Bianchini, C.; Santo, V. D.; Meli, A.; Oberhauser, W.; Psaro, R.; Vizza, F. *Organometallics* **2000**, *19*, 2433.

[62] C. Bianchini, P. Barbaro, V. Dal Santo, R. Gobetto, A. Meli, W. Oberhauser, R. Psaro, F. Vizza, *Adv. Synth. Catal.* 2001, **343**, 41.

[63] F. M. de Rege, D. K. Morita, K. C. Ott, W. Tumas, R. D. Broene. *Chem. Commun,* 2000, 1797.

[64] J. P. Arhancet, D. E. Davis, J. S. Merola, B. E. Hanson, *Nature*, 1989, **339**, 454.

[65] (a) M. E. Davis, *CHEMTECH*, 1992, August, 498. (b) J. P. Archancet, M. E. Davis, B. E. Hanson, *J. Catal.* 1991, **129**, 110. (c) I. Guo, B. E. Hanson, I. Tóth, M. E. Davis, *J. Organomet. Chem.* 1991, **403**, 221. (d) J. P. Arhancet, M. E. Davis, B. E. Hanson, *J.Catal.* 1991, **129**, 94. (e) Y. Yuan, J. Xu, H. Zhang, K. Tsai, *Catal. Lett.* 1994, **29**, 387. (f) G. Fremy, E. Monfier, J.-F. Carpentier, Y. Castanet, A. Mortreux, *J. Catal.* 1996, **162**, 339.

[66] B. B. Bunn, T. Bartik, B. Bartik, W. R. Bebout, T. E. Glass, B. E. Hanson, *J. Mol. Catal.* 1994, **94**, 157.

[67] I. T. Horváth, *Catal. Lett.* 1990, **6**, 43.

[68] A. J. Sandee, V. F. Slagt, J. N. H. Reek, P. C. J. Kamer, P. W. N. M. van Leeuwen, *Chem. Commun.* 1999, 1633.

[69] G. Fremy, E. Monflier, J.-F. Carpentier, Y. Castanet, A. Mortreux, *Angew. Chem. Int. Ed. Engl.* 1995, **34**, 1474.

[70] (a) K. T. Wan, D. E. Davis, *Nature*, 1994, **370**, 449. (b) K. T. Wan, D. E. Davis, *J. Catal.* 1994, **148**, 1.

[71] R. Melhem and M. Aljarrah. Plant design and feasibility study of a heterogeneous hydroformylation process of 1-octene Report of a Process Plant Design. Project supervised by H. van den Berg and A.G.J. van der Ham, University of Twente, Department of Science and Technology, 2004.

CHAPTER 4 SEPARATION BY SIZE-EXCLUSION FILTRATION

Homogeneous Catalysts Applied in Membrane Reactors

N.J. RONDE AND D. VOGT
Laboratory of Catalysis and Inorganic Chemistry, Schuit Institute of Catalysis, Eindhoven University of Technology, Eindhoven, The Netherlands

4.1 Introduction

The high selectivity and activity of homogeneous catalysts under mild reaction conditions combined with the ability to tailor and fine-tune the systems for a given substrate are unbeaten by their heterogeneous counterparts. Unfortunately, the separation of the catalysts from the reaction medium is still an important drawback, which often hampers industrial applications. Only a few processes are applied in industry nowadays, among which are the production of adiponitrile by Koch (formerly Dupont), acetic acid and acetic acid anhydride by Monsanto, butanal by Celanese (former Ruhr Chemie), and linear α-olefins in the SHOP (Shell Chemicals).[1,2] In each case an individual solution was developed for catalyst separation and recovery. A general toolbox for this is required. Table 4.1 summarizes the advantages and disadvantages of homogeneous versus heterogeneous catalysts. In this way the major problem of homogeneous catalysis becomes obvious.

TABLE 4.1. Homogeneous versus heterogeneous catalysis

	Homogeneous	Heterogeneous
Activity	+++	-
Selectivity	+++	+
Catalyst Description	++	-
Catalyst Recycling	-	+++
TON	+	+++
Quantity of Catalyst	++	+++

A number of potential methods for homogeneous catalyst separation and recovery have been discussed in the preceding chapters. This chapter addresses the separation of homogeneous catalysts by means of advanced filtration techniques. Separation of homogeneous catalysts by size exclusion (ultra- or nanofiltration, defined in detail in Section 4.3.1) offers several advantages:

- the catalyst remains homogeneous
  - no to little mass-transfer limitations
  - high activity
- low energy consumption of the separation step
- simultaneous catalyst and product separation

D. J. Cole-Hamilton and R. P. Tooze (eds.), Catalyst Separation, Recovery and Recycling, 73–104.

- potential for continuous homogeneous catalysis

A prerequisite for the application of filtration methods is a significant difference in molecular size of the catalyst and the reactants / products. Molecular enlargement, i.e. binding the homogeneous catalyst to soluble supports, is often the method of choice. These supports can be dendrimers, hyper-branched polymers or even simple polymers, giving the opportunity to tailor the support according to the given process.

All the above-mentioned methods will be discussed in this chapter. A number of filtration and reactor units that have been used in published work will be described and an overview of the different membrane materials will be given. As generally catalytic reactions deal with organic substrates and are usually carried out in an organic solvent, the membranes have to be solvent resistant. Thermal stability can also be an issue depending on the process conditions.

4.2 Reactors

All experimental setups described in the literature for the separation of homogeneous catalysts by membrane filtration technology can be divided into two general classes: Dead-end filtration and cross-flow filtration. The first type of unit is characterized by a product flow perpendicular to the surface of the membrane, while the flow in the case of cross-flow filtration is parallel to the membrane surface (see Figure 4.1).

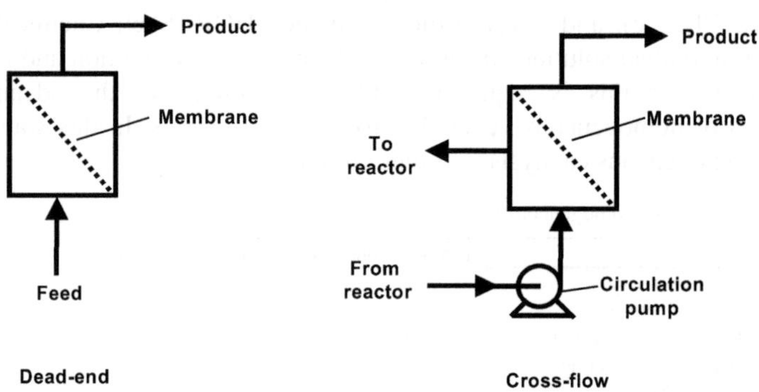

Figure 4.1. Principles of dead-end and cross-flow filtration

The dead-end setup is by far the easiest apparatus both in construction and use. Reactor and separation unit can be combined and only one pump is needed to pump in the feed. A cross-flow setup, on the other hand, needs a separation unit next to the actual reactor and an additional pump to provide a rapid circulation across the membrane. The major disadvantage of the dead-end filtration is the possibility of concentration polarization, which is defined as an accumulation of retained material on the feed side of the membrane. This effect causes non-optimal membrane performance since losses through membrane defects, which are of course always present, will be amplified by a high surface concentration. In extreme cases concentration polarization can also lead to precipitation of material and membrane fouling. A membrane installed in a cross-flow setup, preferably applied with a turbulent flow, will suffer much less from this phenomenon. In the following section different types of reactors that have been reported in literature are discussed.

4.2.1 DEAD-END FILTRATION REACTORS

The separation of homogeneous catalysts by means of membrane filtration has been pioneered by Wandrey and Kragl. Based on the enzyme-membrane-reactor (EMR),[3,4] that Wandrey developed and Degussa nowadays applies for the production of amino acids, they started to use polymer-bound ligands for homogeneous catalysis in a chemical membrane reactor (CMR).[5] For large enzymes, concentration polarization is less of an issue, as the dimension of an enzyme is well above the pore-size of a nanofiltration membrane.

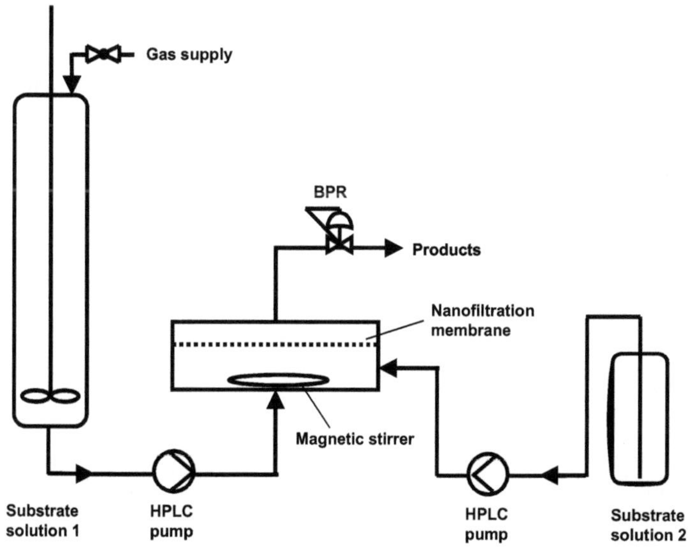

Figure 4.2. Dead-end filtration reactor described by Vogt *et al.*[6,7]

In collaboration with the Wandrey group, Vogt *et al.* developed a dead-end filtration reactor for application at high pressures (Figure 4.2). It was first used for the hydrovinylation of styrene.[6,7] In this reaction, ethylene was used as a solution in dichloromethane, prepared separately in a stainless steal tank (substrate solution 1). Using a second HPLC pump, styrene dissolved in dichloromethane was pumped into the reactor. The catalyst solution was introduced via an HPLC injection valve equipped with a loop of 2 mL. The products and unreacted compounds crossed the membrane and left the reactor on top where the pressure was regulated by a backpressure regulator (BPR). Details about the reaction can be found in Section 4.4.3. This dead-end reactor has also been used for hydrogenation reactions using polymer-stabilized Pd-nanoparticles, applying a solution of H_2 in the solvent.[8] This example clearly shows the limitations of the setup, as an internal gas phase is not permitted.

A similar reactor setup was used by Keurentjes *et al.*[9,10] A Wilkinson catalyst with fluorinated ligands was applied in the hydrogenation of 1-butene in supercritical carbon dioxide. Tubular microporous silica membranes (ECN, Petten, The Netherlands) were used. The process started with the preformation of the catalyst, after which the reactor was pressurized up to 20 MPa. Substrate and hydrogen were fed continuously into the reactor. A trans-membrane pressure was created by opening a needle valve on the permeate side, which started the continuous process (Figure 4.3). The process will be discussed in detail in Section 4.6.1.

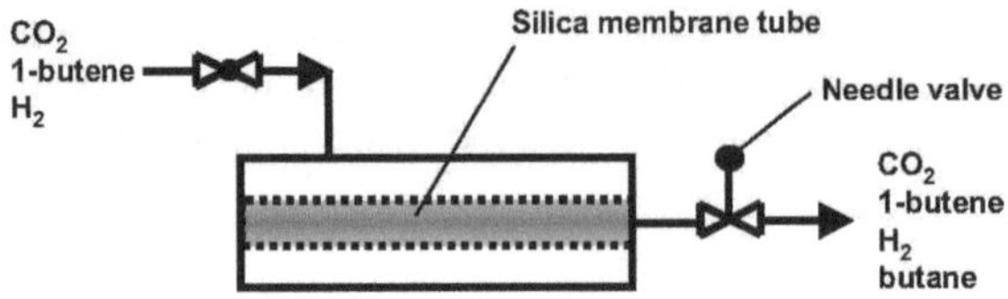

Figure 4.3. Dead-end filtration reactor4 described by Keurentjes *et al.*[9,10]

Figure 4.4. High-Output Stirred Cell from Millipore.[12]

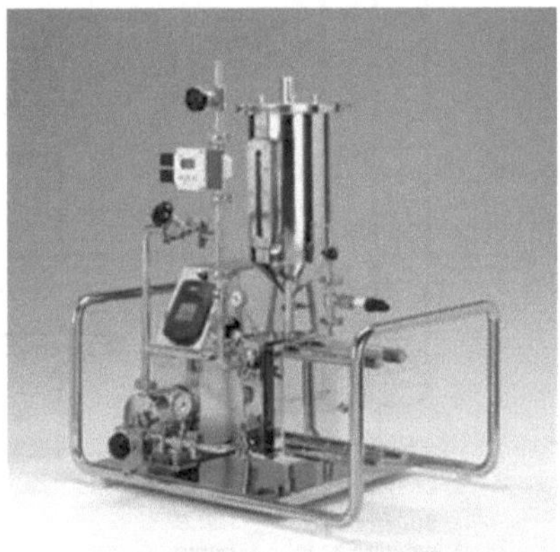

Figure 4.5. ULTRAN® - MaxiFlex from Schleicher & Schuell.[13]

Using unmodified Ru-BINAP and Rh-Et-DUPHOS catalysts Jacobs *et al.* performed hydrogenation reactions of dimethylitaconate (DMI) and methyl-2-acetamidoacrylate (MAA), respectively.[11,47] The continuous hydrogenation reaction was performed in a 100 mL stirred autoclave containing an MPF-60 membrane at the bottom, which also acts as a dead-end membrane reactor. The hydrogenation reactions will be discussed in paragraph 4.6.1.

Commercial dead-end filtration cells are available from Millipore [12] suitable for ultrafiltration (Figure 4.4, e.g. model 8003 and 8010) and from Schleicher & Schuell [13] (Figure 4.5) applicable for ultra- and nanofiltration.

4.2.2 CROSS-FLOW FILTRATION REACTORS

A coupled reactor-separation system (Figure 4.6) was used by Livingston *et al.* in which they were able to perform a Heck reaction and to recycle the catalyst 10 times using a semi batch method.[14,15] To prevent degradation of the membrane by the reaction mixture, the actual catalysis was performed in a separated stainless steel reactor. When the conversion reached more than 95%, the reactor was cooled to room temperature and the reaction mixture was pumped through the solvent resistant nanofiltration membrane (SRNF), where it was filtered until 85% of the initial volume had penetrated. The retained catalyst-rich solution was then transferred back into the reactor and diluted with fresh reactant solution, before a new catalysis cycle was performed. STARMEM membranes were used as well as Koch MPF-60 membranes (discussed in Section 4.3.2). The STARMEM membranes deteriorated in THF and were replaced after each cycle, whereas the Koch membranes showed long-term stability.

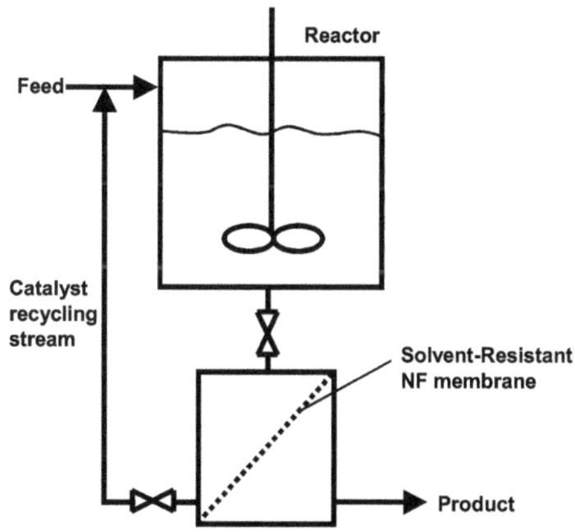

Figure 4.6. Cross-flow filtration reactor described by Livingston *et al.*[14,15]

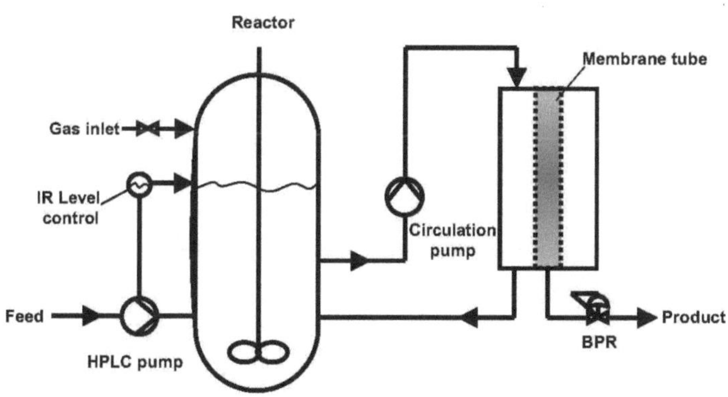

Figure 4.7. Cross-flow filtration reactor described by Vogt *et al.*[16]

A special type of cross-flow reactor was developed in the laboratories of Vogt [16] to handle continuous gas / liquid reactions. The challenge in the reactor design was to combine efficient gas-liquid mixing, liquid level control in the reactor, turbulent flow across the membrane, and efficient gas-liquid separation to avoid gas contacting the membrane, which would lead to a shunt of gas. The total internal volume should not exceed ca. 250 mL. The problems were solved by a combination of an IR level control, steering the feed pump and special geometric arrangements of sieves (Figure 4.7). Ceramic membrane tubes (HITK [26]), showing a higher temperature stability and significantly better solvent resistance, were used.

4.3 Membranes

In the 1970s the first papers were published combining membranes with reaction engineering. Before, membranes were used for water purification, and in the food and dairy industries. Since this new application of membranes started, pioneered by Alan Michaels, membrane reactors were used first in combination with enzymes.[3] Later on, when membranes with smaller pores and resistance against organic solvents became available, membrane technology was also applied for homogeneous catalysis.[17,18] In the following part, the different types of membrane filtration will be discussed. Furthermore, the advantages and disadvantages of the different materials of membranes will be explained. Finally an overview of the membranes used in literature will be given.

4.3.1 CLASSIFICATION OF FILTRATION TYPES

In the field of membrane filtration, a distinction is made based upon the size of the particles, which are retained by the membrane. That is: micro-, ultra-, nanofiltration and reverse osmosis. Figure 4.8 shows a schematic picture of the classification of membrane processes. The areas of importance for application with homogeneous catalysts are ultra- and nanofiltration, depicted in gray.

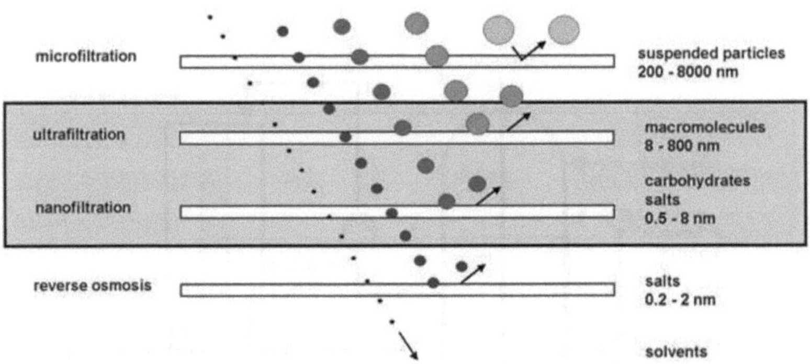

Figure 4.8. Classification of membrane processes

To find a suitable membrane for a certain application, an important parameter is the molecular weight cut-off (MWCO). The MWCO is defined as the molecular weight at which 90% of the solutes are retained by the membrane. It should be taken into account that the pore size of many ultra- and nanofiltration membranes is greatly influenced by the solvent and by the temperature used under experimental conditions. This particularly concerns polymeric membranes as will be discussed in the next paragraph.

4.3.2 MEMBRANE MATERIALS

There are basically two types of membranes, inorganic (ceramic) membranes and organic (polymer) membranes. Both types of membranes have their advantages and disadvantages as will be discussed in this section. The ideal membrane would have a very narrow pore size distribution, which is not influenced by solvent or temperature changes. Furthermore it should be stable under real reaction conditions, i.e. stable towards organic solvents, aggressive reagents and elevated temperature. Additionally a sufficient flow under all circumstances and at a low trans-membrane pressure is required.

The most widely applied commercial nanofiltration membrane is the MPF series of Koch Int.[19] (Figure 4.9). The series contains polymeric silicon-derived membranes in the MWCO range of 400-700 Da.[20] The membranes are supplied in a 50% ethanol / water mixture, and should under no circumstances be allowed to dry out. Before use, the membranes must be conditioned with the solvent of choice.

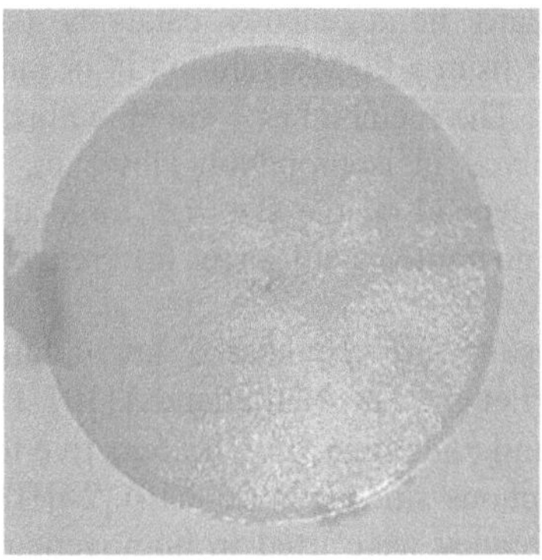

Figure 4.9. A used nanofiltration membrane (Koch MPF-50)

Figure 4.10. TEM picture of a ceramic membrane

Since these membranes are polymer-based, their MWCO is highly dependent upon the solvent used.[21]

Another nanofiltration membrane often applied is the STARMEM™ series of MET.[22] These membranes have an active layer manufactured from polyimides and are available with MWCO's in the range of 200-400 Da. The membranes are supplied in dry form containing a conditioning agent, which can be easily washed out. In contrast to the MPF membranes, these membranes are not stable in chlorinated solvents or THF, but have higher temperature stability. Livingston *et al.* made a comparative study for these two membrane types together with the Desal-5 (Osmonics [23]) membrane.[50] More details about the polymeric membranes are given in Table 4.2.

TABLE 4.2. Commercially available solvent resistant nanofiltration membranes

Membrane	MWCO	Manufacturer	Material	Solvents	Max temp (°C)
STARMEM120	200	MET	polyimide	Tol, Xyl, EtOAc	60
STARMEM220	220	MET	polyimide	Tol, Xyl, EtOAc	60
STARMEM228	280	MET	polyimide	Tol, Xyl, EtOAc, C6	60
STARMEM240	400	MET	polyimide	Tol, Xyl, EtOAc, C6	60
MPF-50	700	Koch	silicon derived	common org solvents, limited stable in: DMF,NMP, DMA, DMSO	40
MPF-60	400	Koch	silicon derived	Common org solvents, limited stable in: DMF,NMP, DMA, DMSO	40
Desal-5	350	Osmonics			50

MWCO: Molecular weight cut-off

Dense membranes are a special type of polymeric membranes. Jacobs *et al.* published on the use of polydimethylsiloxane (PDMS) dense membranes in the hydrogenation of dimethylitaconate and acetophenone using standard homogeneous catalysts (see Section 4.6.1)[48]. The membranes were homemade from a PDMS solution in hexane, which was cross-linked in a vacuum oven at 100°C. The membranes were able almost completely to retain unmodified Ru-BINAP dissolved in isopropanol. However, as mentioned earlier, these applications will strongly depend on the size, i.e. molecular weight, of the substrate to be converted in order to guarantee a sufficient difference in size of the product and the catalyst to be retained.

The only ceramic membranes of which results are published, are tubular microporous silica membranes provided by ECN (Petten, The Netherlands).[10] The membrane consists of several support layers of α- and γ-alumina, and the selective top layer at the outer wall of the tube is made of amorphous silica (Figure 4.10).[24] The pore size lies between 0.5 and 0.8 nm. The membranes were used in homogeneous catalysis in supercritical carbon dioxide (see paragraph 4.6.1). No details about solvent and temperature influences are given but it is expected that these are less important than in the case of polymeric membranes.

In Table 4.3, a selection of inorganic membrane suppliers is given. Very little data are available on applications in catalysis so far. The membranes can be used for separation in the ultra- or nanofiltration range.

TABLE 4.3. Some ceramic ultra- and nanoflitration membranes

Manufacturer	Pore size (nm)	Material support	Material active layer
ECN[25]	0.5 - 0.8	α- and γ-alumina	amorphous silica
HITK[26]	30 - 1000	α-alumina	Al_2O_3, TiO_2, ZrO_2
Corning[27]	MF and UF		

4.4. Dendrimer Supported Catalysts

Since the pioneering work on dendritic structures by Vögtle *et al.*,[28] dendrimers have attracted much attention. The synthesis and investigation of their structural properties became a new field in science. The application of dendrimers as support molecules for homogeneous catalysts was first reported by Van Koten *et al.* in 1994.[29] Dendrimers have the advantage of having perfect structures unlike polymeric structures and are therefore preferred in academic research. Catalytic properties can be investigated using standard high-resolution techniques in solution, whereas heterogeneously supported catalysts require more sophisticated techniques that often give less detailed information. In the next section the applications of dendritic supports for the recycling of homogeneous catalysts will be discussed.

4.4.1 KHARASCH ADDITION REACTION

Building on the experience with enzyme catalysis in membranes reactors, Kragl and Van Koten performed one of the first continuous catalysis experiments using dendrimer supported homogeneous catalysts.[39,40] Zeroth and first generation carbosilane dendrimers were functionalized with 4, respectively 12 NCN pincer ligands (Figure 4.11). Nickel was introduced by lithiation followed by the reaction with $[NiCl_2(PEt_3)_2]$. Both dendritic catalysts were tested for their retention in a nanofiltration cell equipped with a Koch (former SELRO) MPF-50 membrane using CH_2Cl_2 as the solvent. The measured retention for zeroth generation catalyst was found to be 97.4%, while 99.75% of the first generation was retained.

After reaching its maximum productivity (after ca. 8 hours.) the [G1]-Ni_{12} showed a fast deactivation when applied in continuous catalysis performed in a membrane reactor (Figure 4.12). The fast loss of activity cannot be due to a lack of retention of the catalyst. Due to the high retention measured, this process should be much slower. A model study revealed that this deactivation process probably takes place by the formation of insoluble Ni(III) species (see Section 4.5 for further details).

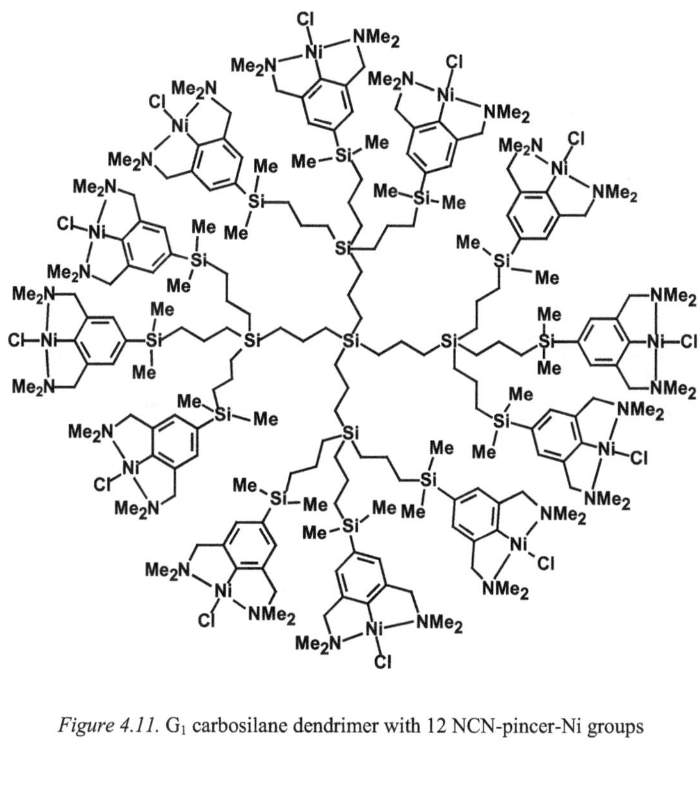

Figure 4.11. G_1 carbosilane dendrimer with 12 NCN-pincer-Ni groups

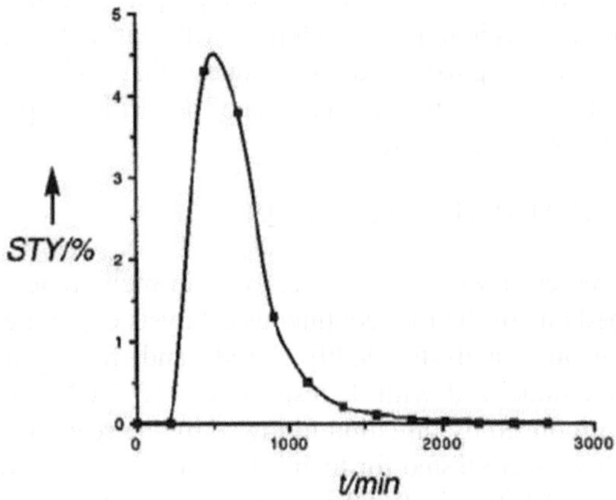

Figure 4.12. Continuous Kharasch addition. (Reprinted with permission from ref. 40. Copyright 2000 American Chemical Society)

4.4.2 ALLYLIC SUBSTITUTION REACTIONS

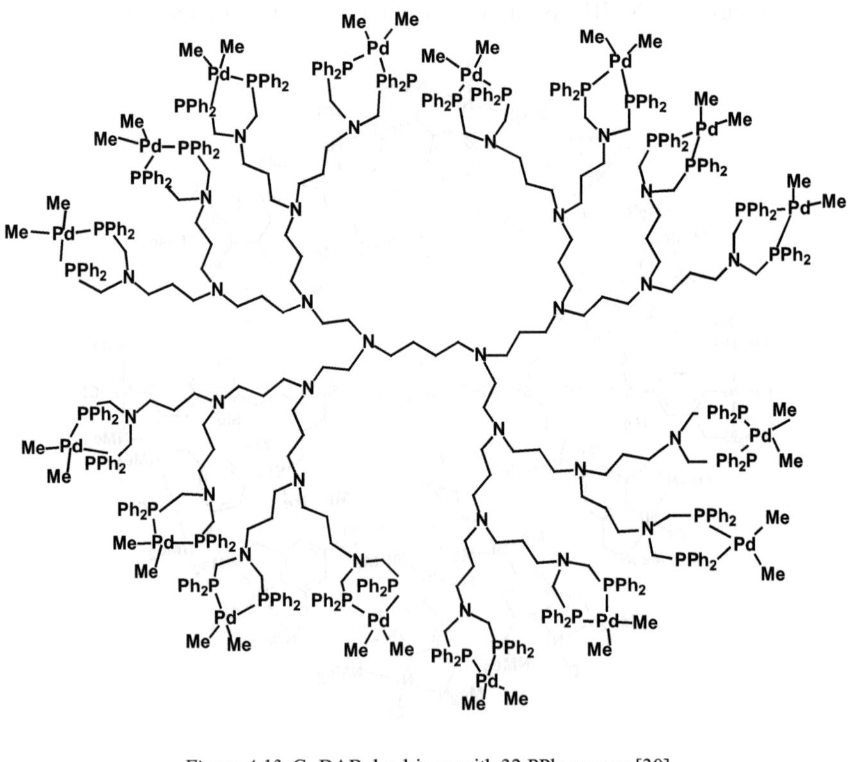

Figure 4.13. G_2 DAB dendrimer with 32 PPh_2 groups.[30]

Another reaction performed in the dead-end reactor discussed before, is the allylic amination of 3-phenyl-2-propenyl-carbonic acid methyl ester with morpholine.[30] First and second generation commercially available DAB-dendrimers were functionalized with diphenylphosphine groups (Figure 4.13). Two different membranes were used, the Nadir UF-PA-5 (ultrafiltration) and the Koch MPF-50 (former SELRO) (nanofiltration), which gave retentions of 99.2% and 99.9% respectively for the second generation functionalized dendrimers.

The third generation dendritic catalyst was applied in continuous catalysis. The catalyst remained active for more than 100 residence times with a conversion ranging from 100% at the beginning to 80% at the end of the run (Figure 4.14). Leaching of palladium of only 0.07-0.14% per residence time was found. The authors give the formation of inactive palladium species as another possible reason for the drop in activity, next to the leaching of palladium. NMR data indeed indicate the formation of inactive palladium species such as [ligand-$PdCl_2$] complexes, possibly resulting from a decomposition of the complex by a reaction with dichloromethane.

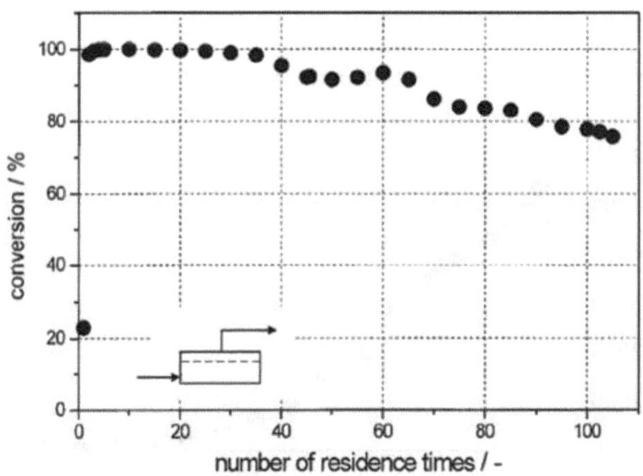

Figure 4.14. Continuous allylic amination. (Reprinted with permission from ref. 30. Copyright 1999 Science Direct)

Van Leeuwen *et al.* used several generations of carbosilane dendrimers with 4, 8, 24, and 36 diphenylphosphine end-groups (Figure 4.15) for the allylic alkylation reaction of allyl trifluoracetate with sodium diethyl 2-methylmalonate.[31]

Application of the largest dendritic catalyst **8** (Figure 4.15) in a continuous process showed activity over 15 exchanged reactor volumes (Figure 4.16). The decrease in activity caused by wash out was calculated to be only 25% (retention of ligand 98.1%). The drop in activity was therefore ascribed to the decomposition of the palladium catalyst. Addition of membrane material to batch catalysis experiments did not change the conversion showing that this was not the cause of decomposition.

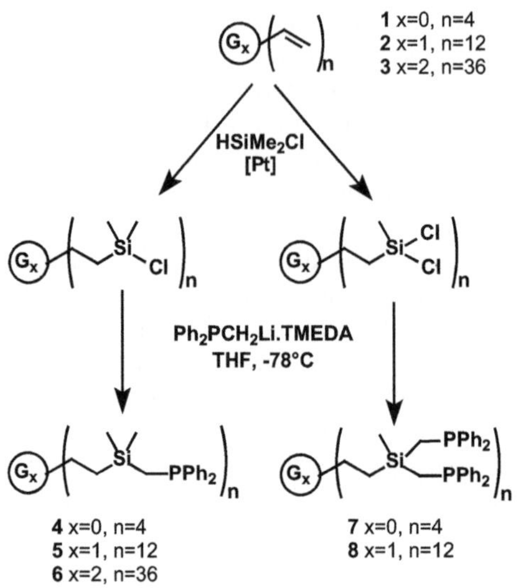

Figure 4.15. Synthesis of phosphine-functionalized carbosilane dendrimers.[31]

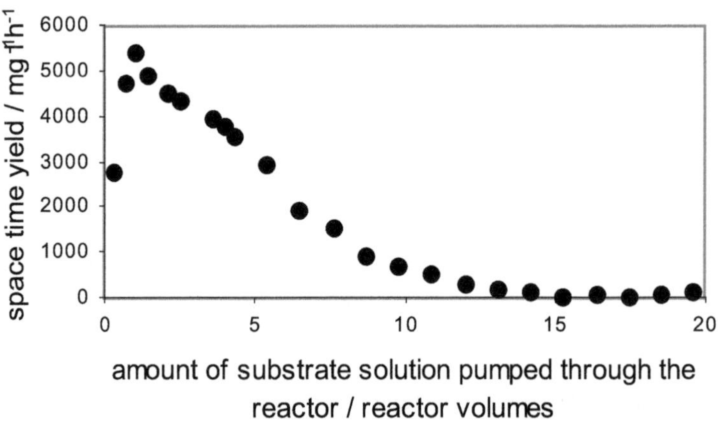

Figure 4.16. Continuous allylic alkylation.[31]

Further investigations using model compounds showed that the formation of $PdCl_2$ by a reaction with the solvent, as suggested by Brinkmann *et al.* [30], was also not responsible for the observed rapid deactivation. Palladium leaching after formation of Pd(0) was also excluded by experiments. The authors concluded therefore that the presence of allyl acetate facilitated the decomposition.

More dendritic ligand systems were synthesized with an ethylene spacer between the terminal silicon atom and the phosphorus atom.[32] This enlarged system showed a higher stability in the continuous allylic amination. During the exchange of 15 reactor volumes, more or less constant conversion was observed, slowly decreasing from its maximum (~70%) after 5 h, to ~50% after 15 reactor volumes. When the P / Pd ratio was increased from 2 to 4, the activity increased significantly giving almost quantitative conversion after about 7 exchanged reactor volumes which slightly decreased to ca. 80% after 15 reactor volumes (Figure 4.17). This finding showed that a relative small change in the backbone of dendritic catalysts could increase the stability significantly.

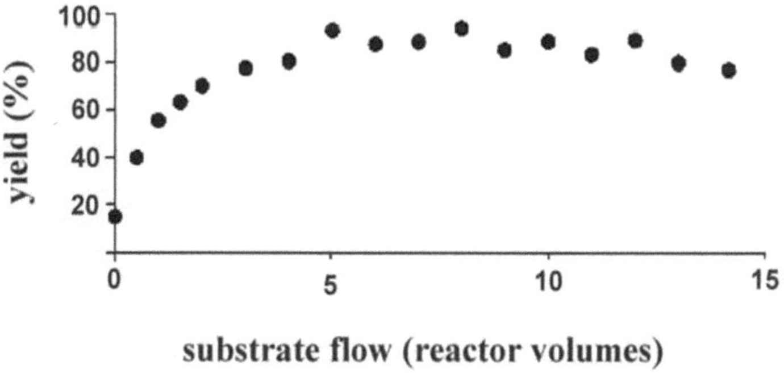

Figure 4.17. Continuous allylic amination with a modified catalyst. (Reprinted with permission from ref. 32. Copyright 2002 Wiley)

A non-covalently functionalized dendrimer was also applied in a continuous allylic amination reaction.[33] PPI dendrimers functionalized with urea adamantyl groups can act as host molecules for phosphorus ligands equipped with acetyl urea groups (Figure 4.18). The so formed supramolecular complex was reacted with a palladium precursor and used as catalyst in the allylic amination reaction.

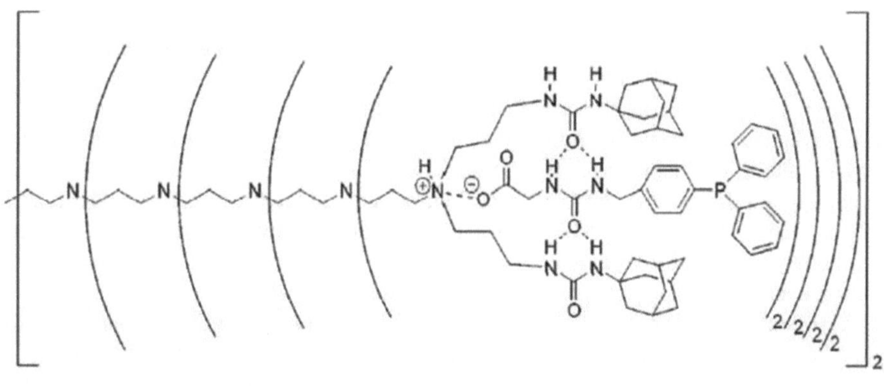

Figure 4.18. G_5 PPI dendrimer with 32 host-guest complexes.[33]

Both the acid and ester were applied in continuous allylic amination. The maximum conversion (ca. 80%) was reached after 1 h in both experiments. Using the acid derivative of the guest, a slight drop in activity was observed (**(a)** in Figure 4.19), which is probably caused by a slow deactivation of the catalyst and has also been observed for covalently functionalized dendrimers (described above). When using the ester-functionalized guest, the activity dropped faster (**(b)** in Figure 4.19). This decrease in activity is caused by lack of retention (99.4% for the acid vs. 97% for the ester) as well as by deactivation.

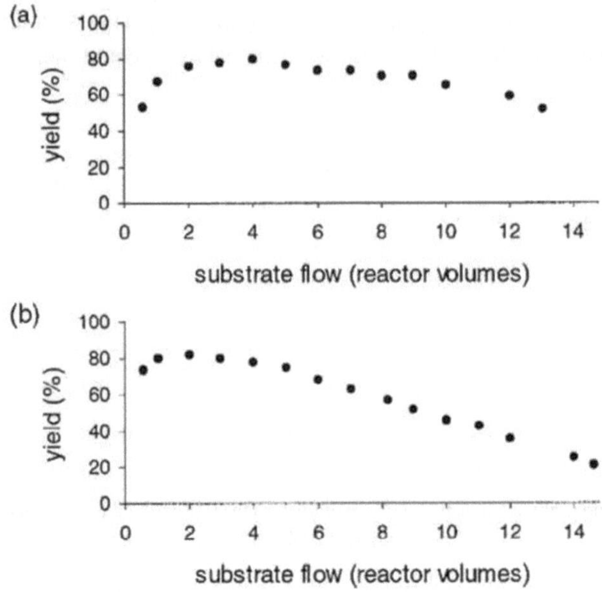

Figure 4.19. Continuous catalysis with non-covalently functionalized dendrimers a) acid-, b) ester-functionalized guest. (Reprinted with permission from ref. 33. Copyright 2001 American Chemical Society)

4.4.3 HYDROVINYLATION REACTION

In the group of Vogt, zeroth and first generation carbosilane dendrimers were functionalized on the periphery with 4 or 12 hemilabile P,O-ligands (Figure 4.20).[6,7] The allyl palladium complexes of these systems were used as catalysts in the hydrovinylation of styrene. At higher conversion the primary codimerization product 2-phenyl-1-butene can isomerize in a consecutive reaction to form the E/Z-mixture of internal 2-phenyl-2-butene, an undesired byproduct, especially in case of asymmetric catalysis. The possibility to perform this homogeneously catalyzed reaction continuously is therefore interesting since it can combine a high productivity with incomplete conversion suppressing the isomerization reaction.

The retention of the free [G_0]-ligand system (without palladium) was 85%, the molecular weight of the actual catalyst is much larger (M_w = 2868 Da versus 1314 Da for the G_0-system), which should be sufficient for initial experiments. The result of this continuous hydrovinylation is shown in Figure 4.21.

After an induction period of ca. 9 hours, the maximum productivity was reached. This was followed by a decrease in activity, which cannot solely be explained by the lack of retention. Calculations showed that at least 20% of the catalyst should still remain in the reactor after 80 h. Additional to this wash-out effect, a deactivation process took place, visible by precipitation of palladium black on the surface of the membrane. Although the catalytic system suffered from deactivation, its selectivity towards 3-phenylbut-1-ene was excellent, being 98% and 85% for the G_0- and G_1-catalysts respectively.

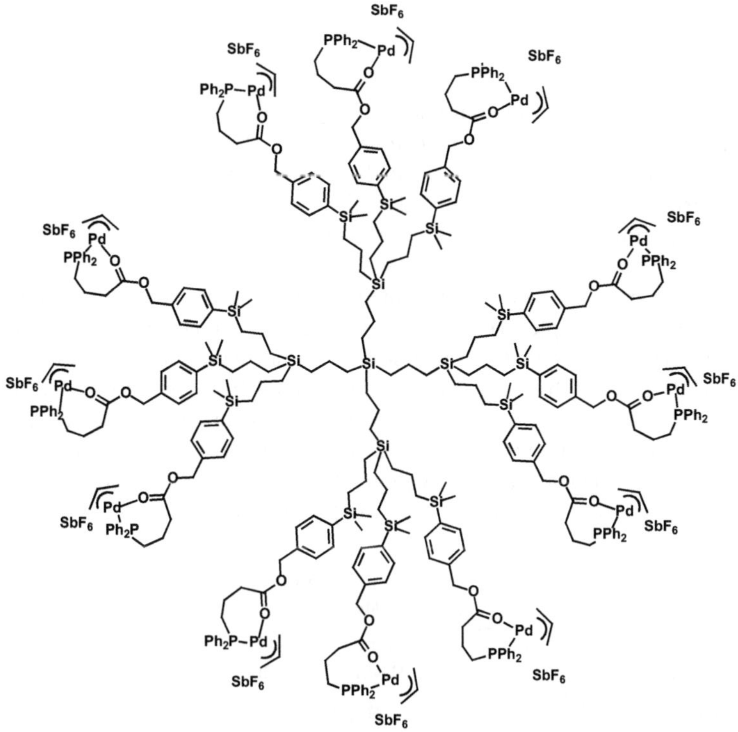

Figure 4.20. G_1 carbosilane dendrimer with 12 Pd-complexes.[6,7]

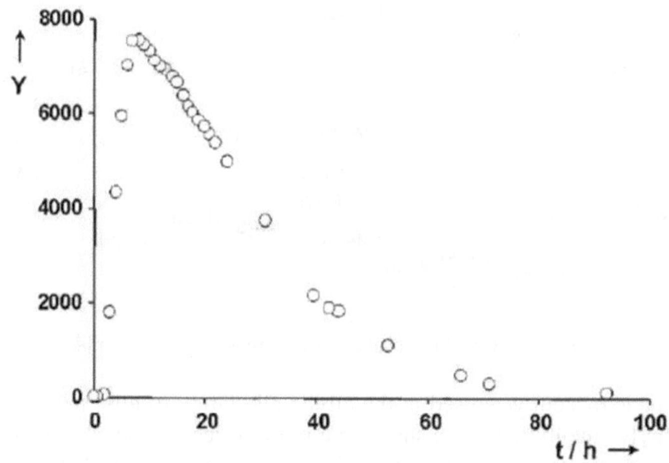

Figure 4.21. Continuous hydrovinylation experiment.[7]

4.4.4 HYDROGENATION REACTION

A set of core-functionalized dendrimers was synthesized by Van Leeuwen *et al.* and one compound was applied in continuous catalysis.[45] The dendritic dppf, Xantphos and triphenylphosphine derivatives (Figures 4.22, 4.30 and 4.31) were active in rhodium-catalyzed hydroformylation and hydrogenation reactions (performed batch-wise). Dendritic effects were observed which are discussed in paragraph 4.5. The dendritic rhodium-dppf complex was applied in a continuous hydrogenation reaction of dimethyl itaconate.

In their experiments, an unsubstituted dppf-complex was compared with the analogous dendritic complex (Figure 4.23). After 35 exchanged reactor volumes the dendritic catalyst still showed a conversion of 77% (maximum: 85% after 10 reactor volumes) while the unsubstituted catalyst deactivated from 70% to 15% at the end of the catalytic run. The drop in activity for both systems can be completely explained by their retention (97% and 99.8% for the unsubstituted and the dendritic complex, respectively). This means, no deactivation of the catalyst occurred during catalysis, which is often the case for palladium-catalyzed continuous catalysis.

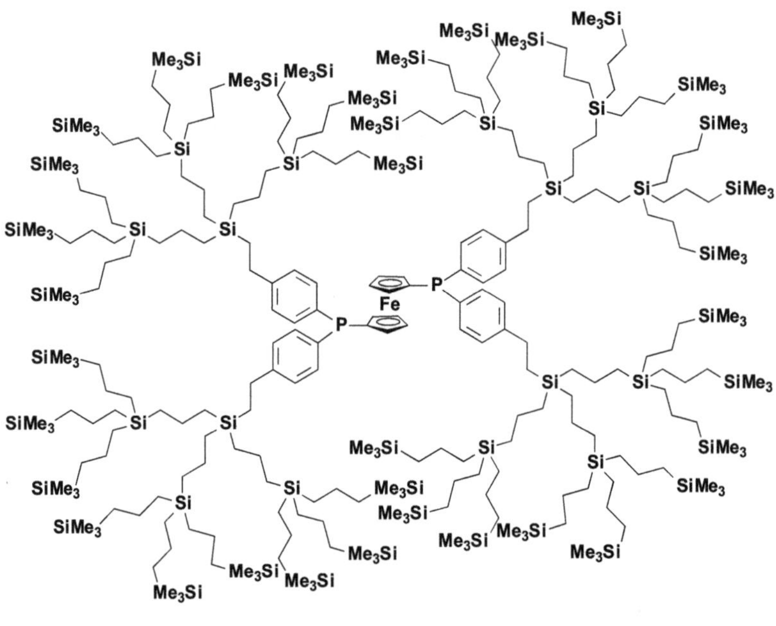

Figure 4.22. Core functionalized dppf-carbosilane dendrimer.[45]

The same ligand system was used in the allylic alkylation of allyl trifluoroacetate with sodium diethyl-2-methylmalonate showing a more or less constant conversion over 8 h (20 exchanged reactor volumes). This is in contrast to peripheral functionalized dendrimers (Section 4.4.2), which deactivated at longer reaction times.

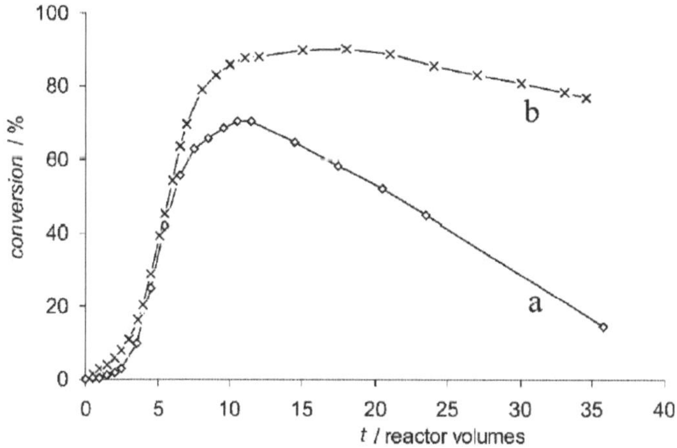

Figure 4.23. Continuous hydrogenation using dppf (a) and the core-functionalized dppf-carbosilane system (b). (Reprinted with permission from ref. 45. Copyright 2002 Kluwer)

4.4.5 MICHAEL ADDITION REACTION

A dodecakis(NCN-Pd^{II}) catalyst, synthesized in the group of Van Koten (Figure 4.24), was applied in the a continuous double Michael addition reaction between methyl vinyl ketone (MVK) and ethyl α-cyanoacetate.[34] The reaction was performed in the dead-end reactor discussed in paragraph 4.2.1. Two catalytic runs were performed differing in the amount of catalyst and in the applied flow (both increased by a factor 2.5). Both runs showed high productivity for more than 24 h (Figure 4.25).

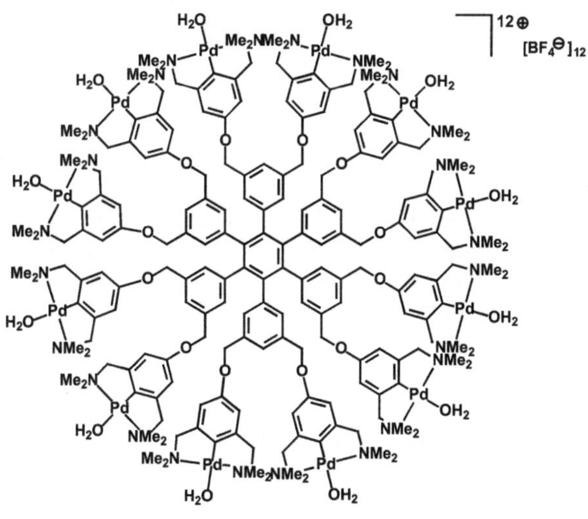

Figure 4.24. Dodecakis(NCN-Pd^{II}) dendrimer for Michael addition reaction.[34]

For run II, the Pd concentration in both the retentate and the permeate was determined by ICP-AAS analysis. A catalyst retention of 99.5% was determined, which is almost identical to the value determined earlier for an isostructural platinum analogue (R = 99,9%).[35] This low leaching of the catalyst completely accounts for the slow decrease in activity after reaching stable conversion.

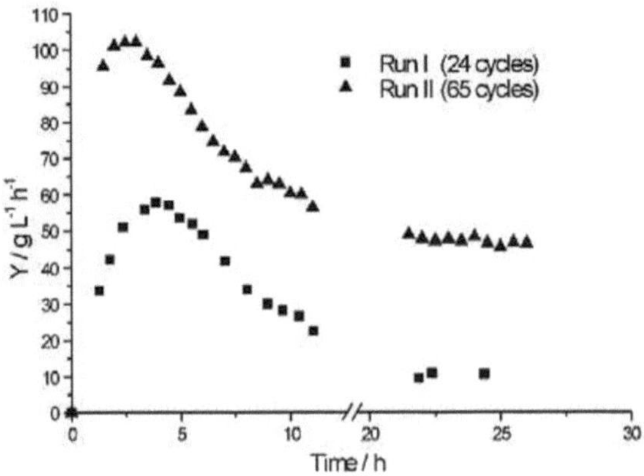

Figure 4.25. Continuous double Michael addition. (Reprinted with permission from ref. 34. Copyright 2003 Wiley)

4.5 Dendritic Effects

Dendrimers are not only unreactive support molecules for homogeneous catalysts, as discussed in the previous paragraph, but they can also have an important influence on the performance of a catalyst. The dendrons of a dendrimer can form a microenvironment in which catalysis shows different results compared to classical homogeneous catalysis while peripheral functionalized dendrimers can enforce cooperative interactions between catalytic sites because of their relative proximity. These effects are called "dendritic effects". Dendritic effects can alter the stability, activity and (enantio)selectivity of the catalyst. In this paragraph, different dendritic effects will be discussed.

Cole-Hamilton *et al.* reported one of the first dendritic effects. Dendrimers based on polyhedral oligomeric silsequioxane (POSS) cores were synthesized (Figure 4.26); the dendrons of this dendrimer were functionalized on the periphery with 8, 24 and 72 PR_2 arms respectively (R = Me, Et, hexyl, Cy, or Ph).[36]

The dendritic ligands were tested in the cascade rhodium-catalyzed hydroformylation of alkenes followed by the hydrogenation of the aldehyde towards alcohols in ethanol as solvent. A slight increase in selectivity was observed using the dendritic ligand systems with PEt_2 bound to 24 arms. A linear to branched ratio (l:b) of 3.1 was observed in the product distribution while unsubstituted PEt_3 has a l:b ratio of 2.4, suggesting that perhaps large dendrimer-based ligands exert some control over the reaction. For the PMe_2 and PPh_2 derivatized dendrimers, the l:b ratio was similar to that obtained using related monodentate phosphines.

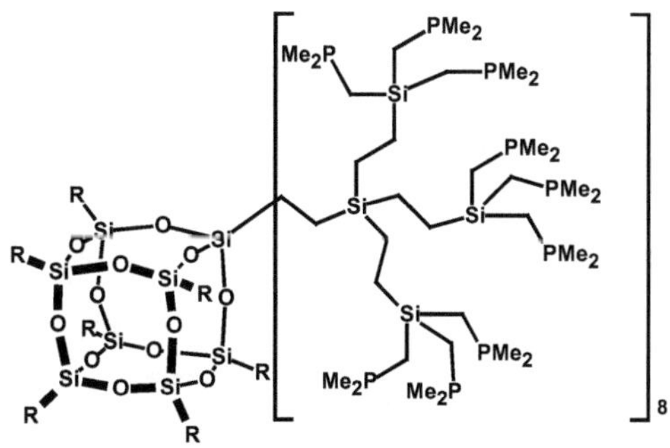

Figure 4.26. POSS core with 72 PMe_2 groups

In a later publication [37] these effects were addressed in more detail using the POSS dendrimers with 8 -$C_2H_4(SiMe(C_2H_4PPh_2)_2$ dendrons. In the hydroformylation of 1-octene, a l:b of 14:1 was observed while $Me_2Si(C_2H_4PPh_2)_2$ dendrons give a l:b ratio of 3-4:1. From this study it was concluded that the dendrimers have sufficient steric crowding to make eight-membered ring bidentate coordination favorable and that these rings enhance the selectivity towards the linear aldehydes in hydroformylation reactions. Different spacer lengths or replacing CH_2 with O gave dendrimers which show little enhancement of selectivity over the small molecule analogues.

Jacobsen *et al.* reported enhanced catalytic activity by cooperative effects in the asymmetric ring opening (ARO) of epoxides.[38] Chiral Co-salen complexes (Figure 4.27) were used, which were bound to different generations of commercial PAMAM dendrimers. As a direct consequence of the second-order kinetic dependence on the [Co(salen)] complex concentration of the hydrolytic kinetic resolution (HKR), reduction of the catalyst loading using monomeric catalyst leads to a sharp decrease in overall reaction rate.

In order to assess whether intramolecular cooperativity occurs, catalysis was performed with very low (dendritic) catalyst loading (0.027 mol% vs. 0.5 mol% for the monomeric catalyst). The dendritic Co complex effected complete kinetic resolution (98% ee, 50% conversion), while the unsubstituted analogue showed no measurable conversion.

To investigate this dendritic effect, a dimeric model compound was synthesized which mimics the tethered relationship of two catalytic units within one branch of the PAMAM dendrimer. All dendritic catalysts were more active in the HKR than the parent complex. Furthermore, the dendritic catalysts also displayed significantly higher activity than the dimeric model compound. The authors proposed that this positive dendritic effect arises from restricted conformation imposed by the dendrimer structure, thereby creating a bigger effective molarity of [Co(salen)] units. Alternatively, the multimeric nature of the dendrimer, may lead to higher order in productive cooperative interactions between the catalytic units.

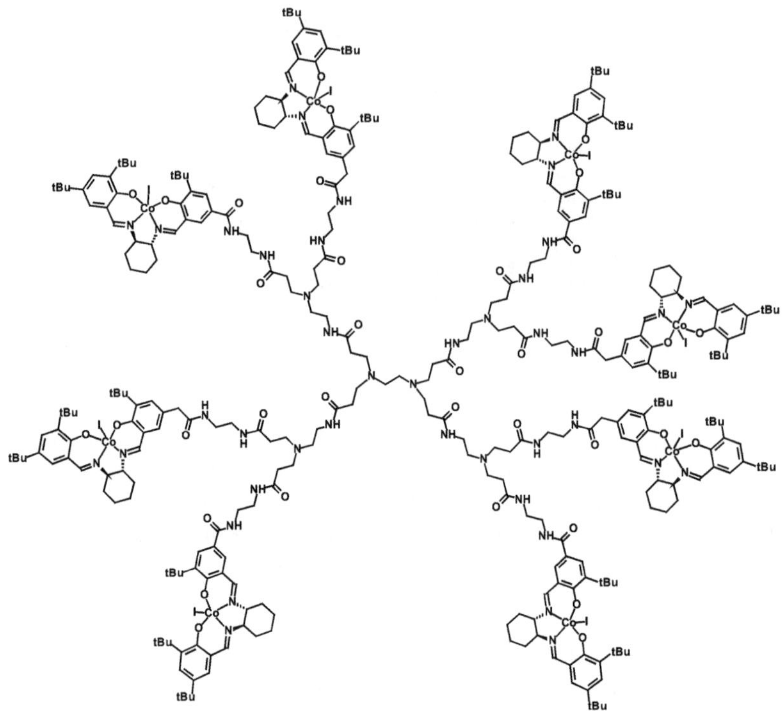

Figure 4.27. G_1 PAMAM dendrimer with 8 [Co(salen)] units.[38]

Van Koten *et al.* reported on a negative dendritic effect in the Kharasch addition reaction.[39,40] A fast deactivation for the carbosilane dendrimer supported NCN pincer catalyst (Figures 4.28 and 4.29) was observed by comparison with a mononuclear analogue. This deactivation is expected to be caused by irreversible formation of inactive Ni(III) sites on the periphery of these dendrimers.

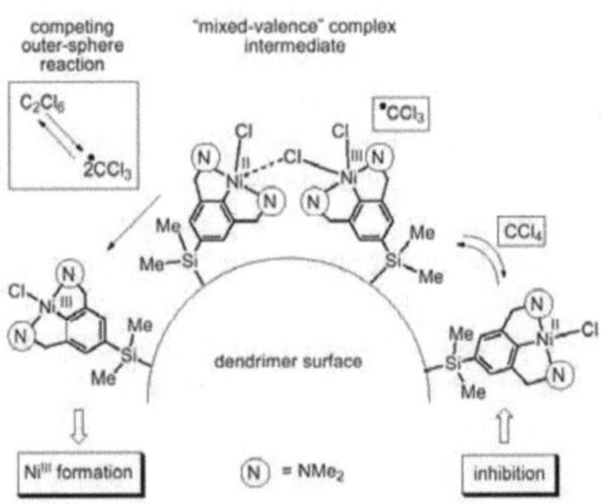

Figure 4.28. Formation of inactive Ni(III) species. (Reprinted with permission from ref. 39. Copyright 2000 American Chemical Society)

This hypothesis was supported by results of model studies as well as ESR spectroscopic investigations. The use of alternative Ni-containing dendrimers in which the distance between the Ni sites was increased (Figure 4.30), led to significantly improved catalytic efficiency.

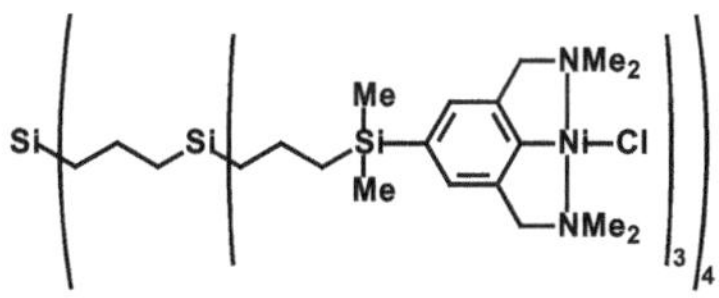

Figure 4.29. Carbosilane dendrimer with 12 Ni-NCN groups.[39]

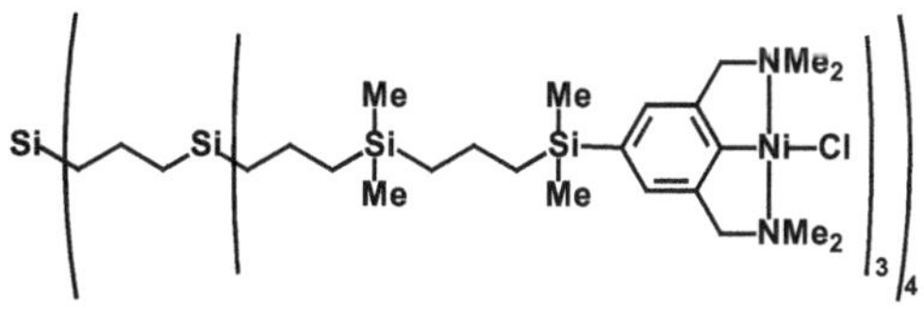

Figure 4.30. Extended G_1 carbosilane dendrimer.[40]

Zeroth to fourth generation poly(propyleneimine) (PPI) dendrimers functionalized with Pyrphos ligands (Figure 4.31) were synthesized by Gade *et al.*[41] In the asymmetric hydrogenation reaction of acetamidocinnamate the activity of the catalysts decreased when going to higher generations. Also the enantioselectivity decreased from 93 to 88% on going from G_2 to G_4. Back folding of the attached rhodium complexes is expected to be the cause of the drop in activity for higher generations of dendrimers. It reduces the accessibility of the catalytic centers and at the same time renders their immediate environment less uniform than originally envisaged. It was stated that the use of more rigid dendrimers might partially suppress this negative effect.

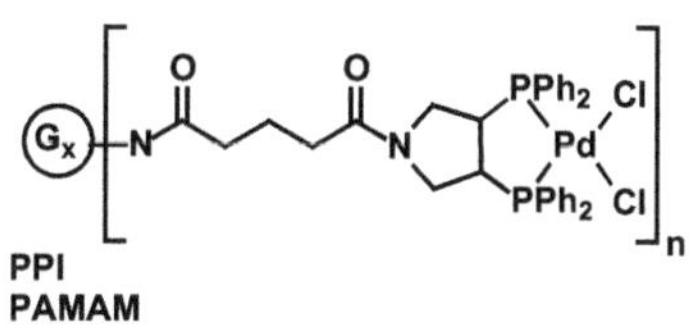

Figure 4.31. PPI and PAMAM dendrimers functionalized with Pyrphos.[41,42]

In 2003 Gade *et* al. reported on poly(amidoamine) (PAMAM) dendrimers functionalized with Pyrphos ligands (Figure 4.31).[42] In the palladium-catalyzed allylic amination reaction, an increase in enantioselectivity was observed on going from the mononuclear BOC-Pyrphos to the fifth generation PAMAM dendrimers bearing 64 Pyrphos ligands. The results with the PAMAM supported catalysts were compared with those from PPI supported catalysts. The latter dendrimers had a smaller, but still positive effect on the enantioselectivity when going to higher generations. In contrast to the highly charged cationic Pyrphos-rhodium dendrimers used in the aforementioned asymmetric hydrogenation, the neutral dichloropalladium derivatives showed no significant tendency to aggregate in solution or upon precipitation. This was concluded from TEM studies and from the determination of the hydrodynamic radii in solution. The aggregation of the cationic dendrimer-metal complexes could contribute to the lower activity and selectivity discussed above.

A slight increase in reactivity and enantioselectivity in Diels-Alder reactions was reported by Chow and Wan.[43] Rigid catalysts bearing three and six (chiral) [1,1'-

binaphthalene]-2,2'-diol (Binol) groups were compared to their unsubstituted analogues. The turn-over-frequency (TOF) increased from 2.0 h^{-1} for the unsubstituted Binol to 15.0 h^{-1} for the catalyst with six Binol groups. The ee increased from 10 to 16%, respectively. The increase in ee was not further investigated.

Simanek *et al.* investigated the structure dependence of the kinetics of thiol-disulfide exchange reactions.[44] In general, the rate of exchange decreased as the size of the dendrimers increased. Dendrimers with disulfides attached near the core undergo exchange more slowly than dendrimers with disulfides at the periphery. No evidence was found for intramolecular macrocyclization (cooperative) exchange.

The last example of a dendritic effect discussed in this chapter is the use of core-functionalized dendritic mono- and diphosphine rhodium complexes by Van Leeuwen *et al.*[45] Carbosilane dendrimers were functionalized in the core with Xantphos, bis(diphenylphosphino)ferrocene (dppf) and triphenylphosphine (Figures 4.22, 4.32 and 4.33).

Rhodium complexes of these mono- and bidentate ligands were tested in the hydroformylation of 1-octene. Depending on the ligand, a small effect of the dendritic encapsulation was observed.

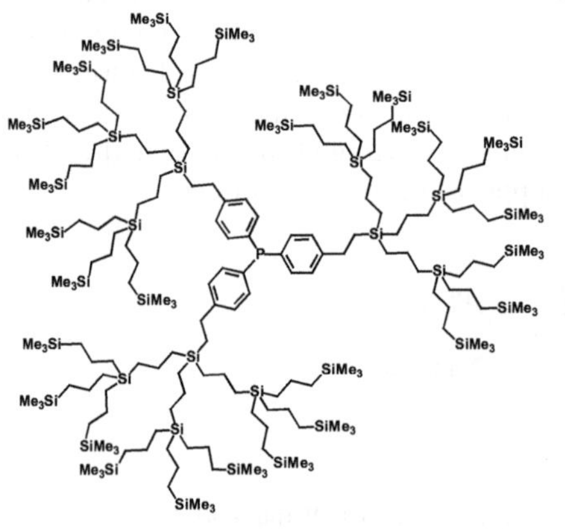

Figure 4.32. Triphenylphosphine derived dendrimer

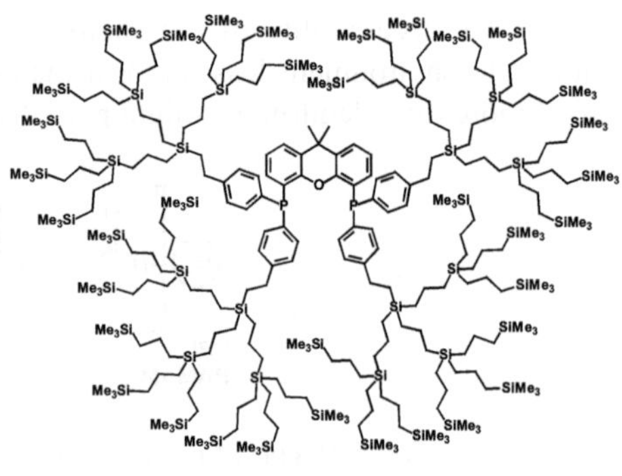

Figure 4.33. Xantphos-derived dendrimer

When bulky substrates were used in the hydroformylation reaction with the dppf-based ligand, a decrease in reactivity was observed on going to higher generations. This is in agreement with the first order dependency in substrate concentration observed for these types of reactions because of the lower diffusion rate of bulky substrates to the active center. The authors suggested taking advantage of this decreased activity for larger substrates as a possibility for substrate selective catalysis.

4.6 Unmodified or Non-dendritic Catalysts

Under carefully adjusted experimental conditions unmodified catalysts can be used in nanofiltration coupled homogeneous catalysis. Also non-dendritic but nanosized rigid catalytic systems can be retained by nanofiltration membranes. In this section, unmodified catalysts and rigid non-dendritic systems applied in continuous catalysis will be discussed.

In 1977, Parshall and co-workers published their work on the separation of various homogeneous catalysts from reaction mixtures.[46] Homemade polyimide membranes, formed from a solution of polyamic acid were used. After reaction the mixture was subjected to reverse osmosis. Depending on the metal complex and the applied pressure, the permeate contained 4-40% of the original amount of metal. This publication was the beginning of research on unmodified or non-dendritic catalysts separated by commercial and homemade membranes.

4.6.1 HYDROGENATION

Using unmodified Ru-BINAP and Rh-Et-DUPHOS catalysts (Figures 4.34 and 4.35) Jacobs *et al.* performed hydrogenation reactions of dimethylitaconate (DMI) and methyl-2-acetamido-acrylate (MAA), respectively.[47]

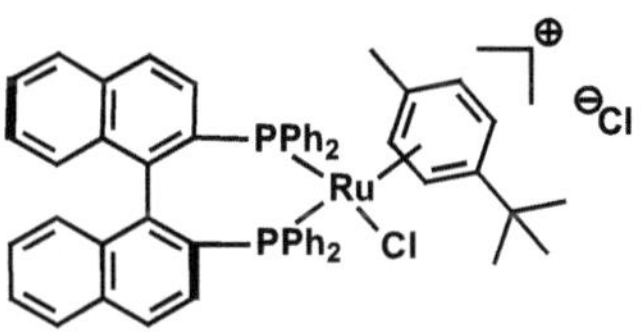

Figure 4.34. Ru-BINAP

Figure 4.35. Rh-Et-DUPHOS

The continuous hydrogenation reactions were performed in a stirred 100 mL autoclave containing a Koch MPF-60 membrane at the bottom. The hydrogenation of DMI was performed for more than 40 h (10 exchanged reactor volumes) without a significant loss in enantiomeric excess. However a slight decrease in activity was observed caused by the incomplete retention (>98%) of the catalyst. In the hydrogenation of MAA with Rh-Et-DUPHOS, a significant drop in activity and selectivity was observed. The retention of 97% alone, cannot account for this effect. A slow deactivation of the catalyst, possibly due to oxidation of the phosphine ligand, was assumed but needs further investigation. The total TON for the hydrogenations with Ru-BINAP and Rh-Et-DUPHOS were 1950 and 930, respectively.

Another setup used for the hydrogenation of DMI with Ru-BINAP was equipped with dense PDMS elastomer membranes (Jacobs *et al.*[48]). The catalyst solution was present in a submerged membrane system, prepared as a sealed "PDMS capsule". The catalytically active complex was retained by the membrane while substrate and products, dissolved in the bulk phase, could cross the membrane under the influence of the concentration difference without the need for mechanical pressure.

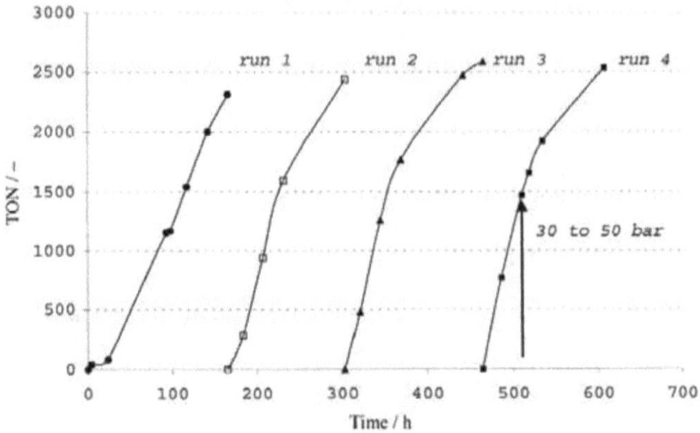

Figure 4.36. Catalysis using a catalyst in a PDMS capsule. (Reprinted with permission from ref. 48. Copyright 2003 Wiley)

The catalysis was performed batch-wise (Figure 4.36). After reaching ca. 90% conversion, the bulk phase was replaced and similar turnover frequencies (TOF) of about 25 h^{-1} were obtained in the following three runs 2, 3 and 4. When the catalyst capsule was removed, no further activity was detected. Furthermore, the Ru content in the bulk phase was always below the detection limit of AAS, which shows good catalyst retention by the membranes used.

Keurentjes *et al.* performed a continuous hydrogenation of 1-butene in supercritical carbon dioxide.[9,10] A fluorous derivative of Wilkinson's catalyst was prepared *in situ* by mixing the ligand with $[(COD)RhCl]_2$ under hydrogen / carbon dioxide pressure (Figure 4.37).

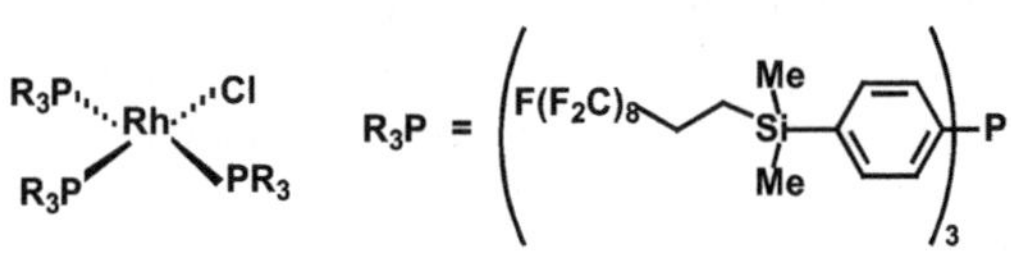

Figure 4.37. Fluorous derivative of Wilkinson's catalyst

After preformation, the substrates and carbon dioxide were supplied continuously. The membrane reactor was pressurized at the feed side up to 20 MPa with the reaction mixture. A trans-membrane pressure was created by opening a needle valve on the permeate side after which the continuous process started.

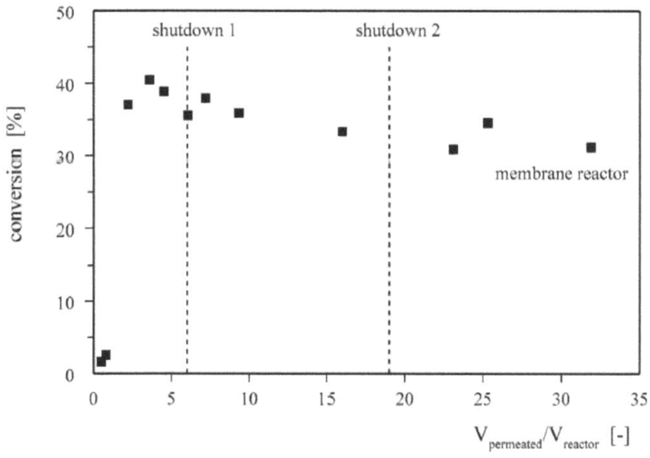

Figure 4.38. Continuous hydrogenation of 1-butene in supercritical CO_2. (Reprinted with permission from ref. 10. Copyright 2003 Science Direct)

After an activation period of 4 h, the conversion showed a maximum of 40% followed by a steady decrease in conversion (Figure 4.38). Overnight, the pressure was decreased to 6 MPa and the needle valve on the permeate side was closed. This shutdown procedure caused the catalyst to precipitate and no reaction occurred anymore. The precipitated catalyst can be used for a new cycle by pressurization of the membrane reactor, redissolving the catalyst. At the end of the third run the conversion had dropped to ~33%. A TON of 1.2×10^5 in 32 h ($\tau \sim 62$ min) was obtained. ICP-AAS analysis of the permeate stream indicated complete retention of the catalyst. The authors propose possible traces of oxygen as the cause of the decrease in activity of the catalyst.

4.6.2 PHASE TRANSFER CATALYSIS

Livingston *et al.* performed phase transfer catalysis (PTC) using membrane technology to separate the catalyst from the products in the post-reaction mixture.[49,50] The PT catalyst (tetraoctylammonium bromide) was re-used after separation in subsequent reactions. The conversion of bromoheptane into iodoheptane was performed in a 100 mL glass vessel with 40 mL aqueous phase (2 M KI) and 40 mL organic phase (0.5 M bromoheptane and 0.05 M catalyst). After complete reaction, the organic phase was transferred into a separation cell (SEPA ST, Osmonics, CA, USA) equipped with a STARMEM™ 122 membrane. The cell was pressurized to 30 bar at room temperature starting the filtration of the post reaction mixture. After filtering 35 mL of the original 40 mL of organic liquid, the catalyst was washed out using toluene or fresh bromoheptane to form a new organic phase for the next catalytic run. Both procedures were applied for the recycling of the catalyst in three consecutive runs. Similar conversions were obtained in all three runs while more than 99% of the catalyst was retained.

In a later publication,[51] Heck reactions (between iodobenzene and styrene) were performed in the same way. In contrast to the PTC, the catalyst for the Heck reaction had a retention lower than 90% (for the first cycle). Nevertheless, this catalyst was recycled 5 times resulting in only 20% of the initial activity in the last run.

The retentions of the catalysts were also measured on a synthetic reaction mixture. The Heck-catalyst showed a retention of 96% while under experimental conditions retentions lower than 90% were obtained. For the PTC the values are both higher than 99%. The authors assume that this big difference for the Heck-catalyst is caused by the formation of smaller Pd species in the catalytic cycle. However, no precipitation or Pd-black formation was observed.

4.7 Soluble Polymer Supported Catalysts

Because of the tedious synthesis involved in the preparation of dendrimers needed to support catalysts in a homogeneous way, many research groups are working on the use of soluble polymers as homogeneous supports. The soluble polymers have the advantage of easy, and thus cheaper synthesis, while the resulting catalyst still acts as a true homogeneous system without the problems related to heterogeneous systems such as mass transfer limitation and lower activity. In this section, catalytic systems using soluble polymers as supports and separation from the reaction mixture using membranes are discussed.

In the early 1970's, Bayer *et al.* reported the first use of soluble polymers as supports for the homogeneous catalysts.[52] They used non-crosslinked linear polystyrene (M_w ca. 100 000), which was chloromethylated and converted by treatment with potassium diphenylphosphide into soluble polydiphenyl(styrylmethyl)phosphines. Soluble macromolecular metal complexes were prepared by addition of various metal precursors e.g. $[Rh(PPh_3)Cl]$ and $[RhH(CO)(PPh_3)_3]$. The first complex was used in the hydrogenation reaction of 1-pentene at 22°C and 1 atm. H_2. After 24 h (50% conversion in 3 h) the reaction solution was filtered through a polyamide membrane [53] and the catalysts could be retained quantitatively in the membrane filtration cell.[54] The catalyst was recycled 5 times. Using the second complex, a hydroformylation reaction of 1-pentene was carried out. After 72 h the reaction mixture was filtered through a polyamide membrane and recycled twice.

Kragl and Dreisbach reported on the asymmetric diethylzinc addition to benzaldehyde using a chiral homogeneous catalyst supported on a soluble polymethacrylate.[55] ʟ,ʟ-Diphenyl-L-prolinol was bound to a copolymer from 2-hydroxyethylmethacrylate and octadecylmethacrylate forming a soluble catalyst with a M_w of 96 000 (Figure 4.39).

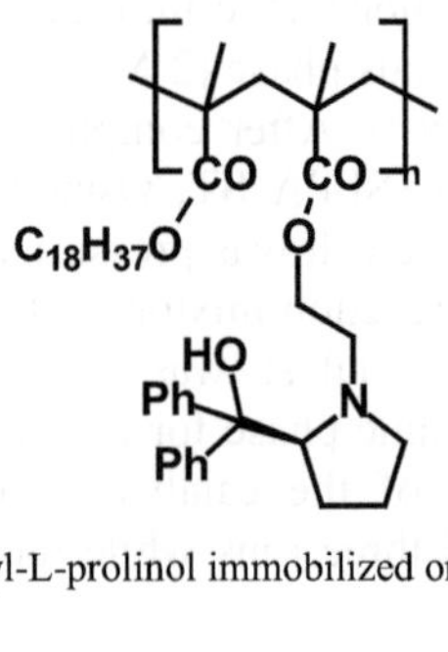

Figure 4.39. ʟ,ʟ-Diphenyl-L-prolinol immobilized on a polymethacrylate.[55]

Membrane filtration using a polyaramide membrane [56] showed a retention of more than 99.8%. Application of this catalyst in a continuously operated membrane reactor showed conversion for more than 150 h. The ee dropped from 80% in the beginning (non-bonded analogue 97%) to about 20% after 150 h. The average ee for the first 80 h was 50%.

Another soluble polymer-enlarged catalyst was synthesized and tested by Wandrey *et al.*[57] The catalyst was prepared by a coupling of an oxazaborolidine via a hydrosilylation reaction to a methyl hydrosiloxane-dimethylsiloxane copolymer (Figure 4.40). The catalyst was used in the enantioselective borane reduction of ketones.

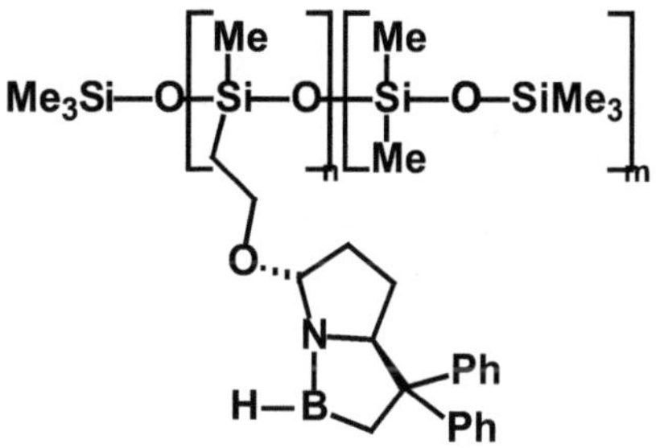

Figure 4.40. Oxazaborolidine ligand on siloxane polymer.[57]

The reduction of acetophenone was carried out at r.t. giving 86% yield with an ee of 97%. This is similar to the ee obtained with unbound analogues. A limited study was conducted on the retention of the catalyst by nanofiltration. It was found that the compound could be retained in the membrane reactor but no specific details were given about these measurements.

The use of such an oxazaborolidine system in a continuously operated membrane reactor was demonstrated by Kragl *et al.*[58] Various oxazaborolidine catalysts were prepared with polystyrene-based soluble supports. The catalysts were tested in a dead-end setup (paragraph 4.2.1) for the reduction of ketones. These experiments showed higher ee's than batch experiments in which the ketone was added in one portion. The ee's vary from 84% for the reduction of propiophenone to up to >99% for the reduction of L-tetralone. The catalyst showed only a slight deactivation under the reaction conditions. The TTON could be increased from 10 for the monomeric system to 560 for the polymer-bound catalyst.

Kragl and Wandrey made a comparison for the asymmetric reduction of acetophenone between oxazaborolidine and alcohol dehydrogenase.[59] The oxazaborolidine catalyst was bound to a soluble polystyrene [58] and used borane as the hydrogen donor. The carbonyl reductase was combined with formate dehydrogenase to recycle the cofactor NADH which acts as the hydrogen donor. Both systems were run for a number of residence times in a continuously operated membrane reactor and were directly comparable. With the chemical system, a space-time yield of 1400 g L^{-1} d^{-1} and an ee of 94% were reached whereas for the enzymatic system the space-time yield was 88 g L^{-1} d^{-1} with an ee of >99%. The catalyst half-life times were

1.2 and 31.1 days respectively for the chemical and the enzymatic system, which lead to a TTON of 560 for the chemical system vs. 78 000 for the enzymatic system.

Liese *et al.* attached a transfer-hydrogenation catalyst to a soluble polymer and applied this system in a continuously operated membrane reactor.[60] A Gao-Noyori catalyst was bound to a soluble polysiloxane polymer via a hydrosilylation reaction (Figure 4.41).

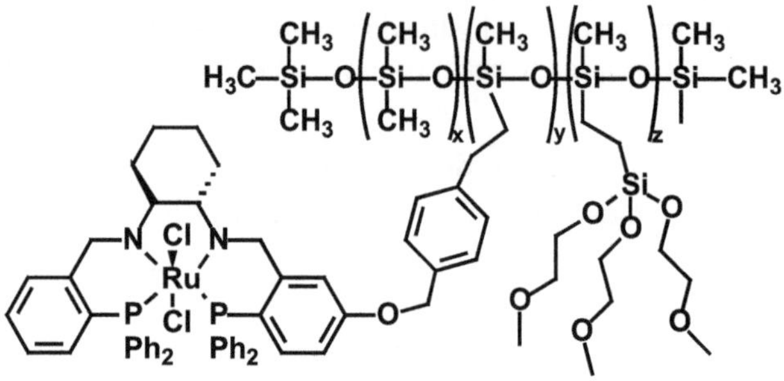

Figure 4.41. Goa-Noyori catalyst bound to a polysiloxane polymer.[60]

Application of this system in the continuous transfer-hydrogenation reaction of acetophenone gave a stable conversion of about 87%, an ee of 94%, and a space-time yield of 255 g L^{-1} d^{-1}. A continuous dosage of isopropoxide was necessary in order to compensate for deactivation caused by traces of water in the feed stream. Under these circumstances a TTON of 2360 was reached. Comparison of this system with an enzymatic process showed that both approaches offer different advantages and are therefore complementary.

A non-covalently bound catalyst was reported by Mecking *et al.*[61] The system was prepared by reacting poly(diallyldimethylammmonium chloride) (PDADMAC) with $NaBAr^F_4$ to form PDADMA-BAr^F_4. Exchanging multiple $NaBAr^F_4$ for [(H)Rh(CO)$(NaTPPTS)_3$] resulted in the formation of the electrostatically polymer-bound complex (Figure 4.42).

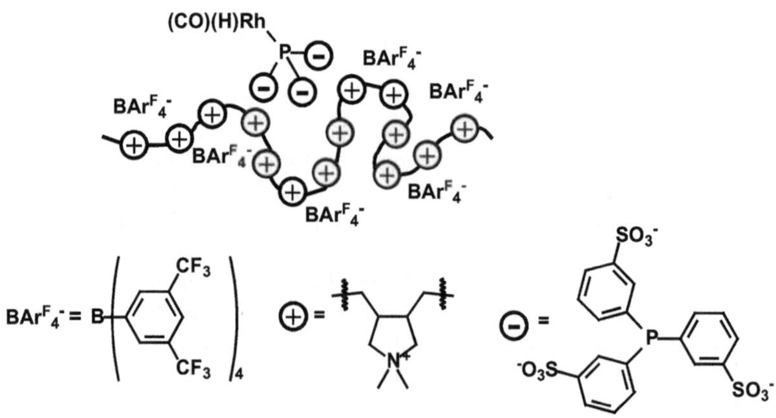

Figure 4.42. Electrostatically polymer-bound rhodium complex.[61]

After a hydroformylation run, the reaction solution was subjected to ultrafiltration using an asymmetric polyethersulfone membrane (MWCO 50 kDa) supplied by Sartorius. A retention of 99.8% was found. When the catalyst solution was recycled, virtually the same catalytic activity was observed again (165 TO h^{-1}). Repetitive recycling experiments resulted in 2-7% loss of rhodium, which was subscribed to partial oxidation of the phosphine ligand.

Plenio *et al.* tested an adamantyl phosphine ligand bound to soluble polystyrene (Figure 4.43) in various palladium-catalyzed C-C coupling reactions.[62] The retention of metal complexes of the polymer-bound phosphine ligand were determined to be higher than 99.95%.

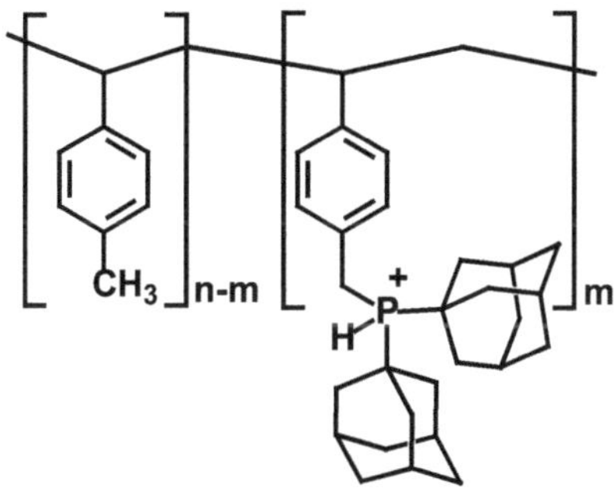

Figure 4.43. Polymer-bound adamantyl phosphine ligands.[62]

A complex prepared from this ligand system and $[Pd(PhCN)_2Cl_2]$ was tested in the Sonogashira coupling showing high yields. After the reaction the solution was filtered through a hydrophobic membrane composed of a dense poly(dimethylsiloxane) layer cast on a porous sublayer of poly(acrylonitrile). The catalyst was recycled seven times, resulting in a slight decrease in conversion from 98% to 80% in the last cycle. The $Pd(OAc)_2$ complexes of the same polymeric ligand were tested in the Suzuki coupling also showing high yields with a slowly decreasing conversion over multiple cycles. Using $[Pd(dba)_2]$ as the metal source, the system was tested in the Heck coupling. Yields between 87% and 80% were obtained. However, on attempting the nanofiltration of the NMP solution, the membrane immediately suffered severe damage. No suitable solvent could be found for this system. Nanofiltration experiments were conducted with the NMP solutions diluted with large amounts of cyclohexane. However, low yields and deactivation of the catalyst resulted.

Van Koten and Frey used a hyperbranched poly(triallylsilane) as the support for palladium- pincer complexes.[63] The supported palladium-pincer complexes were applied in the catalytic aldol condensation of benzaldehyde and methyl isocyanate. Their activity was similar to that of single site Pd catalysts. According to the authors, the complex is suitable for continuous membrane applications, as demonstrated by their purification by means of dialysis.

4.8 Concluding Remarks

Since none of the commercially available nano- or ultrafiltration membranes so far shows real long-term resistance against organic solvents under the reaction conditions needed for a commercially interesting hydroformylation process and since no prices are available for bulk quantities of membranes for larger scale applications, considerations about the feasibility of such processes are difficult and would be highly speculative.

However, with the combined arguments of process intensification due to simultaneous product and catalyst separation, increased product purity, and the reaction engineering advantages provided by a continuous homogeneous catalysis, a rapid development can be foreseen. A lot more fundamental work is needed in order to underline the applicability, together with the development of new membrane materials showing better performance. This field is an ideal example of multidisciplinary research in which contributions from many fields is a prerequisite for success. The availability of larger quantities of solvent resistant membranes will be driven by a strong technology push in the near future.

4.9 References

[1] B. Cornils, W. A. Herrmann, *Applied Homogeneous Catalysis with Organometallic Compounds*, 2nd Ed. Wiley-VCH, **2002**.

[2] D. Vogt, in *Aqueous Phase Organometallic Catalysis – Concepts and Application* (Eds.: B. Cornils, W. A. Herrmann) 2nd compl. rev. and enl. ed., Wiley-VCH **2004,** 541.

[3] U. Kragl, D. Gygax, O. Ghisalba, C. Wandrey, *Angew. Chem., Int. Ed.* **1991**, *30*, 827.

[4] J. Wöltinger, A. S. Bommarius, K. Drauz, C. Wandrey, *Org. Process Res. Dev.* **2001**, *5*, 241.

[5] U. Kragl, C. Dreisbach, *Angew. Chem., Int. Ed.* **1996**, *35*, 642.

[6] N. J. Hovestad, E. B. Eggeling, H. J. Heidbüchel, J. T. B. H. Jastrzebski, U. Kragl, W. Keim, D. Vogt, G. van Koten, *Angew. Chem. Int. Ed.* **1999**, *38*, 1655.

[7] E. B. Eggeling, N. J. Hovestad, J. T. B. H. Jastrzebski, D. Vogt, G. van Koten, *J. Org. Chem.* **2000**, *65*, 8857.

[8] R. Sablong, U. Schlotterbeck, S. Mecking, D. Vogt, *Adv. Synth. Catal.* **2003**, *345*, 333.

[9] L. J. P. van den Broeke, E. L. V. Goetheer, A. W. Verkerk, , E. de Wolf, B.-J. Deelman, G. van Koten, J. T. F. Keurentjes, *Angew. Chem. Int. Ed.* **2001**, *40*, 4473.

[10] E. L. V. Goetheer, A. W. Verkerk, L. J. P. van den Broeke, E. de Wolf, B.-J. Deelman, G. van Koten, J. T. F. Keurentjes, *J. Catal.* **2003,** *219*, 126.

[11] D. E. de Vos, I. F. J. Vankelecom, P. A. Jacobs (Editors), *Chiral Catalyst Immobilization and Recycling*, Wiley-VCH ISBN 3-527-29952-1, **2000**.

[12] Millipore, Billerica, USA, http://www.millipore.com

[13] Schleicher & Schuell MicroScience GmbH, Dassel, Germany, http://www.schleicher-schuell.com.

[14] D. Nair, J. T. Scarpello, I. F. J. Vankelecom, L. M. Freitas Dos Santos, L. S. White, R. J. Kloetzing, T. Welton, A. G. Livingston, *Green Chemistry* **2002**, *4*, 319.

[15] D. Nair, J. T. Scarpello, L. S. White, L. M. Freitas, Dos Santos, I. F. J. Vankelecom, T. Welton, A. G. Livingston, *Tetrahedron Lett.* **2001**, *42*, 8219.

[16] TU Eindhoven, The Netherlands, http://www.catalysis.nl/homogeneous_catalysis.html.

[17] I. F. J. Vankelecom, *Chem. Rev.* **2002**, *102*, 3779.

[18] K. K. Sirkar, P. V. Shanbhag, A. S. Kovvali, *Ind. Eng. Chem. Res.* **1999**, *38*, 3715.

[19] Koch Membrane Systems, Wilmington, USA, http://www.kochmembrane.com.

[20] C. Linder, M. Nemas, M. Perry, R. Katraro, U.S. patent no. 5.265.734.

[21] E. Gibbins, M. D'Antonio, D. Nair, L. S. White, L. M. S. dos Santos, I. F. J. Vankelecom, A. G. Livingston, *Desalination* **2002**, *147*, 307.

[22] Membrane Exctraction Technology, London, U.K., http://www.membrane-extraction-technology.com.

[23] http://www.gewater.com

[24] M. K. Koukou, N. Papayannakos, N. C. Markatos, M. Bracht, H. M. Van Veen, A. Roskam, *J. Membra. Sci.* **1999**, *155*, 241.

[25] http://www.ecn.nl.

[26] Hermsdorfer Institut für Technische Keramik, Hermsdorf, Germany, http://www.hitk.de.

[27] Corning, New York, U.S.A., http://www.corning.com.

[28] E. Buhleier, W. Wehner, F. Vögtle, *Synthesis* **1978**, 155.

[29] J. W. J. Knapen, A. W. van der Made, J. C. de Wilde, P. W. N. M., P. Wijkerns, D. M. Grove, G. van Koten, *Nature*, **1994**, *372*, 659.

[30] N. Brinkmann, D. Giebel, G. Lohmer, M. T. Reetz, U. Kragl, *J. Catal.*, **1999**, *183*, 163.

[31] D. de Groot, E. B. Eggeling, J. C. de Wilde, H. Kooijmans, R. J. van Haaren, A. W. van der Made, A. L. Spek, J. N. H. Reek, D. Vogt, P. C. J. Kamer, P. W. N. M. van Leeuwen, *Chem. Commun.*, **1999**, 1623.

[32] D. de Groot, J. N. H. Reek, P. C. J. Kamer, P. W. N. M. van Leeuwen, *Eur. J. Org. Chem.* **2002**, 1085.

[33] D. de Groot, B. F. M. de Waal, J. N. H. Reek, A. P. H. J. Schenning, P. C. J. Kamer, E.W. Meijer, P. W. N. M. van Leeuwen, *J. Am. Chem. Soc.* **2001**, *123*, 8453.

[34] H. P. Dijkstra, N. J. Ronde, G. P. M. van Klink, D. Vogt, G. van Koten, *Adv. Synth. Catal.* **2003**, *345*, 364.

[35] H. P. Dijkstra, C. A. Kruithof, N. J. Ronde, R. van de Coevering, D. J. Ramón, D. Vogt, G. P. M. van Klink, G. van Koten, *J. Org. Chem.* **2003**, *68*, 675.

[36] L. Ropartz, R. E. Morris, G. P. Schwarz, D. F. Foster, D. J. Cole-Hamilton, *Inorg. Chem. Commun.* **2000**, *3*, 714; L. Ropartz, D. F. Foster, R. E. Morris, A. M. Z. Slawin, and D. J. Cole-Hamilton, *J. Chem. Soc., Dalton Trans.*, **2002**, 1997.

[37] L. Ropartz, R. E. Morris, D. F. Foster, D. J. Cole-Hamilton, *Chem. Commun.* **2001**, 361; L. Ropartz, K. J. Haxton, D. F. Foster, R. E. Morris, A. M. Z. Slawin, and D. J. Cole-Hamilton, *J. Chem. Soc., Dalton Trans.*, 2002, 4323; K. J. Haxton, D. J. Cole-Hamilton, and R. E. Morris, *Dalton Trans.*, 2004, 1665.

[38] R. Beinbauer, E.J. Jacobsen, *Angew. Chem. Int. Ed.* **2000**, *39*, 3604.

[39] A. W. Kleij, R. A. Gossage, J. T. B. H., J. Boersma, G. van Koten, *Angew. Chem. Int. Ed.* **2000**, *39*, 176.

[40] A. W. Kleij, R. A. Gossage, R. J. M. Klein Gebbink, N. Brinkmann, E. J. Reijerse, U. Kragl, M. Lutz, A.L. Spek, J. Boersma, G. van Koten, *J. Am. Chem. Soc.* **2000**, *122*, 12112.

[41] G. D. Engel, L. H. Gade, *Chem. Eur. J.* **2002**, *8*, 4319.

[42] Y. Ribourdouille, G. D. Engel, M. Richard-Plouet, L. H. Gade, *Chem. Commun.* **2003**, 1228.

[43] H.-F. Chow, C.-W. Wan, *Helv. Chim. Acta*, **2002**, *85*, 3444.

[44] W. Zhang, S. E. Tichy, L. M. Pérez, G. C. Maria, P. A. Lindahl, E.E. Simanek, *J. Am. Chem. Soc.* **2003**, *125*, 5086.

[45] G. E. Oosterom, S. Steffens, J. N. H. Reek, P. C. J. Kamer, P. W. N. M. van Leeuwen, *Topics in Catal.* **2002**, *19*, 61.

[46] L. W. Gosser, W. H. Knoth, G. W. Parshall, *J. Mol. Catal.* **1977**, *2*, 253.

[47] K. de Smet, S. Aerts, E. Ceulemans, I. F. J. Vankelecom, P. A. Jacobs, *Chem. Commun.* **2001**, 597.

[48] K. de Smet, A. Pleysier, I. F. J. Vankelecom, P. A. Jacobs, *Chem. Eur. J.* **2003**, *9*, 334.

[49] S. S. Luthra, X. Yang, L. M. Freitas dos Santos, L. S. White, A. G. Livingston, *Chem. Commun.* **2001**, 1468.

[50] S. S. Luthra, X. Yang, L. M. Freitas dos Santos, L. S. White, A. G. Livingston, *J. Membr. Sci.* **2002**, *201*, 65.

[51] D. Nair, S. S. Luthra, J. T. Scarpello, L. S. White, L. M. Freitas dos Santos, A. G. Livingston, *Desalination* **2002**, *147*, 301.
[52] E. Bayer, V. Schurig, *Angew. Chem.*. **1975**, *7*, 493.
[53] Polyamide, MW cut-off limit 10 000, manufactured by Berghof GmbH, 7401 Tübingen-Lustnau.
[54] Homemade from glass and Teflon.
[55] U. Kragl, C. Dreisbach, *Angew. Chem. Int. Ed.* **1996**, *35*, 642.
[56] Hoechst–Nadir UF-PA20, not commercially available anymore.
[57] M. Felder, G. Giffels, C. Wandrey, *Tetrahedron: Asymm.* **1997**, *8*, 1975.
[58] G. Giffels, J. Beliczey, M. Felder, U. Kragl, *Tetrahedron: Asymm.* **1998**, *9*, 691.
[59] S. Risson, J. Beliczey, G. Giffels, U. Kragl, C. Wandrey, *Tetrahedron: Asymm.* **1999**, *10*, 923.
[60] S. Laue, L. Greiner, J. Wöltinger, A. Liese, *Adv. Synth. Catal.* **2001**, *343*, 711.
[61] E. Schwab, Stefan Mecking, *Organometallics* **2001**, *20*, 5504.
[62] A. Datta, K. Ebert, H. Plenio, *Organometallics* **2003**, *22*, 4685.
[63] C. Schlenk, A. W. Kleij, H. Frey, G. Van Koten, *Angew. Chem. Int. Ed.* **2000**, *39*, 3445.

CHAPTER 5 BIPHASIC SYSTEMS: WATER – ORGANIC

ERNST WIEBUS[1], BOY CORNILS[2]

[1] *Celanese Europe GmbH, Werk Ruhrchemie, D-46128 Oberhausen, Germany;*

[2] *D-65719 Hofheim, Germany; e-mail: boy.cornils@t-online.de*

5.1 Introduction

Biphasic techniques for recovery and recycle are among the recent improvements of homogeneous catalysis - and they are the only developments which have been recently and successfully applied in the chemical industry. They are specially introduced into the hydroformylation (or "oxo") reaction, where they form a fourth generation of oxo processes (Figure 5.1 [1]). They are established as the "Ruhrchemie/Rhône-Poulenc process" (RCH/RP) [2]; cf. also Section 5.2.4.1), with annual production rates of approximately 800,000 tonnes y^{-1} (tpy).

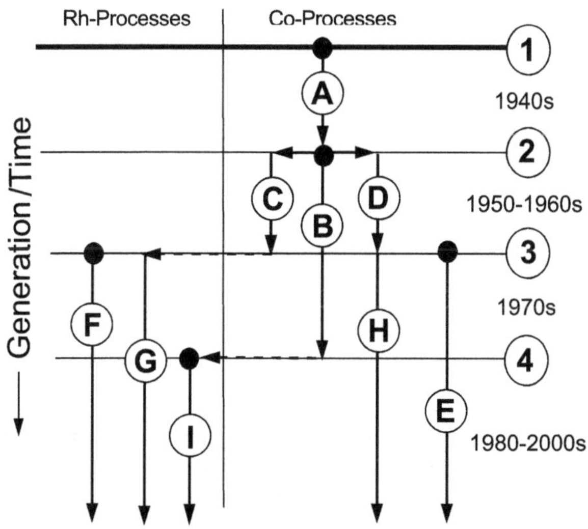

Figure 5.1. The generations of oxo processes [3] (symbolized by full points).A, First generation Ruhrchemie process 1943 (diaden process [4]); B, second generation Ruhrchemie process; C, second generation BASF process; D, second generation Kuhlmann process; E, third generation Shell process; F, third generation LPO (UCC) process; G, third generation BASF process; H, third generation Exxon (Kuhlmann) process; I, fourth generation Ruhrchemie/Rhône-Poulenc process

According to the early scientists developing this process, Manassen/Whitehurst [5a] and

D. J. Cole-Hamilton and R. P. Tooze (eds.), Catalyst Separation, Recovery and Recycling, 105–143.

Joó [5b,c,d], biphasic stands for the use of two immiscible liquid phases, one containing the catalyst and the other containing the unreacted substrate and the reaction products. The tremendous progress in homogeneous catalysis is that both liquids, i.e., phases, can be separated off after the reaction is complete by simply separating the second phase from the catalyst solution, thus making it easy to recirculate the latter without any thermal or chemical stress (which is the case in all classical separation methods, see Chapter 2 and [3]). Therefore, this technique allows full utilization of the inherent advantages of homogeneous catalysis and avoids the costly recycling procedures of traditional homogeneous catalysts. In this respect, biphasic catalysis is the most successful variant of an immobilization technique, i.e., an "anchoring" of the catalyst onto a liquid support [6]. The application of biphasic catalysts simplifies the process of homogeneous catalysis considerably.

Early experimental work on biphasic techniques came up with large scale processes for the above mentioned hydroformylation of alkenes, oligomerization of ethene, telomerization of butadiene, and some small-scale processes for the manufacture of fine chemicals. Whereas the oligomerization proceeds in an organic/organic biphasic system (SHOP process of Shell; [7]), the other reactions use biphasic *aqueous* conditions which offer an especially advantageous mode of operation. Speaking of "Green Chemistry" and "Green Catalysis" and their definitions (Ken Seddon: "Green chemistry is all about reducing the number and amount of harmful chemicals that are used and/or generated in research and industry.... This new field is all about minimizing the amount of waste..."), reacting substrates in aqueous biphasic operation is the most progressive example [8] - far beyond other techniques such ionic liquids where the specialists make encouraging statements such as "You think your process is green, how do you know?" [8c] or "Ionic liquids are not always green!" [8d].

This Chapter will concentrate on the hydroformylation of propene by means of rhodium catalysts, modified by water-soluble ligands such as TPPTS (**tri****p**henyl**p**hosphine *m*-**t**ri**s**ulfonate).

5.2 Immobilization with the Help of Liquid Supports

5.2.1 GENERAL

In Chapter 3 about *Supported Catalysts* the numerous unsuccessful attempts to heterogenize oxo-active transition metal complexes are compiled: the difficulty of catalyst/product separation appeared to have been solved. However, it was discovered during long-term tests to verify the utility of the concept (tests which do simulate industrial conditions in terms of uninterupted long runs and numerous catalyst cycles, TONs and TOFs, changes of temperatures, pressures, loads, etc.) all of these essentially supported oxo catalysts, tended to leach out. This means that the active catalyst metal and also the modifying ligand slowly, but steadily, became detached from the heterogeneous support and were carried away after an uneconomically short time: the problem of homogeneous catalysis may have been delayed but has not been solved. Even hybrid techniques such as

SLPC or SAPC (**s**upported **l**iquid [or **a**queous] **p**hase **c**atalysis [9,10,62,64]; see also Section 5.2.5) provide no improvement, probably because of the tremendous stress on the support/transition metal bond during the repeated change between tetrahedral and trigonal-bipyramidal metal carbonyls over the course of a single catalyst cycle. Only recent publications [11,21,26b,28h] report on successful realization of supported homogeneous hydroformylation catalysts, but so far there is no confirmation by practise-soriented tests - not to mention by commercial applications.

5.2.2 BIPHASIC SYSTEMS

Only the biphasic method, specially of *aqueous*-biphasic catalysis, has provided a fundamental remedy to the problem of stress-free and economical recovery and recycle of homogeneous oxo catalysts [12]. The fact that the catalyst, which still acts homogeneously, is dissolved in water, thus in a polar solvent, and remains dissolved, enables it to be separated from the nonpolar products without problems and with minimal effort after reaction.

The decisive step was the development of ligands with two concurrent properties: they are simultaneously water-soluble and they modify the catalytically actice complex [13]. Pioneering work was carried out by Joó (mainly concerning hydrogenations) [5], Kuntz (then with Rhône-Poulenc, [12g,14]), and the former Ruhrchemie AG [12,77,79] - and thus in industry - which led to the first large-scale utilization of the aqueous, homogeneous catalysis technique at the beginning of the 1980s [7,12]. The generally used embodiment of two-phase catalysis (e.g., as practized in Shell's SHOP method [7]; see also Section 5.2.6) was thus joined by *aqueous* two-phase catalysis.

In classical homogenous catalysis, an organic compound, i.e., a real liquid phase (a solvent) dissolves all of the reactants, catalysts, and products. The role of the solvent is underlined by the fact that it has to be separated from the reaction products by an additional and costly step, for example by distillation.

Olivier-Bourbigou and Hugues [15] define the importance of solvents as follows: "The role of the solvent in organic reactions is of the utmost importance [16b]. Its effect can just be limited to a "physical effect" in making possible the solubilization of the reactants, with no direct interaction with the active center. More interesting are the cases in which the solvent interacts through specific forces, such as hydrogen bonds, thus altering the mechanism, the rate, and eventually the selectivity of the reaction in stabilizing certain reaction intermediates. In homogeneous catalysis, the solvent effect is often difficult to explain due to the large number of reaction intermediates involved. However, for catalytic applications involving coordinatively unsaturated cationic metal centers, the "ideal" solvent should be able to solubilize the metal ion, while maintaining its ionic character, and should create weak and labile metal-solvent bonds. It should he highly polar and non-coordinating for the active metal center. Most of the classical organic solvents are covalent and do not satisfy the above requirements." Especially the necessity for a separate and costly separation step is a heavy burden for bulk chemical processes operating with real solvents, diluents, etc.

For aqueous biphasic systems, the "solvent" is water which shows pronounced solvent effects (cf. Section 5.2.3.1) but, on the other hand, does not have to be separated by means involving thermal or other stresses either from the reactants or from the products. Therefore, the water of the aqueous-phase processes has to be regarded as a "supporting fluid" rather than a real solvent although there are many cases known in which water acts as an accelerator for organic and even organometal-catalyzed reactions [17,18]. The key for this behavior is the role of water in both, the influence on activity and selectivity of the desired reaction **and** the suitability to act as phase separating agent [19] - presumably enhanced by its tendency to form micelles, microemulsions, or other surface-based aggregates [51b,51c,52d, 54,58]. Not recognized by all academian scientists [20], the phase-separating power is the decisive advantage of water.

5.2.3 AQUEOUS BIPHASIC CATALYSIS

5.2.3.1 Water as a Solvent

Aqueous biphasic catalysis is a special case of the two-phase processes of homogeneous catalysis. Despite the academic literature's provocative question "Why water?" [18a,18b], the advantages of water as the second phase and the "liquid support" are numerous. On the one hand, the search for the necessary solubility gap is much easier with water than with various organic-phase liquids (Figure 5.2). Additionally, water has many properties which predestine it as a ideal liquid support in homogeneous catalysis (see Table 5.1) [18c,18d].

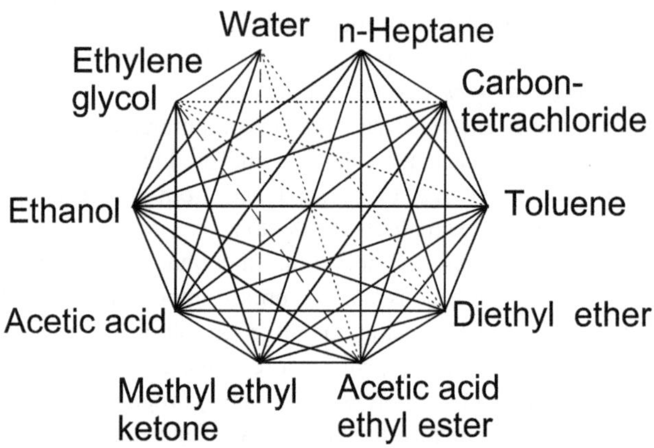

Figure 5.2. Miscibility diagram (and solubility gaps) of water and organic-phase liquids. Solvents not connected by a binding line in Figure 5.2 are immiscible; solvents of unlimited miscibility are connected by a solid line, those of limited miscibility by a dotted line [16]

Water has several anomalous features (e.g., density, being the only nontoxic and liquid "hydride" of the non-metals, melting point varying with pressure, etc.). Of direct importance for the aqueous biphasic process are the physiological (entries 2 and 4 of Table 5.1), economic (1,3,6,9), ecological/safety-related (2,3,4,9), process engineering (1,6,7,9,10,11,12), and chemical and physical properties (1,5,6,8,11,13) of water. The different properties interact and complement each other. Thus water, whose high

Hildebrand parameter and high polarity advantageously influence organic chemical reactions (such as hydroformylation), has sufficiently high polarity and density differences compared to organic (reaction) products to enable separation of the phases after the homogeneously catalyzed reaction is completed [17].

TABLE 5.1. Properties of Water as a Liquid Support [12j]

Entry	Property
1	Water is polar and easy to separate from non-polar solvents or products; its polarity may influence (i.e., improve) reactivity
2	Water is non-flammable and incombustible: A decisive advantage in terms of safety and occupational health
3	Water is ubiquitous and available with suitable quality
4	H_2O is odourless and colourless which makes contamination easily detectable
5	The physical and physico-chemical characteristics (e.g., hexagonal 2D surface structure, tetrahedral 3D molecular network) influence the mutual (in)solubility significantly; chaotropic compounds lower the order by H-bond breaking
6	The Hildebrand parameter as the unit of solubility of non-electrolytes in organic solvents/ reaction products is high
7	Water's density of 1 $g{\cdot}cm^{-3}$ provides a sufficient difference in density from those of most organic substances
8	The dielectric constant, DK, is very high; the refraction index n_D is low
9	The high thermal conductivity, the high specific heat capacity, and the high evaporation enthalpy of water make it suitable as solvent **and** heat removing fluid
10	Water has a high solubility for gases, especially CO_2
11	Water may form hydrates and solvates
12	Water is highly dispersable; it has a high tendency to support micelle and/or microemulsion formation. These tendencies may be enhanced by additives such as surfactants
13	Water has amphoteric behaviour in a Brønsted sense

Compared with this, the high solvent power for many compounds and gases, in some cases boosted by solvate or hydrate formation or by H-bonding, facilitates reactions in the two-phase system.

The chaotropic properties of many chemical compounds prevent the H_2O cage structures necessary for the formation of solvates and thus facilitate the transfer of nonpolar molecules between nonaqueous and aqueous phases. Water is incombustible and non-flammable, odorless and colorless, and is universally available in any quality: important prerequisites for the solvent of choice in catalytic processes. The DK and n_D can be important in particular reactions and are advantageously used for the analysis and control of substrates and products. The favorable thermal properties of water make it highly suitable for its simultaneous dual function as a mobile support *and* heat transfer fluid, a feature that is utilized in the RCH/RP process (see below).

Compared to the inexpensive and ubiquitous solvent and support, water, with its unique combination of properties, other alternative solvents may well remain unimportant. Others make the same comments using different words (Table 5.2).

TABLE 5.2. Pros and cons of water as a reaction fluid [21]

Advantages	Disadvantages
* Not inflammable	* Large heat of evaporation[1]
* Non-toxic	* Detection in case of leakage[2]
* No smell	* Low solubility of many nonpolar substrates
* Good separation with many organics	* Hard to collect in case of spills[3]
* Cheap	
* Unique fluid properties	* No incineration of bleed streams[4]
* Stabilization of certain organometalic complexes	* Decomposition of water-sensitive compounds[5]

[1] Possibly an advantage within the economics of the heat compound; [2] an advantage since contaminated water smells intensively; by the way the danger occuring in case of spills is severely overestimated. Additionally, it must be underlined that the danger in case of spills is identical with biphasic or with homophasic (conventional) operation; [3] True for all "solvents" and all organic liquids and no specific disadvantage of water; [4] Advantageous when containing the water-soluble catalyst - the system is self-extinguishing; [5] A matter of evaluation: nobody will recommend water in case of decomposition of water-sensitive compounds.

5.2.3.2 Aqueous-phase Catalysis as a Unit Operation

Operating homogeneously catalyzed conversions under aqueous-biphasic conditions is certainly a recently introduced technique but does not necessarily require newly designed apparatuses and exorbitant new sets of reaction conditions. To "do the solubility split" [19], the catalyst should only be soluble in one of the liquids (i.e., water), while the products (and occasionally the reactants) should be mainly soluble in the other - preferentially the phase of the reaction products itself. In this case, the separation of the catalyst from the products can simply be achieved by decantation (or temporarily by other means such as extraction, etc.). In all cases, the reactor design ought to maintain biphasic operation by proper mixing of the reactants. After decantation, the phase containing the catalyst can simply be recycled to the reactor, and the products can be separated and/or purified without the need to consider any stability or reactivity problems for the catalyst or the product. Basically, the flow sheet of an aqueous-biphasically and homogeneously catalyzed oxo process is as simple as shown in Figure 5.3 b) in comparison to the scheme of a classical hydroformylation unit Figre 5.3 a). The comparison gives an impression of the tremendous savings in investment made possible by aqueous biphasic operation.

An important feature of biphasic hydroformylation is the separability due to density differences. Because of the differences in density of the polar compound water (1.0 $g \cdot cm^{-1}$) and the hydrophobic oxo products (average 0.8), no problems occur. Additionally, the hydroformylation products are not sensitive to water. Another important question is to what extent water and the reactants are mixed. Therefore, the reactor in Figure 5.3 b), a continuously stirred tank reactor (CSTR) [22], normally contains usual installations to guarantee excellent mixing. For the lower alkenes with their significant water solubility (propene, butene) this is no problem. In these cases, the hydroformylation reaction takes place at the interfacial region [23].

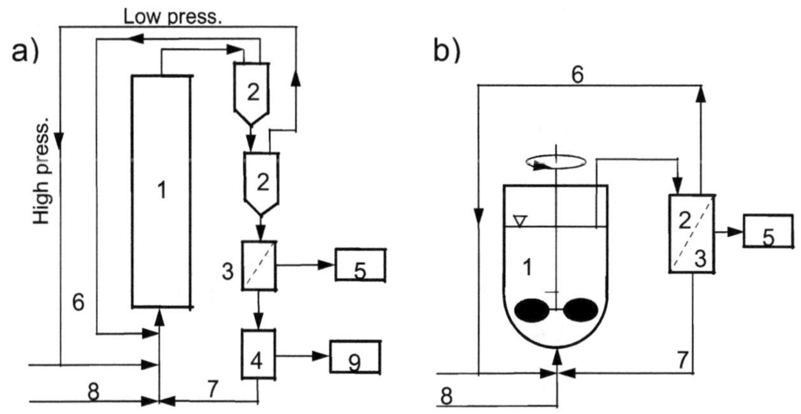

Figure 5.3. Basic flow-sheets of a) a conventional, homogeneously catalyzed process and b) an aqueous-biphasically, homogeneous catalytic process. 1, Reactor; 2 Separator(s); 3, Catalyst separator; 4 , Make-up; 5, Further purification and processing; 6, Gas recycle(s); 7, Catalyst recycle; 8, Reactant feed; 9, Withdrawal of high boilers

So far only propene and butene are hydroformylated commercially using the RCH/RP process. A reason which has been postulated for this is the decreasing solubility in water with increasing number of C atoms in both the starting alkene and the reaction products (Figure 5.4) and the associated mass-transfer problems in the relatively complicated gas-liquid-liquid, three-phase reaction.

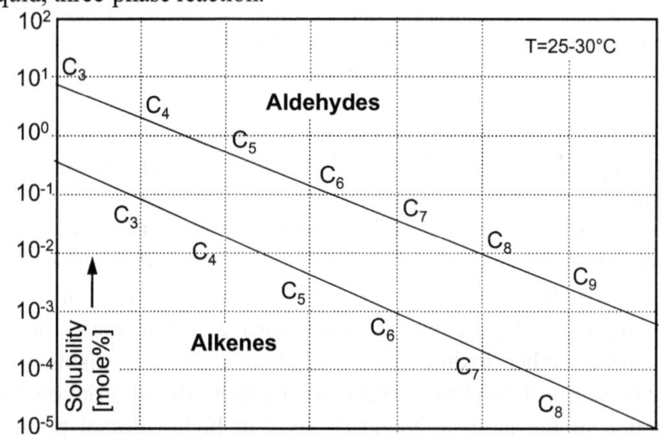

Figure 5.4. Solubility of alkenes and aldehydes in water (calculated. from [24])

Up to now only limited kinetic data and thus rate models (and even mechanistic details) of aqueous phase operation are available. Thus, in many cases only estimates and experimentally found data are at the disposal for reaction engineers' work (e.g.[25]). The state of the art of the hydroformylation of higher alkenes (>C_5) comprises additions of supplementary solvents/diluents or extraction fluids, surface-active agents (detergents), intensity and mode of stirring ([22b], power of agitation (cf. Figure 5.5) operation in micelles [51b,51c,52d,54,56h,58], thermo-regulated systems, etc. [26a,27,54a]. Up to now, it has to be noted that many measures may improve conversion and yield but not sufficiently to cover the additional expenses. The same is true for new and exotic ligands, although this measure would be the ideal solution of the problem when no additional solvents/extracting agents and their costly recycling are permitted [27,28,41e].

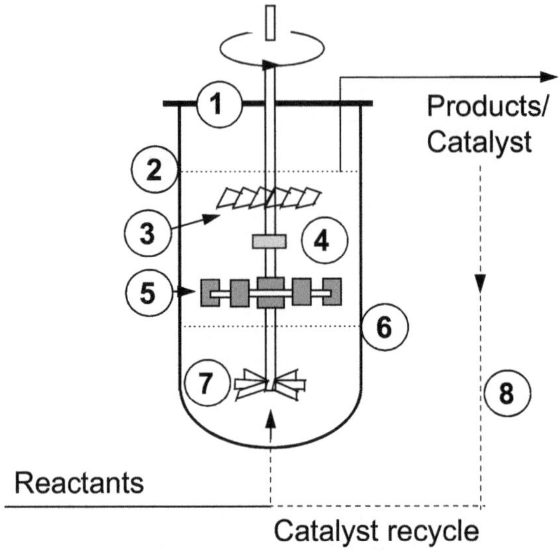

Figure 5.5. Example of a highly sophisticated CTSR [25a; other examples see 25c].
1, CTSR; 2, Interphase gas/liquid; 3, Self-rotating floating baffle; 4, Annulus for position limiting; 5, Rushton disk turbine; 6, Interphase liquid/liquid; 7, Pitched blade turbine upward (mixer/stirrer); 8, Aqueous-phase catalyst recycle

Quite new ideas for the reactor design of aqueous multiphase fluid/fluid reactions have been reported by researchers from Oxeno. In packed tubular reactors and under unconventional reaction conditions they observed very high space-time yields which increased the rate compared with conventional operation by a factor of 10 due to a combination of mass transfer area and kinetics [29]. Thus the old question of aqueous-biphase hydroformylation "Where does the reaction takes place?" - i.e., at the interphase or the bulk of the liquid phase [23,56h] - is again questionable, at least under the conditions (packed tubular reactors, other hydrodynamic conditions, in mini plants, and in the unusual,and costly presence of ethylene glycol) and not in harsh industrial operation. The considerable reduction of the laminar boundary layer in highly loaded packed tubular reactors increases the mass transfer coefficients, thus Oxeno claim the successful hydroformylation of 1-octene [25a,26,29c,49a,49e,58d,58f]. The search for a new reactor design may also include operation in microreactors [59].

The hydroformylation reaction is highly exothermic, which makes temperature control and the use of the reaction heat potentially productive and profitable (e.g, steam generation). The standard installation of Ruhrchemie/Rhône-Poulenc's aqueous-phase processes is heat recovery by heat exchangers done in a way that the reboiler of the distillation column for work-up of the oxo products is a falling film evaporator incorporated in the oxo reactor itself [2,12]. The heat of the oxo reaction is thus recovered as the reboiler heat source. This is a great advantage over the classical (non-aqueous) hydroformylation, which simply discard parts of the oxo reaction heat. The RCH/RP process is a net steam exporter. In the case of the above mentioned Oxeno developments the large catalyst flow is said to serve as a heat transfer medium itself thus making the process nearly isothermal [29g].

There are many proposals for addressing the problem of proper phase separation (and reducing the leaching of the catalyst) of aqueous-phase processes when partial mixing of the two phases occurs (or is necessary because of reaction engineering reasons). In this case the simple, rapid, and perfect phase separation without any leaching as in the RCH/RP process (the losses are of the order of 1 ppb [12e]) needs considerable effort; for example extraction with an appropriate additional solvent which is either added after the catalytic conversion or is already present during reaction [30]. In all cases, the reaction loop has to be completed by additional stages such as extraction devices and separation, recovery, and recycle of the extraction fluid. Two examples are given in Figure 5.6; they should be compared with the simple flowsheet of RCH/RP's oxo process (cf. Section 5.2.4.1) to recognize the expense which ia associateded with those solutions [21,30a,31].

Theoretically, it is possible to control chemical reactions by the different solubilities of the reactants or of special intermediates (e.g., according to their nucleophilicity) in different solvents or various catalyst fluids. The principle has been proven but is not yet used for special applications in industry [32].

The search for new ligands (which in a restricted sense is no unit operation) which would combine various properties such as modification of the central atoms of the complex catalyst, creating modifying and surfactant properties, offering chiral properties, etc. is still going on. There are some results but no ligand better than TPPTS has been developed so far ("better" includes simultaneously water-solubility, activity, selectivity, and price [17h,33]). On the contrary, all proposed new ligands are more expensive and shift the level of their costs close to those of the precious metals such as rhodium, palladium, or platinum. The search for "novel catalysts" for multiphase reactions must be done as an integrated task in close relationship with reactor design [34,48] - colloidal suspensions included [35a].

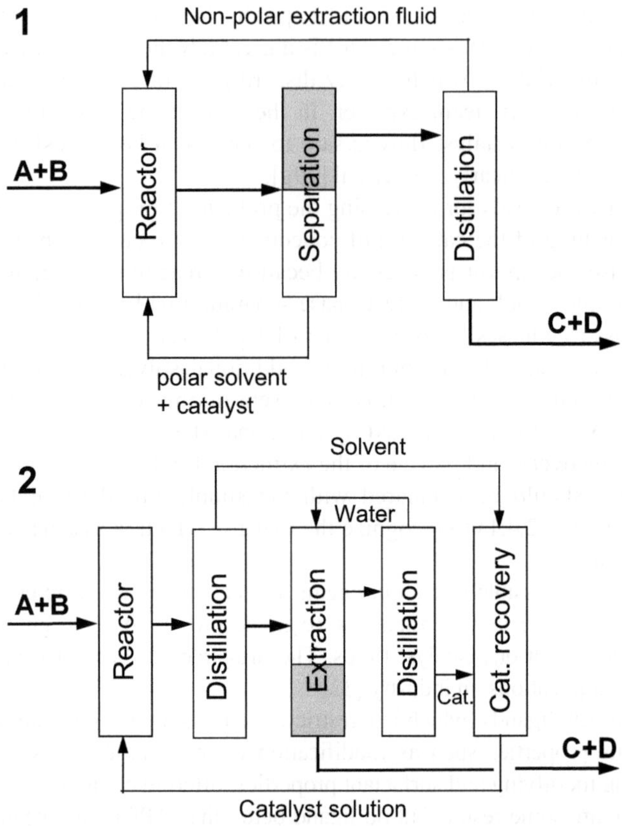

Figure 5.6. Process flowsheets of biphasic reaction A+B → C+D and extraction
1, Simultaneous reaction and extraction within the reactor; 2, Separate reaction and catalyst extraction

5.2.4 EXAMPLES OF AQUEOUS BIPHASIC CATALYSIS

5.2.4.1 Hydroformylation (Ruhrchemie/Rhône-Poulenc[RCH/RP] process)

The RCH/RP process converts propylene to *n*- and *iso*-butyraldehydes (or butenes to valeraldehydes) in the presence [HRh(CO)(TPPTS)] (with TPPTS = tris[sodium-*m*-sulfonatophenyl]phosphine as water soluble ligand) according to Equation 5.1.

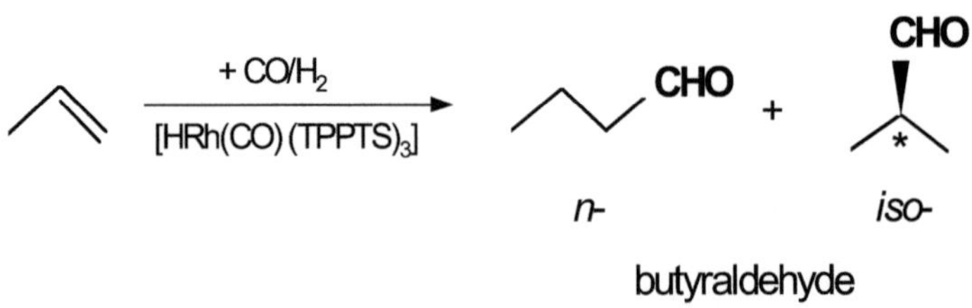

Equation 5.1. Hydroformylation of propylene in the RCH/RP

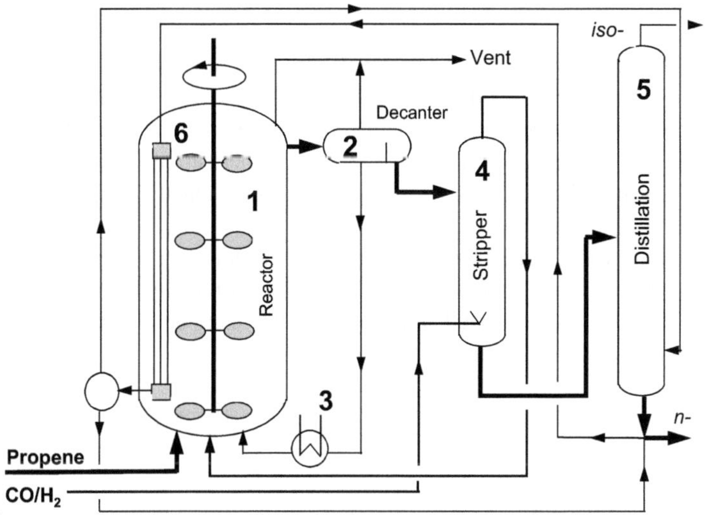

process

Figure 5.7. Process design for the Ruhrchemie/Rhône-Poulenc oxo process

The process scheme is shown in Figure 5.7. The reactor **1** is stirred and supplied with reactants and the catalyst. The catalyst [HRh(CO)(TPPTS)$_3$] is prepared by mixing a Rh salt and aqueous TPPTS solution (which may partly be recycled; see Section 5.3) in a simple “preforming” step (much simpler than the preforming in other oxo processes [35b-e]). Temperature is controlled *via* the heat recovery system with the reaction heat from the exothermic hydroformylation being used in the reboiler of the distillation column **5**. The reboiler is the internal cooler **6** of the reactor. The cooling medium is the reaction product, *n*-butanal. From the reactor, the reaction products pass through a downstream phase separator (decanter **2)** and stripping column **4**. The major part of the catalyst solution remains in the reactor; only a smaller part is separated in the decanter and returns directly to the reactor. In the phase separator, the crude aldehyde formed is freed of gases and further separated into mutually insoluble phases. This decanter ensures the essentially spontaneous phase separation. The reaction heat that is retained in the catalyst solution is recovered by a heat exchanger **3**, and the catalyst solution, supplemented by an amount of water equivalent to that carried off with the *iso*-butanal (since the *n*-aldehyde is practically anhydrous), is recirculated to the reactor. The crude aldehyde is freed of any unreacted alkene in the stripper **4** by means of syngas flowing countercurrently. It is essential that this stripping is accomplished in the absence of the oxo catalyst and, therefore, no selectivity- and/or yield-reducing secondary reactions of the crude aldehyde occur. Typical for the RCH/RP process are high yields and selectivities (99% butanals with an *n/i* ratio of up to 98:2; selectivity toward C_4 products: >99.5%) under mild conditions (120°C at 5 Mpa) [2,12].

The process is highly effective and, as the process data illustrate, represents an economic advantage as well (cf. Section 5.3.3). In ecological terms, the RCH/RP process also provides a considerable improvement (Section 5.3.4) over conventional methodologies.

Recovery and recycle of the catalyst are described in Section 5.3.

5.2.4.2 Other Industrially Used Aqueous-biphasic Processes

Hydroformylation comprises the state-of-the-art of bulk chemical production via aqueous-biphasic processes. At present five plants produce worldwide some 800,000 tpy of oxo products [1]. Another bulk process - the hydrodimerization of butadiene and water, a variant of telomerization – is run by Kururay with a capacity of 5000 tpy (Equation 5.2 [31b,36]).

The reaction product of telomerization is 2,7-octadiene-1-ol. In subsequent steps this dienol may be converted to 1-octanol by hydrogenation or hydrogenated/dehydrogenated to 1-octenal. This unsaturated aldehyde again can be hydroformylated to yield nonadialdehyde and then hydrogenated to nonadiol.

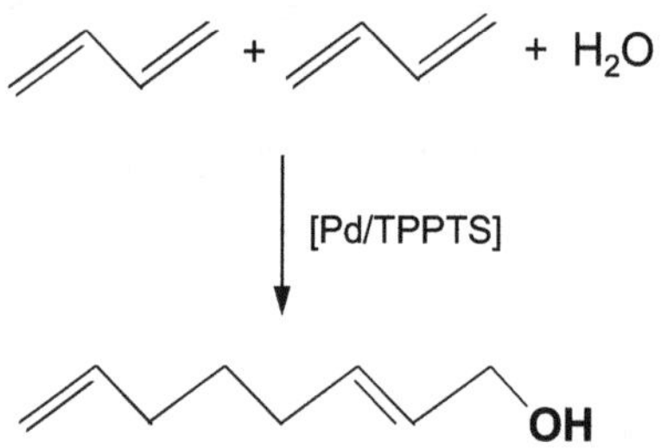

Equation 5.2. Hydrodimerization of butadiene and water

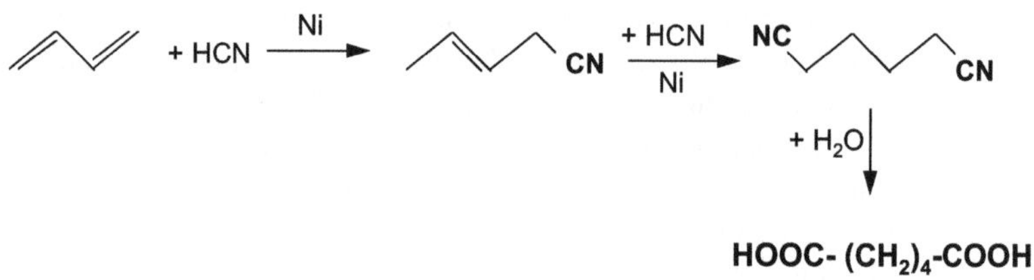

Equation 5.3. Hydrocyanation of butadiene

The remarkably versatile C_1 building block HCN may be used in the aqueous two-phased hydrocyanation, too (Equation 5.3; [37]). Also, some fine chemicals such as intermediates for vitamins, phenyl acetic acid, etc. are manufactured on an industrial scale using this technology (Equations 5.4 and 5.5; [12e,31b,38]).

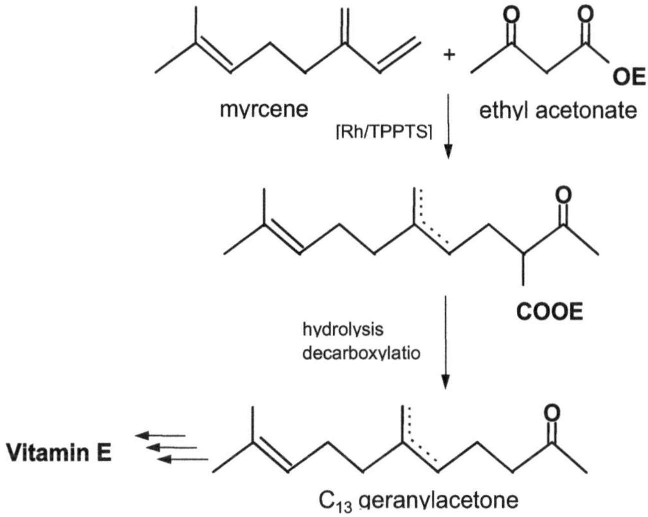

Equation 5.4. Manufacture of vitamin E precursors

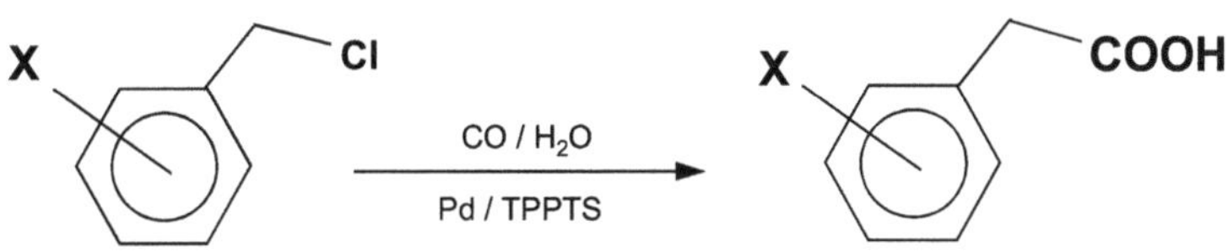

Equation 5.5. Manufacture of phenyl acetic acid

Today, the Suzuki cross coupling of aryl halides and arylboronic acids is also carried out in aqueous-biphasic operation starting from chlorinated derivatives instead of their more costly bromo or iodo equivalents (Equation 5.6, [39]).

The commercially applied biphasic processes are compiled in Table 5.3. Tests to produce economically interesting profens or other analgesics by two-phase hydrocarboxylation [40] remain industrially unsuccessful.

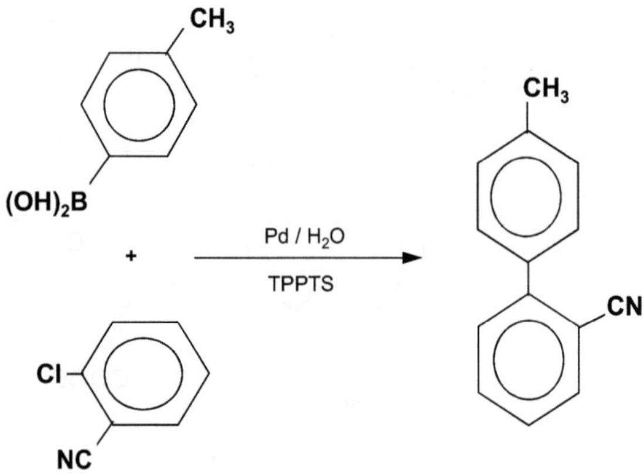

Equation 5.6. Suzuki coupling to yield aromatic biphenyls

TABLE 5.3. Commercial biphasic processes

Process/Catalyst	Products	Capacity / [tpy]
Shell SHOP/ Ni-P,O ligand	∟-Olefins and internal olefins	900 000
Ruhrchemie-RP (now Celanese)/Rh-TPPTS	*n*-Butanal	800 000
Kururay Co Ltd / Pd-TPPMS	*n*-Octanol, nonadiol	5 000
Clariant AG/Pd-TPPTS	Substituted biphenyls	<1 000
Rhodia (former Rhône-Poulenc)/Rh-TPPTS	Vitamin precursors	*

* Exact production figures not known

5.2.4.3 Short Overview of Other Reactions

A multitude of other reactions are compiled in (Table 5.4,[12m]). A proper choice of ligands and reaction conditions will make many other reactions available *via* aqueous-biphasic operation.

TABLE 5.4. Recently described examples for aqueous-biphasically operated reactions

Type of reaction/Catalytic metal involved	Refs.
Aldolization/Sc,In,Cu,Ln	[41a-d]
Alkylation/K,Pd, organocatalysts	[41e-g]
Allylation/Ru,Pd	[41d,h]
Alternate copolymerisation of CO-ethylene/Pd	[42]
Carbonylation	[43]
Claisen rearrangement/Al	[41i]
Diels-Alder reaction/Si,Cu	[41d,j,k]
Epoxidation/W,Re,Mn	[41l-p]
Friedel-Crafts reaction/Sc	[41q]
Heck reaction/Pd	[41r-v]
Hydrogenation/Ru,Rh,Pd,Ir,Pt	[5b],[41w-ii]
Hydrosilylation/Pt	[41jj]
Mannich reaction/Zn	[41d,vv]
Metathesis/Ru	[44]
Michael reaction/Ln	[41d,45]
Oxidation/Ru,Pd,Os,W,Mn	[46]
Polymerization/Cu,Ti,Ni,Pd,Rh,Co	[41kk-pp]
Reformatsky reaction/Zn	[57]
Sharpless dihydroxylation/Os	[41qq-ss]
Ullmann reaction/Pd,Rh,Ni	[41tt,uu]

5.2.5 OTHER PROPOSALS FOR WATER - BIPHASIC SYSTEMS

With the RCH/RP process, it is possible to hydroformylate propene up to pentenes with satisfying space time yields. On the other hand, heavier aldehydes such as C_{10} (*iso*-decanal) or higher from the hydroformylation of nonene(s), decenes, etc. can not be separated from the oxo catalysts by conventional means such as distillation due to thermal instability at the required temperatures and thus especially needs the careful aqueous-biphasic separation technique. There are numerous attempts to overcome the problem of low reactivity of higher alkenes which is due to low miscibility of the alkenes in water [26,27b, 50a,58d]. These proposals can briefly be summarized as:

(1) The use of amphiphilic water-soluble ligands which influence the alkene solubility or increase the catalyst concentration at the interface area of the phases [48,49];

(2) The use of co-solvents such as alcohols or detergents to improve interface exchange of the feedstock and the reaction products [11,21,26b,28h,50,51c,56g, 58i];

(3) Substitution of the water-biphasic procedure by supported aqueous phase catalysis (SAPC, see below) [39,10,26b,64];

(4) The separation of the reaction products from the catalyst solution by membrane techniques (Figure 5.8, [52]);

(5) In continuation of Bayer/Schurig's work from 1975/1976 [28a,b], the use of

polymeric water-soluble supports as “smart ligands” (see also the work of Bergbreiter and Mecking [48,53,56m,60,61]);

(6) The addition of surfactants to improve the formation of micelles, microemulsions, etc. [26e, 51,58];

(7) The introduction of specially developed ligands [11,26h,27b,28,33,49a,54a,60,61];

(8) Measures for the controlled switch of the catalyst system from the two-phase system (suitable for the *separation* of products from the catalyst) to a monophasic system which supports the *reaction* itself (Figure 5.9, [27,54]).

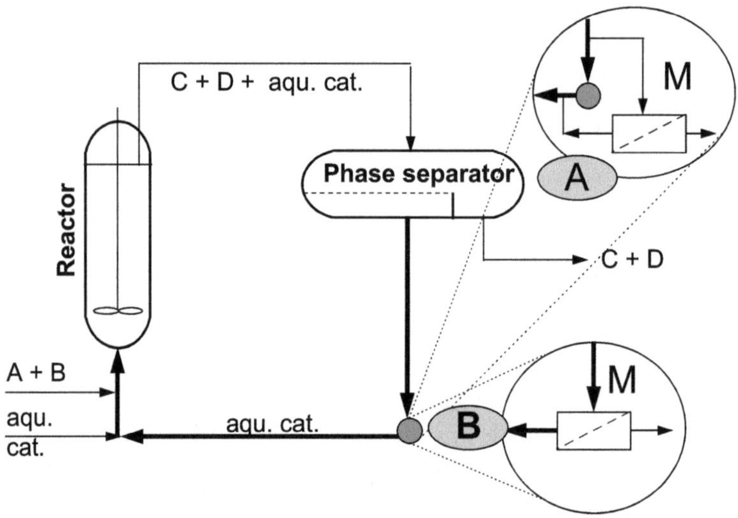

Figure 5.8. Membrane steps as a constituent part of aqueous-biphasic hydroformylation A+B->C+D

The method outlined in 8) above may be applied by the use of thermo-regulating or -separating systems. These take advantage of a temperature-dependent "cloud-point" associated with P-bound poly(alkylene glycol ether) as ligands, i.e., oxo catalysts which combine water solubility together with phase separating properties depending upon temperature. Developments of Fell and Jin based on ethoxylated phosphines give the first pointers to such a procedure.

As shown in Figure 5.9, at the cloud point the ligand (and thus the catalyst complex) loses its hydration shell, just as in the case of other compounds of this type, causing the two-phase reaction mixture, normally obtained when (higher) alkene is added to the catalyst solution, to merge into a single phase, thereby initiating a rapid conversion that is no longer impeded by mass transport problems. Subsequent lowering of the temperature causes the hydration shell to be reversibly restored, inducing the catalyst complex solution once again to separate out as an independent phase, this time from the reaction products, viz. the desired higher aldehydes. The disadvantage of this higher cost technique is the resulting lower linear/branched (*n/iso*) ratio. Jin refined this technique to a highly sophisticated method, including transitions to other variants (e.g., water-soluble polymers, pH-dependent operation, etc.)[55].

The membrane process, incorporated into the catalyst recycle either in the main stream or in the side stream, may also separate the reaction products from the remaining catalyst (Figure 5.8).

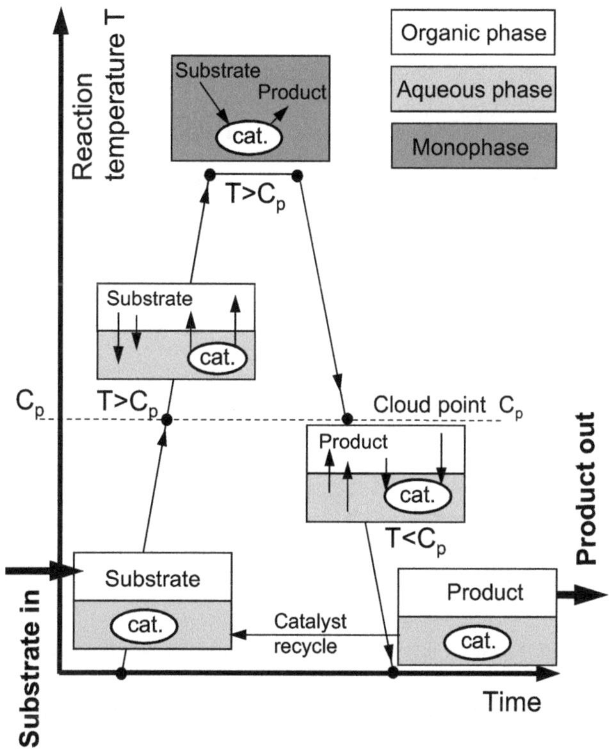

Figure 5.9. Phase change of thermoregulating ligands

As an alternative to the heterogenization of homogenous catalysis, there are some proposals to realize a solid catalyst with an immobilized species in aqueous-organic media. This concept, a continuation of the "**s**upported (or solid)-**l**iquid **p**hase **c**atalysis" (SLPC) has mainly published and highlighted by Scholten et. al [62], consists of a thin film of catalytic material that resides on a high-surface-area support such as controlled-pore glass, silica, zeolites, etc. Thus this concept of a "**s**upported **a**queos-**p**hase **c**atalysis" (SAPC) contains both a hydrophilic liquid and a hydrophilic organometallic catalytic complex on a solid support as shown in Figure 5.10 [63].

Reactions take place at the water film - organic phase interface and are catalyzed by the phase-immobilized complex catalyst which is usually $[HRh(CO)(TPPTS)_3]$. Other hydrophilic solvents, e.g., glycols or suitable liquids, adjusted to the requirements of the respective reaction, can be used instead of water for the formation of the immobilized liquid layer. As in all SPC variants, the catalytic process proceeds homogeneously in the supported film (SLPC) or at the interface_(SAPC), thus avoiding the problem of the separation of the reaction products from the catalyst. It is believed that the hydrophilicity of the ligands and the support creates interaction energies suffient to maintain the immobilization.

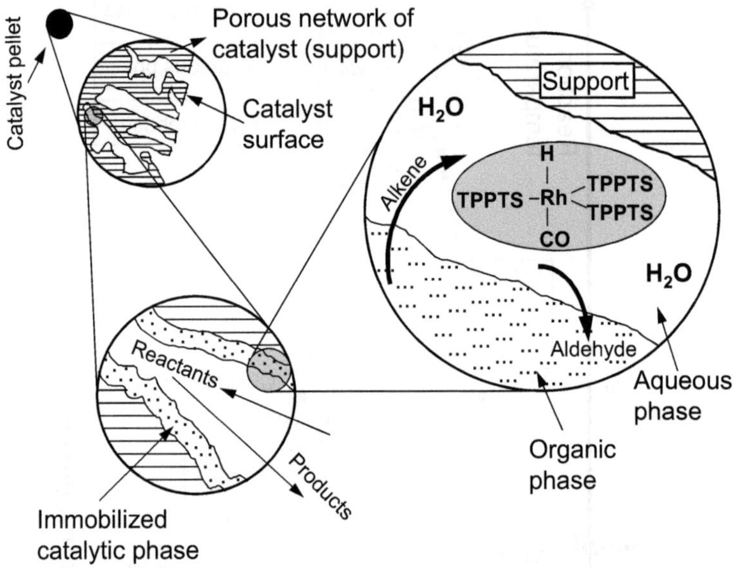

Figure 5.10. Schematic illustration of a supported aqueous-phase catalytic system (SAPC)

So far, various reactions have successfully been accomplished with SAPC such as hydrogenation (including asymmetric hydrogenation), hydroformylation, alkylation, Wacker oxidation, etc.[64]. However, no continous long-term runs nor commercial tests have been carried out. As a process requiring relatively complex processing, the SAPC technique is most likely to find applications in the manufacture of sophisticated and high-priced products (e.g., pharmaceuticals or chiral intermediates). It is likely that rare metals and unusual and specially designed ligands are needed to fully meet the demands of those processes and products. The lifetime of SAPS catalysts under continuous operation and industrial load is as yet unknown.

Advantageously, SAPC as a technique with immobilized catalysts does not need devices for catalyst separation and recycling, since the reactions can in principle be carried out using standard flow reactors commonly used in heterogeneous catalysis. On the other hand, the presumed processes for the work-up of the constituents of the catalyst, the ligand (and - may be - the support) will be demanding and expensive, too.

5.2.6 INTERLUDE - BIPHASIC SYSTEMS: ORGANIC-ORGANIC

The first biphasic process which has been commercialized uses Ziegler's *Aufbaureaktion* ("growing reaction") of ethene yielding oligomeric alkenes (Equation 5.7, [65]).

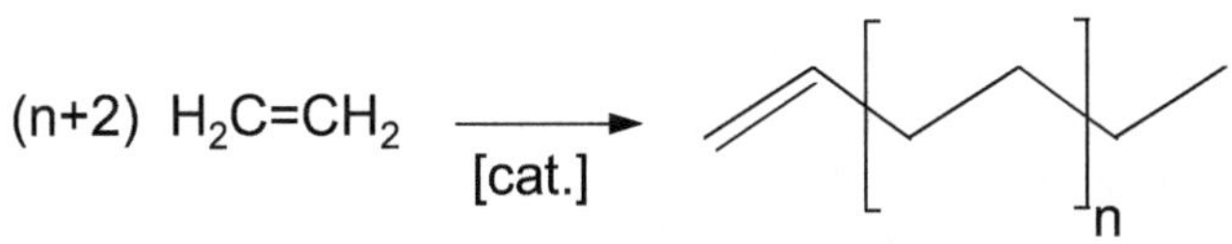

Equation 5.7. "Aufbaureaktion" according to Ziegler

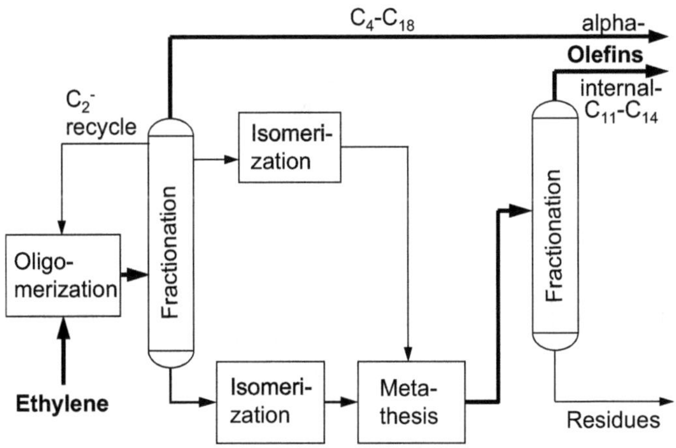

Figure 5.11. Scheme of Shell's SHOP process

Based on this reaction, commercial processes have been developed by different companies such as Chevron, Ethyl, Mitsubishi, Idemitsu, and Shell [66]. The processes use Ziegler-type catalysts (Chevron, Ethyl, Mitsubishi), Zr compounds (Idemitsu), or specially developed, P,O bidentate Ni complexes (Shell). Shell runs the largest capacities with approximately 900,000 tpy of "SHOP" (**S**hell **H**igher **O**lefin **P**rocess) products, consisting of valuable terminal alkenes and of special internal alkenes (see Table 5.3). Combining the oligomerization (Aufbaureaktion) with subsequently operated isomerization and metathesis stages, the overall product range of SHOP plants are linear terminal alkenes C_4-C_{18} *and* internal alkenes C_{11}-C_{14} [67]. The biphasic operation of the process is a consequence of the fact that the chelating catalyst is dissolved in polar 1,4-butandiol and the nonpolar products, the alkenes, are insoluble in the diol. Thus, the catalyst-containing solution can easily be separated from the insoluble products in a decanter. In a series of distillation towers, the various terminal alkene cuts (preferred are C_4-C_{18}) are isolated. The lower and the C_{18+} fractions are combined to be isomerized to internal linear alkenes by catalysts such as Na/K on alumina or MgO in the liquid phase. The resulting linear alkenes are subsequently subject to a metathesis reaction which combines the lower and the higher internal alkenes to give a mixture of alkenes with odd and even carbon chain lengths. The process is depicted in Figure 5.11. The catalytic cycle is shown with Figure 5.12.

Only a small amount of information has been released by Shell about the recovery and recycle of the special P,O bound Ni catalyst. Obviously, the overall operation of the SHOP plants is economical but one has to keep in mind that the non-homogeneously catalyzed parts of Shell's integrated system of processes use inexpensive, heterogeneous catalysts. It is to be expected that a large part of these catalysts can be disposed off without any recycling – neither of the metal content nor of the supports.

According to Figure 5.2 and to chemical experience, the selection of other pairs of non-miscible organic liquids is difficult and yields mainly unusual (not to say, exotic) solvents or pairs of solvents [68] such as fluorous liquids (cf. Chapter 6). This is the reason why no other organic-organic biphasic catalytic processes have yet been commercialised.

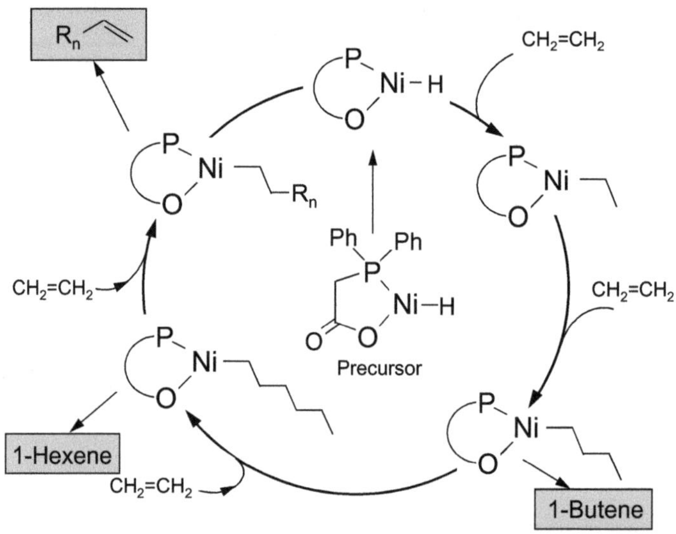

Figure 5.12. Postulated mechanism for ethylene oligomerization according to Shell's SHOP process with a P,O-stabilized catalyst

5.3 Recycle and Recovery of Aqueous Catalysts

Recovery and recycle of homogeneous catalysts are said to be the focal point of any new generation of, for example, hydroformylation technologies. Additionally, the costs of the new process are determined by the losses, which may be a particular problem when using high-price rare metal catalysts such as rhodium or palladium. Obviously this is true because a new mode of recyling charaterizes a new process. But on the other hand, metal losses (and supplementary ligand costs) are only a (minor) part of the overall cost as will be demonstrated in Section 5.3.3.

In recent years, the evaluation of environmentally benign (so-called "green" processes) has been a focal point of oxo developments, according to the definitions and targets of the

OECD Workshop on Sustainable Chemistry (1998) [8c]:

> "Within the broad framework of sustainable development, we should strive to maximize resource efficiency through activities such as energy and non-renewable resource conservation, risk minimization, pollution prevention, minimization of waste at all stages of a product life-cycle, and the development of products that are durable and can be re-used and recycled. Sustainable chemistry strives to accomplish these ends through the design, manufacture and use of efficient and effective, more environmentally benign chemical products and processes".

In this respect, risk minimization and life cycle assessments play a major role.

Figure 5.13. Large-scale oxo plant using the water-soluble Ruhrchemie/Rhône-Poulenc catalyst [$HRh(CO)(TPPTS)_3$]. Note the horizontal decanter in front of the reactor

5.3.1 RECYCLING

As mentioned in Section 5.2.4.1, the part of the aqueous catalyst solution which leaves the oxo reactor accompanying (but not dissolved in) the reaction products passes through a phase separator (decanter **2** in Figure 5.7) which is a characteristic part of the plant shown in Figure 5.13. In this decanter, which ensures essentially spontaneous phase separation, the crude aldehyde formed by hydroformylation according to Equation 5.1 is freed of gases and separated into mutually insoluble phases. The catalyst solution, supplemented by an amount of water equivalent to the water content of the crude aldehyde, is recirculated to the reactor. During its active life, the Rh catalyst is mainly situated in the oxo reactor, it is not moved in bulk and no aliquots are withdrawn as in other processes. For this reason, rhodium losses are low - in the range of parts-per-billion (ppb) - and thus the background for the high economy (see Section 5.3.3) [12e,121]. Like every technically rused and thus "real" catalyst, the complex [HRh(CO)(TPPTS)$_3$] and the excess ligand TPPTS undergo a degree of decomposition that determines the catalyst's lifetime as measured in years. The catalyst deactivation mechanism has been clarified in detail (Figures 5.14 and 5.15; Equation 5.8) [69].

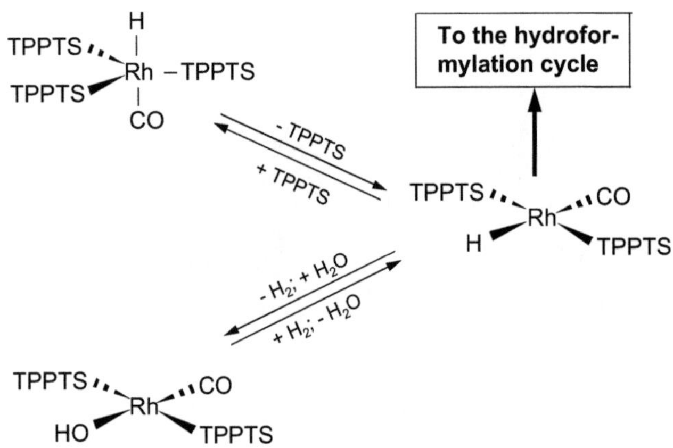

Figure 5.14. From the precatalyst to the oxo-active species

The compound [HRh(CO)(TPPTS)$_3$] is a "precatalyst" and dissociates to the 16e species [HRh(CO)(TPPTS)$_2$]. This oxo-active complex initiates the hydroformylation cycle. Under oxo conditions (presence of CO/H$_2$, H$_2$O, and a surplus of TPPTS) the hydroxo complex [(HO)Rh(CO)(TPPTS)$_2$] may be formed and again reversibly converted to [HRh(CO)(TPPTS)$_3$] (equilibrium lies almost completely towards the hydride). However, higher carbon monoxide partial pressures may cause the displacement of TPPTS by CO according to Equation 5.8.

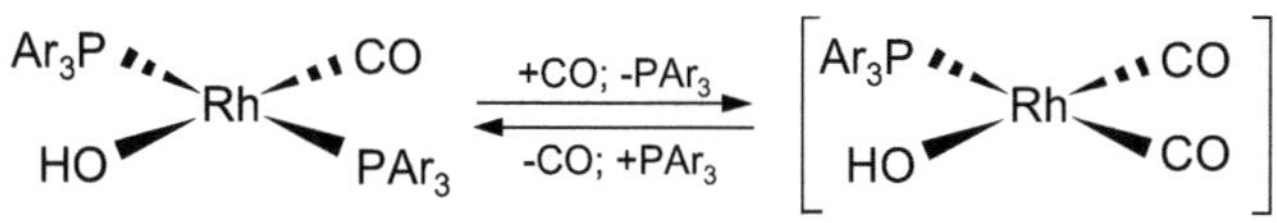

Equation 5.8. The Influence of CO

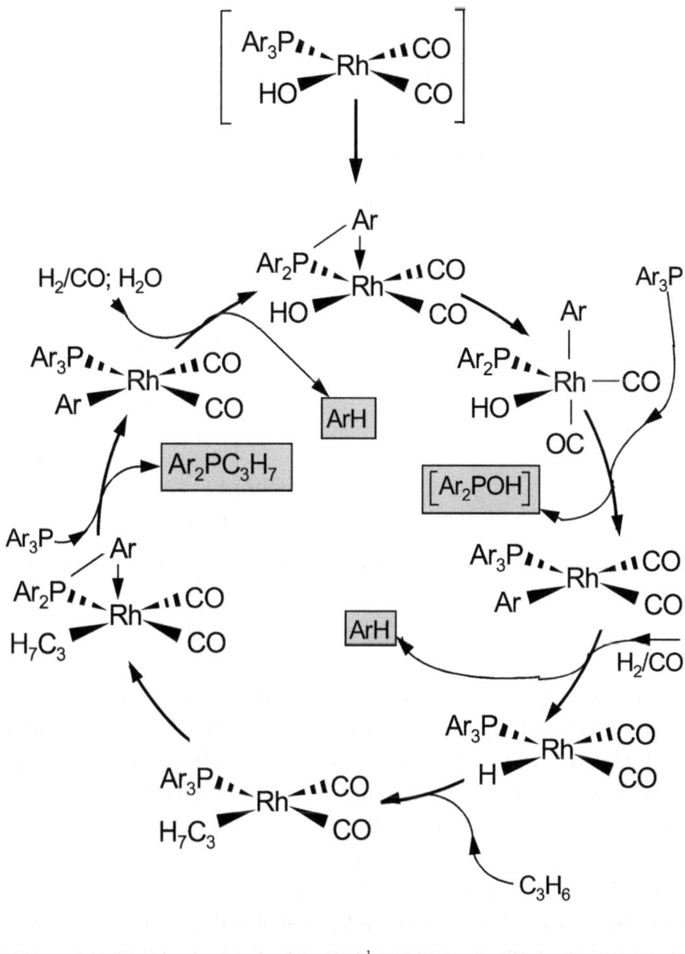

Figure 5.15. Deactivation Mechanism of Rh[I]-TPPTS Catalyst (Ar = 3-$C_6H_4SO_3Na$)

The hydroxo complex [(HO)Rh$(CO)_2$TPPTS] starts the deactivation cycle as shown in Figure 5.15 [69] [70a,b]. The single steps will not be discussed here. Other decomposition products such as the reductively eliminated bis(*meta*-sulfophenyl)phosphinous acid Ar_2POH (cf. Figure 5.15) and the phosphine oxides $Ar_2P(OH)(=O)$ and $Ar_2P(=O)(CH[OH]C_3H_7)$ (Ar = 3-$C_6H_4SO_3Na$) have been identified. According to Figure 5.16, the intermittent addition of excess ligands extends the catalyst's lifetime in a saw-tooth curve. This addition of ligand compensates for the system-immanent formation of deactivating substances which are brought into the system by the feedstocks. Most of the gaseous poisons are removed with the heavy ends of the stripper **4** in Figure 5.7. Filters, guard beds, or special precautions to avoid larger sulfur or oxygen inputs and their concentrations (as in other processes, [70c,d]) are not necessary. Additionally, other activity-lowering oxo poisons may be separated with the organic product phase of the decanter and are thus continuously removed at the very point of their formation from the system: any accumulation of activity-decreasing poisons in the catalyst solution is prevented. It might be worth mentioning that the Ruhrchemie plant has been supplied over longer periods with syngas manufactured from coal by the TCGP (Texaco coal gasification process) [12e].

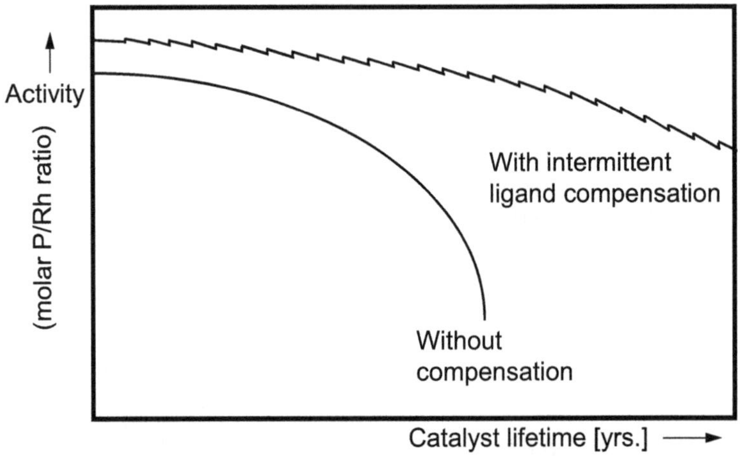

Figure 5.16. The effect of excess ligand on the catalyst's lifetime

Catalyst deactivation includes (among other reactions) the formation of inactive Rh species, ligand decomposition, or P-C cleavage by direct oxidative insertion of the rhodium metal for formation of PDSPP (propyldi[*m*-sulfophenyl]phosphine) acting as strong electron donor reducing the amount of active Rh catalyst. It turned out to be beneficial to control the P^{III}/Rh ratio and the CO partial pressure very carefully.

After years of use catalyst solutions typically contain 20 mg·L^{-1} iron and 0.7 mg·L^{-1} of nickel, thus showing no corrosivity. The Rh content of crude aldehyde is in the ppb range; this corresponds to losses of less than 10^{-9} g kg^{-1} *n*-butanal, totalling some kg rhodium over a twenty-year period and a production of approximately 5 million metric tons of *n*-butanal.

5.3.2 RECOVERY

Eventually, the spent catalyst solution has to leave the oxo loop for work-up. The Ruhrchemie works of Celanese AG in Oberhausen (Germany) operate several rhodium-based oxo processes: besides the well-known Ruhrchemie/Rhône-Poulenc process (RCH/RP, the described low pressure oxo process with TPPTS-modified Rh catalyst), there are the "Ruhrchemie process" with an unmodified Rh catalyst at high pressure (comparable to the late ICI process [76]; this variant is for the benefit of a high *iso/n* ratio in cases where this strange ratio is required), and a high pressure technique using ligand-modified rhodium catalysts. The various processes deliver aldehydes from C_4 up to C_{10} and additionally fine chemical aldehydes, starting from bicyclic alkenes, functionalized alkenes, etc.

The compound of the distinct three oxo processes, all rhodium-based, enables a highly efficient recovery system to be achieved. Figure 5.17 describes the TPPTS manufacture and its use for the preparation of the rhodium catalyst, using either freshly introduced Rh acetate or recycled Rh 2-ethylhexanoate. The recycle technique of the RCH/RP process and its performance is depicted earlier. Spent Rh-TPPTS solutions are worked-up (see Figures 5.18 and 5.19), the resulting TPPTS returns to the RCH/RP process. The rhodium portion passes also a work-up stage and is reformulated as Rh 2-ethylhexanoate. This Rh salt may serve all various oxo processes of the oxo loop and will compensate for possible Rh losses as mentioned earlier.

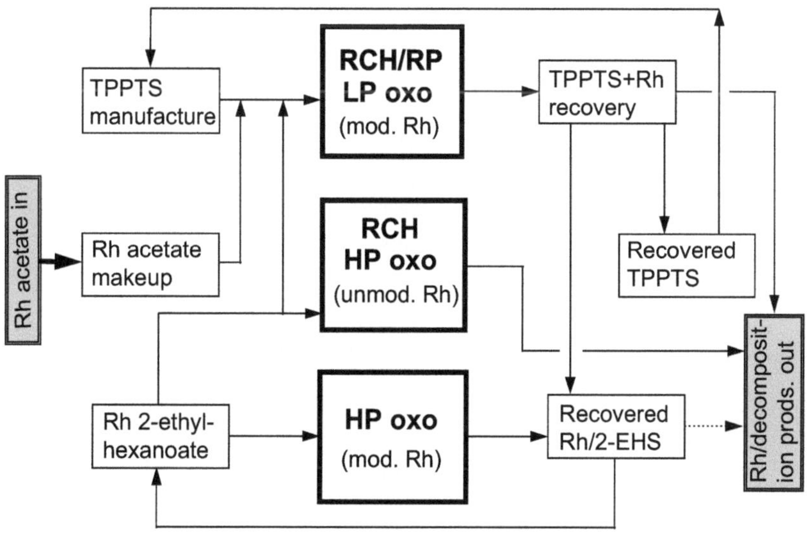

Figure 5.17. Compound of various oxo processes at the Ruhchemie works

According to Figure 5.17, the only external input is rhodium(III)acetate besides TPPTS from its own manufacturing unit. Rhodium compounds which are too difficult to be worked-up (catalyst poisons, metal aggregates, clusters, etc.) together with decomposition products, residues, etc. leave the recovery steps to be processed by external precious-metals refiners. These have their own expertise and trade (and production) secrets as far as the technical know-how is concerned. Depending upon the status and the "quality" of these fractions (if necessary, following a referee's check, various samplings, etc.), the credit for the rhodium is >95% of its content [80].

The work-up scheme as described in Figure 5.17 takes into account the decomposition rate of TPPTS (and thus the ligand losses), the different volume of the consumed Rh streams and their residual activity, and – last but not least – the Rh losses. Advantageously, these losses may be minimized by proper use of the different sources from the various oxo stages. Most steps are covered by patents [77], and the same is true for similar processes developed by competitors [78].

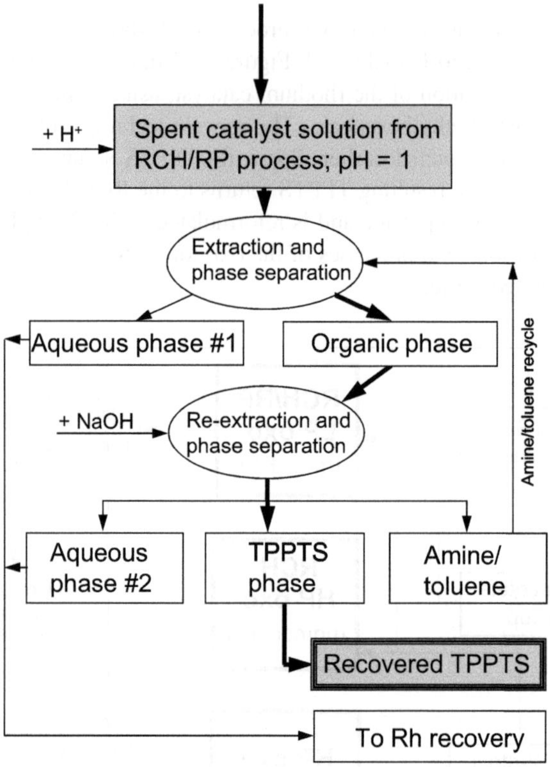

Figure 5.18. TPPTS recovery

The economics of the TPPTS recovery (cf. Figure 5.18) depend on the costs and its relation to the Rh price. Since the procedure of *work-up* is identical with some steps within the *manufacture* of fresh TPPTS [78], existing devices of the TPPTS manufacture unit can be used advantageously. According to Figure 5.18, the work-up of rhodium follows an acidification of the spent catalyst and extraction with a mixture of a tertiary alkylamine and toluene. The resulting two phases (aqueous phase #1, containing rhodium, and the organic phase with amine/toluene and TPPTS) are further processed. The organic phase is reextracted with aqueous NaOH (which purifies existent Rh(III) compounds and TPPTS), yielding three pH-dependent fractions: An aqueous solution #2 of rhodium compounds, the TPPTS fraction (TPPTS as sodium salt, ready for recycle and re-use; recovery rate: approximately 80%), and the amine/toluene phase which is recycled either to the manufacture of fresh TPPTS or to the TPPTS recovery. Both rhodium-containing fractions of the TPPTS recovery (#1 and #2) are subject to the rhodium recovery as depicted in Figure 5.19.

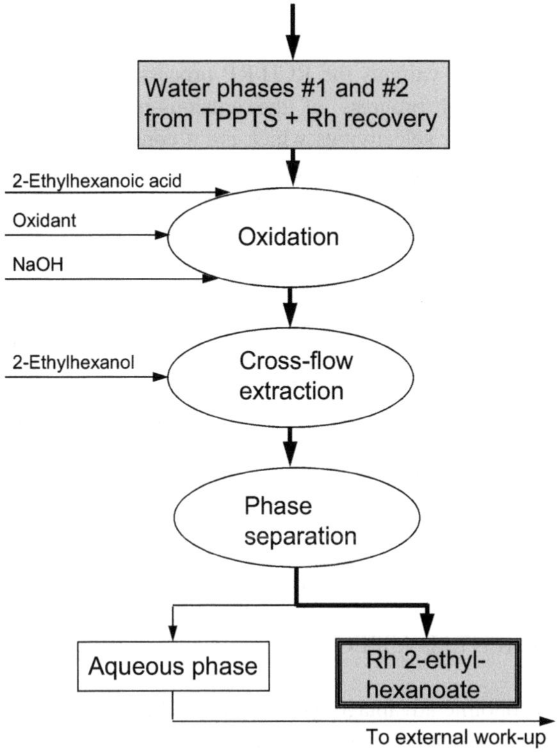

Figure 5.19. Rhodium recovery

These Rh-containing phases are mixed with NaOH and 2-ethylhexanoic acid and subsequently oxidized with air, H_2O_2, or other oxidizing agents. The Rh compounds in phase #1 and #2 are thus converted to Rh 2-ethylhexanoate and are extracted in a cross-flow manner by means of 2-ethylhexanol. The recovery rate of rhodium is 97-98%. The residual water phase ends up in the fraction for external work-up as described above.

It may be expected that higher-valued ligands (which might be used for the conversion of long-chain alkenes) or even co-ligands, co-solvents, modifiers, surfactant promoters,etc. can be processed the same way. There are already small-scale processes employing aqueous-phase catalysis which use "exotic" ligands such as di-, tri-, or multidentate phosphines where the situation demands drastic measures including a high-sophisticated management for rhodium and ligand recycle and recovery (e.g., [73]).

The "real" oxo precatalyst [HRh(CO)$(TPPTS)_3$] is easily made in the oxo reactor by reacting suitable Rh salts (e.g., rhodium acetate or rhodium 2-ethylhexanoate) with TPPTS – both components freshly prepared or recovered and recycled - without any additional preformation step. The reaction starts after formation of the active species and adjustment of the whole system with water to the desired P/Rh ratio (ensuring the stability of the catalyst and the desired *n/iso* ratio).

5.3.3 ECONOMICS OF THE PROCESS

In comparison to other oxo variants the RCH/RP process is highly economic, *inter alia* reflecting the technological progress of the aqueous-phase operation, better energy management, and the higher selectivities which are a consequence of the TPPTS ligand. Under equal conditions (basing on market prices without internal clearing prices) the RCH/RP process offers clear advantages of approximately 10% over other ligand-modified processes (Table 5.5) [12e].

TABLE 5.5. Manufactering costs of 100 kg *n*-butanal (RCH/RP=100)

Costs	RCH/RP Process	Other rhodium catalysed process
1. Raw materials	88.9	89.0
2. Energy	1.5	9.7
3. Credits (*iso*-butanal, *n*-butanol, others)	-7.7	-11.4
4. Costs for materials (sum 1+2+3)	82.7	87.3
5. Fix cost plus license fee[1)]	17.3	22.1
6. Sum 4+5 = Manufacturing costs[1)]	100.0	109.4

[1)] For updating see the *Chemical Engineering Plant Cost Index* (CEPCI, [12n]).

5.3.4 ENVIRONMENTAL ASPECTS

In addition to their economic advantages (see Section 5.3.3), catalyzed organic reactions in aqueous media have received signifcant attention as a result of environmental considerations [6,15]. The fundamental advance represented by the RCH/RP process in terms of the environment, conservation of resources, and minimization of risks such as enviromental pollution - aiming at sustainable “green” processes - can be demonstrated by various criteria and proved by means of the analysis of *atom economy* (according to Trost [71]) and, more convincingly and constructively, by the *environmental E factor* according to Sheldon [72]. The latter defined the *E* factor as the ratio of the amount of waste ("waste" is everything except the desired product) produced per kilogram of "target" products and he specified the *E* factor for every segment of the chemical industry (Table 5.6; [74]).

TABLE 5.6. The Environmental factor *E*

Industry segment	Product tonnage [tpy]	*E* factor
Oil refining	10^6-10^8	approx. 0.1
Bulk Chemicals	10^4-10^6	<1 - 5
Fine chemicals	10^2-10^4	5 - 50+
Pharmaceuticals	10-10^3	25 - 100+

As expected from Table 5.7, this environmental quotient for conventional oxo processes (basis: Co catalysts) and for the manufacture of the bulk chemical *n*-butanal is actually about 0.6-0.9, depending on the definition of the term "target" product. The range 0.6-0.9

indicates that the by-product *iso*-butanal occuring with conventional oxo processes is further processed by some producers (to *iso*-butyric acid, neopentylglycol, etc.) so that the *iso*-aldehyde thus becomes a target product and the E factor falls from 0.9 to 0.6. Strictly speaking, this observation is included in Sheldon's wider assessment, according to which the E factor is refined and becomes the environmental quotient EQ, depending on the nature of the waste. Since such quotients are "debatable and will vary from one company to another and even from one production to another" (Sheldon) they will not be discussed here. The crucial point is that on the same basis, taking into account all by-products including those produced in ligand manufacture, etc., which exhibit for conventional oxo processesve an *E* factor of 0.6-0.9, this factor falls to below 0.1 in the RCH/RP process: an important pointer to the environmental friendliness of the RCH/RP process (Table 5.7).

TABLE 5.7. *E* factors of different processes for the hydroformylation of propene

	E factor	
	iso-butanal as value product	*iso*-butanal as by-product
Co catalyst	>0.6	>0.9
Rh catalyst (RCH/RP)	<0.04	<0.1

Whereas this important quotient is calculated solely from the product spectrum, process simplifications are a consequence of combining the rhodium catalyst with the special two-phase process. Compared with the conventional oxo process and with other variants (which, for example, include disadvantegeously thermal separation of the oxo reaction products from the catalyst) the procedure is considerably simplified (as shown in several papers, e.g., [2,12]).

The conservation of energy resources with the RCH/RP process is dramatic. Note should be taken of the much milder reaction conditions and of the fact that the RCH/RP process is an energy exporter because of an intelligent, integrated heat network - an unusual occurence for conventional oxo processes, competing processes with PPh_3-modified Rh catalysts included. Furthermore, the steam consumption figures for the older, cobalt-based processes are very much higher than those for the Rh process and power consumption was twice as high as that of the RCH/RP process: both of theses factors represent an environmental burden. The relative compression costs alone for the required syngas are 1.7:1 (Co versus Rh process). The volume of waste water for the RCH/RP process is 70 times lower than that from the cobalt-based high pressure process - convincing evidence of an environmental benign and "green" process and one of the reasons for the advantageous results from the life cycle assesssment. It may be added that the process thus meets the requirements of the "production-integrated environmental protection", sometimes known as "cleaner production" which has the aim of avoiding and reducing residues and using resources carefully [8c,75].

The solvent water reliably averts the risk of fire, which was inherent in the old cobalt-based process as a result of leaking highly flammable, metal carbonyls. The technique with its "built-in extinguishing system" reliably prevents such fires, and the painstaking measuring and monitoring procedure necessitated by the valuable rhodium-based catalyst, accompanied by constant simultaneous balancing of the RCH/RP process, permits any leaks from the aqueous system to be detected much earlier than was ever possible with the expensive mass and liquid balance of the old Co process. This also applies for the cooling system, in which any leak from the falling film evaporator would be noticed immediately. For the other advantages of using water as the solvent, see Section 5.2.3.1.

5.4 Concluding Remarks

Taking all criteria into consideration, aqueous two-phase techniques are very sound methods for homogeneously catalyzed processes such as hydrogenations or hydroformylations. Of the various alternatives to the conventional (and solvent-free) processes most progress in terms of ecological impact and economics has been attained by the aqueous biphasic approach (Figure 5.20).

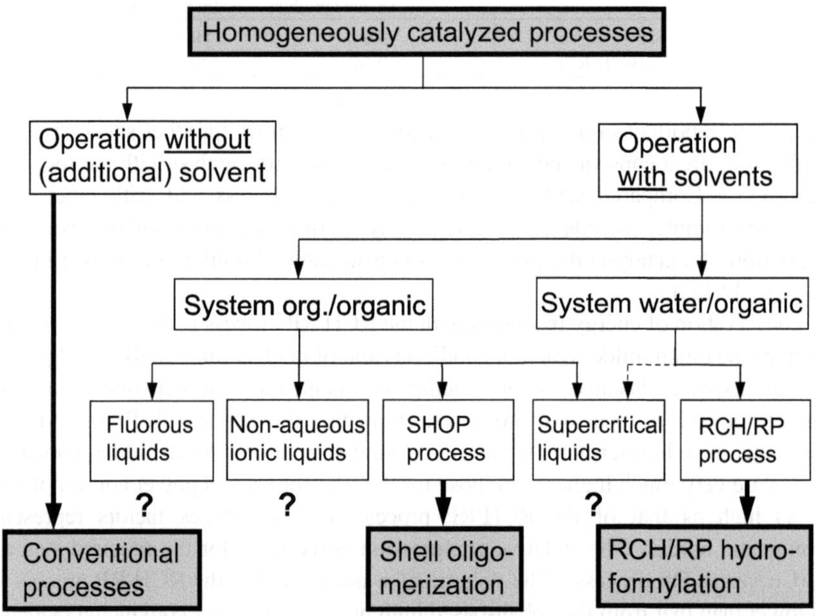

Figure 5.20. The various possibilities for biphasic operation of homogeneous catalysis

So far, only the systems organic/organic and water/organic have produced cost-effective, commercial processes. It is believed that aqueous biphasic systems have larger potential than organic/organic systems because their solubility patterns are more promising. The other possibilities such as fluorous liquids, supercritical fluids and non-aqueous ionic liquids are discussed in detail in later chapters of this book, but none have yet been commercialised They are labeled with a question mark in Figure 5.17 to underline the uncertainty of their development. The same is true for recent work using water-soluble polymers [61]. All of these processes need costly media (fluorous solvents, non-aqueous ionic liquids), high-priced ligands (fluorous processes or processes with supercritical CO_2), and /or additional means such as phase separating fluids, co-solvents, etc.

The aqueous-biphasic processes are easy to handle thus emphasizing their great environmental compatibility. They have significant potential for further development whether in terms of varying the ligand TPPTS (and thus the activity and selectivity of the conversions chosen, e.g., the hydroformylation; [74]) as well as the possibility of achieving asymmetric conversions, or in terms of other feedstocks. The adoption of the aqueous two-phase reaction for hydroformylation in particular (but also for other homogeneously catalyzed conversions in general) shows the striking and unequalled advantages of this special neat type of "immobile" catalyst and its "heterogenization" by the catalyst carrier, water as a result of which it is handled in similar fashion to a heterogeneous catalyst. This affects the costs as well as the enviromental compatibility

The advantages of successful waste management and avoiding by-products are further good points of the process, which ultimately lead to the higher cost-effectiveness as a result of the mentioned advantages together with less downtime: further proof of the likelihood that even in the medium term only environmentally sound processes will remain the most effective. The fact that basicly Sheldon's *E* factor is inversely proportional to a successful life cycle assessment underlines the superior quality of aqueous-organic biphasic systems.

5.5 References

[1] H. -W. Bohnen, B. Cornils, *Adv. Catal.* 2003, **47**, 1.

[2] (a) B. Cornils, C. W. Kohlpaintner, E. Wiebus in J. J. McKetta, G. E. Weismantel (Eds.), *Encyclopedia of Chemical Processing and Design*, Dekker, New York 1999, Vol. 66, p. 273; (b) B. Cornils in I. T. Horváth et al. (Eds.), *Encyclopedia of Catalysis*, Wiley, New York 2003; Vol. 1, p. 805; (c) B. Cornils in P. Knochel (Ed.), *Topics in Current Chemistry*, Vol. 206, Springer, Berlin/Heidelberg 1999, p. 133; (d) B. Cornils, *Ind. Catal. News* 1998, No. 2; (e) R. V. Chaudhari, A. Bhattacharya, B. M. Bhanage, *Catal. Today* 1995, **24**, 123.

[3] J. Falbe(Ed.), *New Syntheses with Carbon Monoxide*, Springer, Berlin 1980.

[4] Field Information Agency, Technical (FIAT) Report No. 1000 (1945), (Reports of the Occupational Forces in Germany).

[5] (a) J. Manassen, in F. Basolo and R. L. Burwell (Eds.), *Catalysis Progress in Research*, Plenum Press, London 1973, p. 183 and YEDA, US 4,415,500 (1983); (b) F. Joó, M. T. Beck, *Magy. Kém. Folyóirat* 1973, **79**, 189; (c) F. Joó, *Aqueous Organometallic Catalysis*, Kluwer, Dordrecht 2001, p. 3; (d) F. Joó, E. Papp, A. Kathó, *Top. Catal.* 1998, **5**, 113.

[6] B. Cornils, W. A. Herrmann (Eds.), *Aqueous-Phase Organometallic Synthesis*, Wiley-VCH, Weinheim 1998; 2nd Ed. 2003,

[7] W. Keim, F. -H. Kowaldt, R. Goddard, C. Krüger, *Angew. Chem.* 1978, **90**, 493.

[8] (a) R. T. Baker, W. Tumas, *Science* 1999, **284**, 1477; (b) S. K. Ritter, *Chem. & Eng. News* 2002, May 20, 38; (c) A. D. Curzons, D. J. C. Constable, D. N. Mortimer, V. L. Cunningham, *Green Chem.* 2001, **3**, 1; (d)

R. P. Swatloski, J. D. Holbrey, R. D. Rogers, *Green Chem.* 2003, **5**, 361; (e) G. Centi, S. Perathoner, *Catal. Today* 2003, **77**, 287.

[9] B. Cornils, W. A. Herrmann, R. Schlögl, C. -H. Wong (Eds.), *Catalysis from A to Z*, Wiley-VCH, Weinheim, 2nd ed. , 2003.

[10] (a) Z. Li, Q. Peng, Y. Yuan, *Appl. Catal. A:* 2003, **239**,79; (b) M. Benaissa, U. J. Jáuregui-Haza, I. Nikov, A. M. Wilhelm, H. Delmas, *Catal. Today* 2003, **79/80**, 419; (c) U. J. Jáuregui-Haza, E. Pardillo-Fontdevila, Ph. Kalck, H. Delmas, *Catal. Today* 2003, **79/80**, 409; (d) A. Riisager, K. M. Eriksen, J. Hjortkjaer, R. Fehrmann, *J. Mol. Catal. A:* 2003, **193**, 259; (e) M. Dessoudeix, U. J. Jáuregui-Haza, M. Heughebaert, A. M. Wilhelm, H. Delmas, A. Lebugle, Ph. Kalck, *Adv. Synth. Catal.* 2002, **344**, 406; (f) Z. Li, Q. Peng, Y. Yuan, *Appl. Catal. A:* 2003, **239**, 79; (g) Y. Yuan, H. Zhang, Y. Yang, Y. Zhang, K. Tsai, *Catal. Today* 2002, **74**, 5; (h) Z. Li, Q. Peng, Y. Yuan, *Appl. Catal. A:* 2003, **239**, 79; (i) G. Fremy, E. Monflier, J. -F. Carpentier, Y. Castanet, A. Mortreux, *Angew. Chem.* 1995, **107**, 1608.

[11] C. Botteghi et al. , *J. Mol. Catal. A:* 2003, **200**, 147.

[12] (a) B. Cornils, J. Falbe, *4th Int. Symp. on Homogeneous Catalysis*, Leningrad, Sept. 1984, Proceedings p. 487; (b) H. Bach, W. Gick, E. Wiebus, B. Cornils, *Int. Symp. High-Pressure Chem. Eng.* , Erlangen/Germany, Sept. 1984, Preprints p. 129; (c) H. Bach, W. Gick, E. Wiebus, B. Cornils, *8th ICC*, Berlin 1984, Preprints Vol. V, p. 417; [C. A. 1987, 106, 198,051; cited also in A. Behr, M. Röper, *Erdgas und Kohle* 1984, **37**, 485]; (d) H. Bach, W. Gick, E. Wiebus, B. Cornils, *1st IUPAC Symp. Org. Chemistry*, Jerusalem 1986, Abstracts p. 295; (e) E. Wiebus, B. Cornils, *Chem. Ing. Techn.* 1994, **66**, 916; (f) B. Cornils, E. Wiebus, *CHEMTECH* 1995, **25**, 33; (g) E. G. Kuntz, *CHEMTECH* 1987, **17**, 570; (h) C. W. Kohlpaintner, R. W. Fischer, B. Cornils, *Appl. Catal. A:* 2001, **221**, 219; (i) B. Cornils, *J. Mol. Catal. A:* 1999, **143**, 1; (j) B. Cornils, *Org. Proc. Res. & Dev.* 1998, **2**, 121; (k) C. -D. Frohning, C. W. Kohlpaintner in [6], p. 294; (l) E. Wiebus, B. Cornils, *Hydrocarb. Proc.* 1996, March, 63; (m) B. Cornils, W. A. Herrmann in [6], Introduction; (n) *Chem. Eng.* 2002, Jan., 62.

[13] (a) P. Kalck, M. Dessoudeix, S. Schwarz, *J. Mol. Catal. A:* 1999, **143**, 41; (b) R. M. Deshpande, S. S. Divekar, B. M. Bhanage, R. V. Chaudhari, *J. Mol. Catal. A:* 1992, **75**, L19.

[14] (a) Rhône-Poulenc Recherche (E. Kuntz), FR 2,314,910 (1975); FR 2,349,562 (1976); FR 2,338,253 (1976); FR 2,366,237 (1976) and other patent applications; (b) B. Cornils, E. G. Kuntz, *J. Organomet. Chem.* 1995, 505, 177.

[15] (a) H. Olivier-Bourbigou, F. Hugues in R. D. Rogers, K. R. Seddon, S. Volkov (Eds.), *Green Industrial Applications of Ionic Liquids*, NATO Science Series, Vol. 92, Kluwer, Dordrecht 2002.

[16] (a) G. Duve, O. Fuchs, H. Overbeck, *Lösemittel Hoechst*, 6th ed. , Hoechst AG, Frankfurt/M. 1976, p. 49; (b) C. Reichardt, *Solvents and Solvent Effects in Organic Chemistry*, 3rd ed. , Wiley-VCH, Weinheim/Germany 2002.

[17] (a) P. A. Grieco (Ed.), *Organic Synthesis in Water*, Blacky Academic, London 1998; (b) H. Fuhrmann, T. Dwars, G. Oehme, *Chem. Unserer Zeit* 2003, **37**, 40; (c) C. J. Li, T. H. Chan, *Organic Reactions in Organic Media*, Wiley, New York 1997; (d) C. J. Li, *Chem. Rev.* 1993, **93**, 2023; (d) C. J. Li, T. H. Chan, *Tetrahedron* 1996, **52**, 5643; (e) M. Groß, *Chem. Unserer Zeit* 2003, **37**, 170; (f) K. Nomura, *J. Mol. Catal. A:* 1998, **130**, 1; (g) N. J. Buurma, M. J. Blandamer, J. B. F. N. Engberts, *Adv. Synth. Catal.* 2002, **344**, 413; (h) F. Reynal, R. Barhdadi, J. Périchon, A. Savall, M. Toupel, *Adv. Synth. Catal.* 2002, **344**, 45; (i) S. Ribe, P. Wipf, *Chem. Commun.* 2001, 299.

[18] (a) A. Lubineau, *Chem. Ind.* 1996, February, 123 (not p. 125, as printed); (b) H. W. Roesky, M. G. Walawalkar, R. Murugavel, *Acc. Chem. Res.* 2001, **34**, 201; K. R. Asmis et al. , *Science* 2003, **299**, 1375 ; T. Köddermann, F. Schulte, M. Huelsekopf, R. Ludwig, *Angew. Chem. Int. Ed.* 2003, **42**, 4904; (c) N. J.

Buurma, M. J. Blandamer, J. B. F. N. Engberts, *Adv. Synth. Catal.* 2002, **344**, 413; (d) K. Itami, T. Nokami, J. Yoshida, *Adv. Synth . Catal.* 2002, **344**, 441.

[19] E. G. Hope, A. M. Stuart, *Chem. &Ind.* 1999, 4 Oct. , 755.

[20] C. C. Tzschucke, C. Markert, W. Bannwarth, S. Roller, A. Hebel, R. Haag, *Angew. Chem. Int. Ed.* 2002, **41**, 3964.

[21] P. W. N. M. van Leeuwen, P. C. J. Kamer, J. N. H. Reek, *Cattech* 1999, **3**, 64.

[22] (a) R. J. Stokes, D. F. Evans, *Fundamentals of Interfacial Engineering*, Wiley-VCH, Mew York/Weinheim 1997; (b) BASF, DE 3,220,858 (1983).

[23] (a) O. Wachsen, K. Himmler, B. Cornils, *Catal. Today* 1998, **42**, 373; (b)E. Lindner, T. Schneller, F. Auer, H. A. Meyer, *Angew. Chem. Int. Ed.* 1999, **38**, 2155; (c) Ph. Kalck, M. Dessoudeix, S. Schwarz, *J. Mol. Catal. A:* 1999, **143**, 41.

[24] (a) R. M. Stephenson, *J. Chem. Eng. Data* 1993, **38**, 630; (b) C. McAuliffe, *J. Phys. Chem.* 1966, **70**, 1267; (c) P. L. Davis, *Gas Chromatogr.* 1968, **6**, 518.

[25] (a) C. Yang, X. Bi, Z. -S. Mao, *J. Mol. Catal. A:* 2002, **187**, 35; (b) Y. Önal, M. Baerns, P. Claus in [6], p. 151; (c) *Chem. Ing. Tech. plus.* 2003, Issue Sept. , 46; (d) Y. Zhang, Z. -S. Mao, J. Chen, *Catal. Today* 2002, **74**, 23; (e) C. Yang, Z. -S. Mao, Y. Wang, J. Chen, *Catal. Today* 2002, **74**, 111.

[26] (a) E. A. Karakhanov, Y. S. Kardesheva, E. A. Runova, V. A. Semermina, *J. Mol. Catal. A:* 1999, **142**, 339; (b) J. N. H. Reek, A. J. Sandee, M. Schreuder Goedheijt, P. C. J. Kamer, P. W. N. M. van Leeuwen, *Erdöl, Erdgas, Kohle* 2001, **117**, 134; (c) H. Chen, Y. Li, R. Li, P. Cheng, X. Li, *J. Mol. Catal. A:* 2003, **198**, 1; (d) U. Ritter, N. Winkhofer, H. -G. Schmidt, H. W. Roesky, *Angew. Chem. Int. Ed.* 1996, **35**, 524; (e) M. Li, Y. Li, H. Chen, Y. He, X. Li, *J. Mol. Catal. A:* 194, **13**; (f) Y. Huang, L. Min, Y. Li, H. Chen, P. Cheng, X. Li, *J. Mol. Catal. A:* 2002, **185**, 41; (g) Y. Zhang, Z. -S. Mao, J. Chen, *Catal. Today* 2002, **74**, 23; (h) H. Chen, Y. Li, J. Chen, P. Cheng, X. Li, *Catal. Today* 2002, **74**, 131; (i) Z. Li, Q. Peng, Y. Yuan, *Appl. Catal. A:* 2003, **239**, 79; (k) F. Monteil, R. Queau, Ph. Kalck, *J. Organomet. Chem.* 1994, **480**, 177; (l) M. Giménez-Pedrós, A. Aghmiz, C. Claver, A. M. Masdeu-Bultó, D. Sinou, *J. Mol. Catal. A:* 2003, **200**, 157; L. Wang, H. Chen, Y. He, Y. Li, M. Li, X. Li, *Appl. Catal. A :* 2003, **242**, 85.

[27] Among a multitude of publications: (a) B. Fell et al. , *J. Prakt. Chem. /Chem. Ztg.* 1996, **338**, 124; *J. Mol. Catal.* 1997; **116**, 55; *Fett/Lipid* 1996, 98, 393; (b) Z. Jin et al. , *J. Mol. Catal. A:* 1999, **147**, 131 and **149**, 113; 2000, **157**, 111 and **159**, 37; 2001, **171**, 85; 2002, **188**, 79; 2003, **195**, 133; 2003, **198**, 23; *Catal. Today* 2002, **74**, 85; *Catal. Commun.* 2003, **4**, 129; (c) E. Paetzold, G. Oehme, C. Fischer, M. Frank, *J. Mol. Catal. A:* 2003, **200**, 95; (d) J. A. Loch, C. Borgmann, R. H. Crabtree, *J. Mol. Catal. A:* 2001, **170**, 75; (e) B. M. Bhanage, M. Shirai, M. Arai, *J. Mol. Catal. A:* 1999, **145**, 69; (f) U. Ritter, N. Winkhofer, H. -G. Schmidt, H. W. Roesky, *Angew. Chem. Int. Ed. Engl.* 1996, **35**, 524.

[28] (a) E. Bayer, V. Schurig, *CHEMTECH* 1976, (March), 212; (b) E. Bayer, V. Schurig, *Angew. Chem. Int. Ed. Engl.* 1975, **14**, 493; (c) M. Giménez-Pedrós, A. Aghmiz, C. Claver, A. M. Masdeu-Bultó, D. Sinou, *J. Mol. Catal . A:*2003, **200**, 157; (d) E. Paetzold, M. Michalik, G. Oehme, *J. Prakt. Chem.* 1997, **339**, 38; (e) A. Kumar, G. Oehme, J. P. Roque, M. Schwarze, R. Selke, *Angew. Chem.* 1994, **106**, 2272; (f) S. Ganguly, J. T. Mague, D. M. Roundhill, *Inorg. Chem.* 1992, **31**, 3500; (g) S. Shimizu, S. Shirakawa, Y. Sasaki, C. Hirai, *Angew. Chem.* 2000, **112**, 1313; (h) M. Schreuder Goedheijt, P. C. J. Kamer, P. W. N. M. van Leeuwen, *J. Mol. Catal. A:* 1998, **134**, 243; (i) K. V. Katti, H. Gall, C. J. Smith, D. E. Berning, *Acc. Chem. Res.* 1999, **32**, 9; (k) M. Beller, J. G. E. Krauter, A. Zapf, S. Bogdanovic, *Catal. Today* 1999, **48**, 279; (l) S. Shimizu, T. Suzuki, S. Shirakawa, Y. Sasaki, C. Hirai, *Adv. Synth. Catal.* 2002, **344**, 370; (m) C. Bertucci, C. Botteghi, D. Giunta, M. Marchetti, S. Paganelli, *Adv. Synth. Catal.* 2002, **344**, 556; (n) E. Monflier, G. Fremy, Y. Castanet, A. Mortreux, *Angew. Chem.* 1995, **107**, 2450; (o) G. T. Baxley, T. J. R. Weakly, W. K.

Miller, D. K. Lyon, D. R. Tyler, *J. Mol. Catal. A:* 1997, **116**, 191; (p) A. Gong, Q. Fan, Y. Chen, H. Liu, C. Chen, F. Xi, *J. Mol. Catal. A:* 2000, **159**, 225; (q) Q. Peng, Y. Yang, C. Wang, X. Liao, Y. Yuan, *Catal. Lett.* 20023, **88**, 219; (r) A. Aghmiz, A. Orejón, M. Diéguez, M. D. Miquel-Serrano, C. Claver, A. M. Masdeu-Bultó, D. Sinou, G. Laurenczy, *J. Mol. Catal. A:* 2003, **195**, 113; (s) X. Z. Liu, F. Kong, X. Zheng, Z. Jin, *Catal. Commun.* 2003, **4**, 129; (t) D. C. Mudalige, G. L. Rempel, *J. Mol. Catal. A:* 1997, **116**, 309; (u) A. Riisager, K. M. Erksen, J. Hjortkaer, R. Fehrmann, *J. Mol. Catal. A:* 2003, **193**, 259; (v) S. Ganguly, J. T. Mague, D. M. Roundhill, *Inorg. Chem.* 1992, **31**, 3500; (w) S. Shimizu, S. Shirakawa, Y. Sasaki, C. Hirai, *Angew. Chem.* 2000, **112**, 1313; (x) R. Amengual, E. Genin, V. Michelet, M. Savignac, *Adv. Synth. Catal.* 2002, **344**, 393 ; (y) E. A. Karakhanov, Y. S. Kardasheva, E. A. Runova, M. V. Terenina, *EuropaCat-VI*, 31,Aug-4, Sept. ,Preprints; (z) F. Chécot, S. Lecommandoux, Y. Gnanou, H-A. Klok, *Angew. Chem. Int. Ed.* 2002, **41**, 1340.

[29] (a) G. Protzmann, K-D. Wiese, *Erdöl,Erdgas,Kohle* 2001, **117**, 235; (b) Oxeno (G. Protzmann, K-D. Wiese, J. Koch, M. Trocha, D. Röttger), DE 199 25 384 (2000); (c) Oxeno (K-D. Wiese, G. Protzmann, J. Koch, D. Röttger, M. Trocha), EP 1,057,524 (2000); (d) Oxeno (G. Protzmann, K-D. Wiese, W. Büschken, D. Röttger), DE 199 57 528 and 199 57 522 (1999); (e) Oxeno (G. Protzmann, K-D. Wiese, W. Büschken), EP 1,057,538 (2001); (f) Oxeno (K-D. Wiese, P. Olbrich, J. Gabriel), DE 199 08 320 (2000); (g) K. -D. Wiese, O. Möller, G. Protzmann, M. Trocha, *Catal. Today* 2003, **79/80**, 97.

[30] (a) B. Driessen-Hölscher, P. Wasserscheid, W. Keim, *Cattech* 1998, June, 47; (b) A. Behr, *Chem. Ing. Tech.* 1998, **70**, 685; (c) A. Behr, C. Fängewisch, *Chem. Ing. Tech.* 2002, **73**, 874.

[31] (a) A. Behr, *Chem. Ing. Tech.* 1998, **70**, 685; (b) B. Auch-Schwelk, C. Kohlpaintner, *Chem. Unserer Zeit* 2001, **35**, 306.

[32] T. Prinz, W. Keim, B. Driessen-Hölscher, *Angew. Chem.* 1996, **108**, 1835.

[33] (a) H. Bahrmann, H. Bach, C. D. Frohning, H. J. Kleiner, P. Lappe, D. Peters, D. Regnat, W. A. Herrmann, *J. Mol. Catal. A:* 1997, **116**, 49; (b) M. Beller, B. Cornils, C. D. Frohning, C. W. Kohlpaintner, *J. Mol. Catal. A:* 1995, **104**, 17; (c) B. Cornils, W. A. Herrmann, R. W. Eckl, *J. Mol. Catal. A:* 1997, **116**, 27; (d) M. Yuan, H. Chen, R. Li, Y. Li, X. Li, *Appl. Catal. A:* 2003, **251**, 181.

[34] G. Centi, S. Perathoner, *Catal. Today* 2003, **79/80**, 3.

[35] (a) C. Larpent, E. Bernard, F. Brisse-le Menn, H. Patin, *J. Mol. Catal. A:* 1997, **116**, 277 ; (b) Davy/McKee (Union Carbide), GB 2,075,857 (1981) and EP 096,987 (1983) ; (c) Union Carbide, EP 055,487 (1982) and EP 068,480 (1082); (d) Daicel, JP 5-8186-443A (1983); (e) BASF EP 053,688 (1982).

[36] (a) Elf Atochem, US 5,345,007 (1994, Monflier); (b) A. Behr, M. Urschey, *J. Mol. Catal. A:* 2003, **197**, 101; (c) N. Yoshimura in [6], p. 408; (d) T. Prinz, W. Keim, B. Driessen-Hölscher, *Angew. Chem.* 1996, **108**, 1835; (e) Kuraray, JP 080642 (1981), JP 5-8131-994° (1982) and JP 58216138 (1984); (f) M. Matsumoto, M. Tamura, *J. Mol. Catal.* 1982, 16, 209; (g) Kuraray, EP 088,955 (1983); IFP (H. Chauvin), US 6,525,228 (2001).

[37] (a) DuPont (W. D. Drinkard et al.), US 3,536,748 (1970), 3,579,560 (1971), 3,676,481 (1972), and 3,739,011 (1973); (b) Rhône-Poulenc (E. G. Kuntz), US 4,087,452 (1978); (c) H. E. Bryndza, J. A. Harrelson in [6], p. 393.

[38] (a) R. Chansarkar, K. Mukhopadhyay, A. A. Kelkar, R. V. Chaudhari, *Catal. Today* 2003, **79/80**, 51; (b) C. W. Kohlpaintner, M. Beller, *J. Mol. Catal. A:* 1997, **116**, 259; (c) A. L. Lapidus, K. B. Petrovskii, I. P. Beletskaya, *Russ. J. Org. Chem.* 1997, **33**, 1083; (d) P. Jayasree, S. P. Gupta, R. V. Chaudhari, *Stud. Surf. Sci. Catal.* 1998, **113**, 875; E. Paetzold, G. Oehme, *J. Mol. Catal. A:* 2000, **152**, 69.

[39] (a) C. Dupuis, K. Adiey, L. Charruault, V. Michelet, M. Savignac, J-P. Gênet, *Tetrahedron Lett.* 2001, **42**, 6523; (b) N. E. Leadbeater, M. Marco, *Angew. Chem. Int. Ed.* 2003, **42**, 1407; (c) E. Paetzold, G. Oehme,

J. Mol. Catal. A: 2000, **152**, 69; (d) S. Haber in [6], p. 440.

[40] F. Bertoux, S. Tilloy, E. Monflier, Y. Castanet, A. Mortreux, *J. Mol. Catal. A:* 1999, **138**, 53.

[41] (a) S. Nagayama, S. Kobayashi, *Angew. Chem. Int. Ed.* 2002, **39**, 567; (b) Y. Mori, K. Manabe, S. Kobayashi, *Angew. Chem.* 2001, **113**, 2898; (c) K. Inoue, T. Ishida, I. Shibata, A. Baba, *Adv. Synth. Catal.* 2002, **344**, 283; (d) K. Manabe, S. Kobayashi, *Chem. Eur. J.* 2002, **8**, 4095; (e) T. Ooi, Y. Uematsu, K. Maruoka, *Adv. Synth Catal.* 2002, **344**, 288; (f) S. Shimizu, T. Suzuki, S. Shirikawa, Y. Sasaki, C. Hirai, *Adv. Synth. Catal.* 2002, **344**, 370; (g) (l) D. Sinou, C. Rabeyrin, C. Nguefack, *Adv. Synth. Catal.* 2003, **345**, 357; (h) S. Sigismondi, D. Sinou, *J. Mol. Catal. A:* 1997, **116**, 289; (i) P. Wipf, S. Rodriguez, *Adv. Synth. Catal.* 2002, **344**, 434; (j) K. Itami, T. Nokami, J. Yoshida, *Adv. Synth. Catal.* 2002, **344**, 441; (k) C. Loncaric, K. Manabe, S. Kobayashi, *Adv. Synth. Catal.* 2003, **345**, 475; (l) Nagoya University, *Bull. SocSoc. Jpn.* 1997, **70**, 905 [ref. in *CHEMTECH* 1997, (Nov.), 5]; (m) I. V. Kozhevnikov, G. P. Mulder, M. C. Steverink-de Zoete, M. G. Oostwal, *J. Mol. Catal. A:* 1998, **134**, 223; (n) H. Rudler, J. Ribeiro Grigorio, B. Denise, J. -M. Bregeault, A. Deloffre, *J. Mol. Catal. A:* 1998, **133**, 255; (o) G. Pozzi, I. Colombani, M. Miglioli, F. Montanari, S. Quici, *Tetrahedron* 1997, **53**, 6145; (p) B. Boyer, A. Hambardzoumian, G. Lamaty, A. Leydet, J. -P. Roque, P. Bouchet, *New J. Chem.* 1996, **20**, 985; (q) K. Manabe, N. Aoyama, S. Kobayashi, *Adv. Synth. Catal.* 2001, **343**, 174;(r) S. Mukhopadhyay, G. Rothenberg, A. Joshi, M. Baidossi, Y. Sasson, *Adv. Synth. Catal.* 2002, **344**, 348; (s) R. Amengual, E. Genin, V. Michelet, M. Savignac, J. -P. Genêt, *Adv. Synth. Catal.* 2002, **344**, 393; (t) B. M. Bhanage, M. Shirai, M. Arai, *J. Mol. Catal. A:* 1999, **145**, 69; (u) M. Beller, J. G. E. Krauter, A. Zapf, S. Bogdanovic, *Catal. Today* 1999, **48**, 279; (v) T. Koike, X. Du, T. Sanada, Y. Danda, A. Mori, *Angew. Chem. Int. Ed.* 2003, **42**, 89; (w) Z. Yang, M. Ebihara, T. Kawamura, *J. Mol. Catal. A:* 2000, **158**, 509; (x) P. J. Baricelli, J. Lopez, E. Lujano, F. Lopez-Linares, *J. Mol. Catal. A:* 2002, **186**, 57; (y) T. Malmström, C. Andersson, *J. Mol. Catal. A:* 2000, **157**, 79; (z) A. Andriollo, J. Carrasquel, J. Marino, F. A. López, D. E. Páez, I. Rojas, N. Valencia, *J. Mol. Catal. A:* 1997, **116**, 157; (aa) G. Süss-Fink, M. Faure, T. R. Ward, *Angew. Chem. Int. Ed.* 2002, **41**, 99 and *J. Mol. Catal. A:* 1998, **132**, 5; (bb) G. Papp, J. Elek, L. Nádasdi, G. Laurenczy, F. Joó, *Adv. Synth. Catal.* 2003, **345**, 172; (cc) G. Laurenczy et al. , *Adv. Synth. Catal.* 2003, **345**, 211,216; (dd) J. A. Loch, C. Borgmann, R. H. Crabtree, *J. Mol. Catal. A:* 2001, **170**, 75; (ee) H. Jiang, Y. Xu, S. Liao, D. Yu, H. Chen, X. Li, *J. Mol. Catal. A:* 1999, **142**, 147; (ff) D. Sinou, *Adv. Synth. Catal.* 2002, **344**, 221; (gg) F. López-Linares, M. G. Gonzalez, D. E. Páez, *J. Mol. Catal. A:* 1999, **145**, 61; (hh) K. Nomura, *J. Mol. Catal. A:* 1998, **130**, 1; (ii) A. W. Heinen, G. Papadogianakis, R. A. Sheldon, J. A. Peters, H. van Bekkum, *J. Mol. Catal. A:* 1999, **142**, 17; (jj) A. Behr, N. Toslu, *Chem. Ing. Tech.* 1999, **71**, 490; (kk) S. Mecking, A. Held, F. M. Bauers, *Angew. Chem. Int. Ed.* 2002, **41**, 545; (ll) B. Manders, L. Sciandrone, G. Hauck, M. O. Kristen, *Angew. Chem. Ind. Ed.* 2001, **40**, 4006; (mm) D. M. Jones, W. T. S. Huck, *Adv. Mater.* 2001, **13**, 1256; (nn) M. A. Zuideveld, F. M. Bauers, R. Thomann, S. Mecking, 36th Meeting of German Catalysis, Weimar 19-21, March, 2003, *Preprints*, p. 385 and *Chem. Commun.* 2000, 301; (oo) A. Held, S. Mecking, *Chem. Eur. J.* 2000, **6**, 4623; (pp) P. Dupau, R. Epple, A. A. Thomas, V. V. Fokin, K. B. Sharpless, *Adv. Synth. Catal.* 2002, **344**, 421; (qq) G. M. Mehltretter, C. Döbler, U. Sundermeier, M. Beller, *Tetrahedron Lett.* 2000, **41**, 8083, *J. Am. Chem. Soc.* 2000, **122**, 10289 and *Angew. Chem. Int. Ed.* 1999, **38**, 3026; (rr) T. Wirth, *Angew. Chem. Int. Ed.* 2000, **39**, 334; (ss) T. Ishida, R. Akiyama, S. Kobayashi, *Adv. Synth. Catal.* 2003, **345**, 576; (tt) S. Venkatraman, T. Huang, C. -J. Li, *Adv. Synth. Catal.* 2002, **344**, 399; (uu) F. Raynal, R. Barhdadi, J. Périchon, A. Savall, M. Troupel, *Adv. Synth. Catal.* 2002, **344**, 45; (vv) T. Akiyama, J. Takaya, H. Kagoshima, *Adv. Synth. Catal.* 2002, **344**, 338.

[42] (a) G. Verspui, F. Schanssema, R. A. Sheldon, *Angew. Chem. Int. Ed.* 2000, **39**, 804; (b) G. Verspui, G. Papadogianakis, R. A. Sheldon, *Chem. Commun.* 1998, 401; (c) Z. Jiang, A. Sen, *Macromol.* 1994, **27**,

7215; (d) A. Sen (Ed.), *Catalytic Synthesis of Alkene-Carbon Monoxide Copolymers and Cooliogomers*, Kluwer, Dordrecht 2003.

[43] (a) R. G. da Rosa, J. D. R. de Campos, R. Buffon, *J. Mol. Catal. A:* 2000, 153, 19; (b) G. Papadogianakis, L. Maart, R. A. Sheldon, *J. Mol. Catal. A:* 1997, **116**, 179; (c) G. Verspui, G. Papadogianakis, R. A. Sheldon, *Catal. Today* 1998, **42**, 449.

[44] (a) R. H. Grubbs et al. , *J. Am. Chem. Soc.* 1998, **120**, 1627, 3187; (b) M. C. Schuster, K. H. Mortell, A. D. Hegeman, L. L. Kiessling, *J. Mol. Catal. A:* 1997, **116**, 209; (c) A. Romerosa, M. Peruzzini, *Organometallics* 2000, **19**, 4005; (d) R. H. Grubbs (Ed.), *Handbook of Metathesis*, Wiley-VCH, Weinheim 2003.

[45] S. Kobayashi et al. , *Tetrahedron Letters* 2000, **418**, 3107.

[46] (a) C. Venturello, M. Gambaro, *J. Org. Chem.* 1991, **56**, 5924; (b) P. L. Mills, R. V. Chaudhari, *Catal. Today* 1999, **48**, 17; (c) G. -J. ten Brink, I. W. C. E. Arends, R. A. Sheldon, *Adv. Synth. Catal.* 2002, **344**, 355 and *Science* 2000, **287**, 1636; (d) Y. Kita, *Angew. Chem. Int. Ed.* , 2000, **39**, 1306; (e) J. Cornely, L. M. Su Ham, D. E. Meade, V. Dragojlovic, *Green Chem.* 2003, **5**, 34; (f) L. Cammarota, S. Campestrini, M. Carrieri, F. Di Furia, P. Ghiotti, *J. Mol. Catal. A:* 1999, **137**, 155; (g) C. Dobler, G. M. Mehltretter, U. Sundermeier, M. Eckert, H. -C. Militzer, M. Beller, *Tetrahedron Lett.* 2001, **42**, 8447; (h) R. A. Sheldon, I. W. C. E. Arends, G. -J. ten Brink, A. Dijksman, *Acc. Chem. Res.* 2002, **35**, 774 and *Adv. Synth. Catal.* 2003, 345, 497; (i) Y. Uozumi, R. Nakao, *Angew. Chem. Int. Ed.* 2003, **42**, 194,; (j) W. Sun, H. Wang, C. Xia, J. Li, P. Zhao, *Angew. Chem. Int. Ed.* 2003, **42**, 1042; (k) G. -J. ten Brink, I. W. C. E. Arends, R. A. Sheldon, *Science* 2000, **287**, 1636; (l) G. J. ten Brink, I. W. C. E. Arends, R. A. Sheldon, *Adv. Synth. Catal.* 2002, **344**, 355; (m) E. Garcia Fidalgo, L. Plasseraud, G. Süss-Fink, *J. Mol. Catal. A*: 1998, **132**, 5; (n) S. Trinkhaus, R. Kadyrov, J. Holz, L. Götze, A. Börner, *J. Mol. Catal. A:* 1999, **144**, 15.

[47] (a) A. Behr, R. Ott, B. Turkowski, *36th Meeting of German Catalysis*, Weimar 2003, Preprints, p. 449.

[48] B. Cornils, W. A. Herrmann, I. T. Horváth, W. Leitner, S. Mecking, H. Olivier-Boubigou, D. Vogt (Eds.), *Multiphase Homogeneous Catalysis*, Wiley-VCH, Weinheim 2004.

[49] (a) A. Buhling, P. C. J. Kamer, P. W. N. M. van Leeuwen, J. W. Elgersma, *J. Mol. Chem. A:* 1997, **116**, 297; (b) A. Buhling, P. C. J. Kamer, P. W. N. M. van Leeuwen, *J. Mol. Catal. A:* 1995, 98, 69; (c) B. E. Hanson, H. Ding, C. W. Kohlpaintner, *Catal. Today* 1998, **42**, 421; (d) F. Robert, G. Oehme, I. Grassert, D. Sinou, *J. Mol . Catal. A:* 2000, **156**, 127; (e) Q. Peng, Y. Yang, C. Wang, X. Liao, Y. Yuan, *Catal. Letters* 2003, **88**, 219; (f) Y. Uozumi, M. Nakazono, *Adv. Synth. Catal.* 2002, **344**, 274.

[50] (a) F. Monteil, R. Queau, P. Kalck, *J. Organomet. Chem.* 1994, **480**, 177; (b) S. Trinkhaus, R. Kadyrov, R. Selke, J. Holz, L. Götze, A. Börner, *J. Mol. Catal. A. :* 1999, **144**, 15; (c) L. Lavenot, A. Roucoux, H,. Patin, *J. Mol. Catal. A:* 1997, **118**, 153; (d) K. C. Tin, N. -B. Wong, R. -X. Li, Y. -Z. Li, X. -J. Li, *J. Mol. Catal. A:* 1999, **137**, 113; (e) B. E. Hanson in [6], p. 181; 2nd edition 2003; (f) R. M Deshpande, Purwanto, H. Delmas, R. V. Chaudhari, *Ind. Eng. Chem. Res.* 1996, **35**, 3927; (g) F. Bertoux, E. Monflier, Y. Castanet, A. Mortreux, *J. Mol. Catal. A:* 1999, **143**, 23; (h) E. Paetzold, M. Michalik, G. Oehme, *J. Prakt. Chem.* 1997, **339**, 38; (i) H. Chen, Y. Li, J. Chen, P. Cheng, X. Li, *Catal. Today* 2002, **74**, 131; (j) M. Deshpande, S. S. Divekar, B. M. Bhanage, R. V. Chaudhari, *J. Mol. Catal.* 1992, **75**, L19; (k) A. Roucoux, J. Schulz, H. Patin, *Adv. Synth. Catal.* 2003, **345**, 222; (l) Ph. Kalck, L. Miquel, M. Dessoudeix, *Catal. Today* 1998, **42**, 431; (m) S. Paganelli, M. Zanchet, M,. Marchetti, G. Mangano, *J. Mol. Catal. A:* 2000, **157**, 1.

[51] (a) L. Wang, H. Chen, Y. He, M. Li, X. Li, *Appl. Catal. A. :* 2003, **242**, 85; (b) A. Riisager, B. E. Hanson, *J. Mol. Catal. A:* 2002, **189**, 195; (c) H. Chen, Y. Li, J. Chen, P. Chen, Y. He, X. Li, *J. Mol. Catal. A:* 1999, **149**,1; (d) A. Roucoux, J. Schulz, H. Patin, *Adv. Synth. Catal.* 2003, **345**, 222; (e) J. Texter, *Reactions and*

Synthesis in Surfactant Systems, Dekker, New York 2001; (f) T. Ooi, Y. Uematsu, K. Maruoka, *Adv. Synth. Catal.* 2002, **344**, 288; (g) J. Texter (Ed.), *Reactions and Synthesis in Surfactant Systems, Marcel* Dekker, New York 2001.

[52] (a) Exxon (J. R. Livingston, E. J. Mozeleski, G. Sartori), US 5,288,818 (1994) and PCT WO 93/04,029 (1993); (b) Th. Müller, H. Bahrmann, *J. Mol. Catal. A:* 1997, **116**, 39; (c) L. W. Gosser, W. H. Knoth, G. W. Parshall, *J. Mol. Catal.* 1977, **2**, 253; (d) T. Dwars, J. Haberland, I. Grassert, G. Oehme, U. Kragl, *J. Mol. Catal. A:* 2001, **168**, 81; (e) H. Fuhrmann, G. Oehme et al. , *Adv. Synth. Catal.* 2003, **345**, 202; (f) X. Liu, M. Jiang, S. Yang, M. Chen, C. Yang, K. Wu, *Angew. Chem. Int. Ed.* 2002, **41**, 2950; (g) A. G. Dixon in J. J. Spivey (ed.), *Catalysis*, Vol. 14, Chapter 2, p. 40, RSC, London 1999; (h) H. Bahrmann, M. Haubs, Th. Müller, N. Schöpper, B. Cornils, *J. Organomet. Chem.* 1997, **545/546**, 139; (i) H. Fuhrmann, T. Dwars, D. Michalik, G. Holzhüter, C. Grüttner, U. Kragl, G. Oehme, *Adv. Synth. Catal.* 2003, **345**, 202.

[53] (a) K. Kuroda, T. Swager, *Chem. Commun.* 2003, 26; (b) M. R. Buchmeiser, *Polymeric Materials in Organic Synthesis and Catalysis*, Wiley-VCH, Weinheim 2003; (c) X. Huang, K. L. Witte, D. E. Bergbreiter, C. -H. Wong, *Adv. Synth. Catal.* 2001, **343**, 675; (d) S. Mecking, A. Held, F. M. Bauers, *Angew. Chem. Int. Ed.* 2002, **41**, 545; (e) D. E. Bergbreiter, *CHEMTECH* 1987, (Nov.), 686; (f) D. E. Bergbreiter, J. G. Franchina, B. L. Case, *Org. Lett.* 2000, **2**, 393; (g) T. Malmström, C. Andersson, *J. Mol. Catal. A:* 2000, **157**, 79 and *Chem. Commun.* 1996, 1135; (h) D. E. Bergbreiter, *Catal. Today* 1998, **42**, 389; (i) T. J. Dickerson, N. N. Reed, K. D. Janda, *Chem. Rev.* 2002, **102**, 3325; (j) Z. Lu, E. Lindner, H. A. Mayer, *Chem. Rev.* 2002, **102**, 3543.

[54] (a) T. Borrmann, H. W. Roesky, U. Ritter, *J. Mol. Catal. A:* 2000, **153**, 31; (b) J. A. Loch, C. Borgmann, R. H. Crabtree, *J. Mol. Catal. A:* 2001, **170**, 75,

[55] E. Paetzold, G. Oehme, C. Fischer, M. Frank, *J. Mol. Catal. A:* 2003, **200**, 95.

[56] (a) P. J. Baricelli, G. Rodríguez, M. Rodríguez, E. Lujano, F. López-Linares, *Appl. Catal. A:* 2003, **239**, 25; (b) D. U. Parmar, S. D. Bhatt, H. C. Bajaj, R. V. Jasra, *J. Mol. Catal. A:* 2003, **202**, 9; (c) F. López-Linares, M. G. Gonzalez, D. E. Páez, *J. Mol. Catal. A:* 1999, **145**, 61; (d) D. C. Mudalige, G. L. Rempel, *J. Mol. Catal. A:* 1997, **116**, 309 and 123, 15; (e) S. Kolaríc, V. Sunjic, *J. Mol. Catal. A:* 1996, **111**, 239; (f) E. Garcia Fidalgo, L. Plasseraud, G. Süss-Fink, *J. Mol. Catal. A:* 1998, **132**, 5; (g) F. Robert, G. Oehme, I. Grassert, D. Sinou, *J. Mol. Catal. A:* 2000, 156, **127**; (h) C. Larpent, E. Bernard, F. Brisse-le Menn, H. Patin, *J. Mol. Catal. A:* 1997, **116**, 277; (i) T. Dwars, G. Oehme, *Adv. Synth. Catal.* 2002, **344**, 239; (j) F. Joó, *Acc. Chem. Res.* 2002, **35**, 738; Jóo, *Aqueous Organometallic Catalys*, Kluwer, Dordrecht 2001; (l) D. Sinou, *Adv. Synth. Catal.* 2002, **344**, 221; (m) Q. -H. Fan, G. -J. Deng, X. -M. Chen, W. -C. Xie, D. -Z. Jiang, D. -S. Liu, A. S. C. Chan, *J. Mol. Catal. A:*, 2000, **159**, 37; (n) T. Malmström, C. Andersson, *J. Mol. Catal. A:* 2000, **157**, 79; (o) Z. Yang, M. Ebihara, T. Kawamura, *J. Mol. Catal. A:* 2000, 158, 509; (p) K. Nomura, *J. Mol. Catal. A:* 1998, **130**, 1; (q) G. Papp, J. Elek, L. Nádashdi, G. Laurenczy, F. Jóo, *Adv. Synth. Catal.* 2003, **345**, 172.

[57] A. Chattopadhyay, A. Salaskar, *Synthesis* 2000, 561.

[58] (a) D. L. Reger, M. M. Habib, *J. Mol. Catal.* 1978, **4**, 315; (b) I. Grassert, J. Kovács, H. Fuhrmann, G. Oehme, *Adv. Synth. Catal.* 2002, **344**, 312; (c) V. Nardello, M. Hervé, P. L. Alsters, J. -M. Aubry, *Adv. Synth. Catal.* 2002, **344**, 184; (d) M. Haumann, H. Yildiz, H. Koch, R. Schomäcker, *Appl. Catal. A:* 2002, **236**, 173; (e) M. Haumann, H. Koch, R. Schomäcker, *Catal. Today* 2003, **79/80**, 43; (f) M. Haumann, H. Koch, P. Hugo, R. Schomäcker, *Appl. Catal. A:* 2002, **225**, 239; (g) M. Giménez-Pedrós, A. Aghmiz, C. Claver, A. M. Masdeu-Bultó, D. Sinou, *J. Mol. Catal. A:* 2003, **200**, 157; (h) X. Liu, M. Jiang, S. Yang, M. Chen, D. Chen, C. Yang, K. Wu, *Angew. Chem. Int. Ed.* . 2002, **41**, 2950; (i) A. Riisager, B. F. Hanson, *J. Mol. Catal. A:* 2002, **189**, 195; (k) H. Chen, Y. Li, J. Chen, P. Cheng, Y. He, X. Li, *J. Mol. Catal. A:* 1999, **149**, 1,

[59] (a) C. de Bellefon, N. Tanchoux, S. Caravieilhes, P. Grenouillet, V. Hessel, *Angew. Chem.* 2000, 112, 3584; (b) R. Hodiamont, W. Leitner, B. Driessen-Hölscher, *36th Meeting of German Catalysis*, Weimar 2003, 19-21, March, Preprints, p. 77.

[60] (a) X. Huang, K. L. Witte, D. E. Bergbreiter, C. -H. Wong, *Adv. Synth. Catal.* 2001, **343**, 675; (b) D. E. Bergbreiter, *CHEMTECH* 1987, Nov. , 686; (c) D. E. Bergbreiter, J. G. Franchina, B. L. Case, *Org. Lett.* 2000, **2**, 393; (d) D. E. Bergbreiter, *Catal. Today* 1998, **42**, 389.

[61] (a) M. A. Zuideveld, F. M. Bauers, R. Thomann, S. Mecking, *36th Meeting of German Catalysis,* Weimar, 19,-21, March 2003, Preprints, p. 385; (b) S. Mecking, A. Held, F. M. Bauers, *Angew. Chem. Int. Ed.* 2002, **41**, 545; (c) B. Xing, M. -. F. Choi, B. Xu, *Chem. Eur. J.* 2002, **8**, 5028.

[62] (a) J. J. F. Scholten, L. A. Gerritsen, J. M. Herman et al. , *J. Mol. Catal.* 1980, 9, special issue; (2) C. D. Frohning, C. W. Kohlpaintner, H. -W. Bohnen in B. Cornils, W. S. Herrmann (Eds.), *Applied Homogeneous Catalysis with Organometallic Compounds*, 2nd ed. , Wiley-VCH, Weinheim 2002, p. 31; W. Strohmeier et al. , *J. Organomet. Chem.* 1981, **221**, 361,

[63] M. E. Davies in B. Cornils, W. A. Herrmann (Eds.), *Aqueous-Phase Organometallic Catalysis*, Wiley-VCH, 1st ed. . ,Weinheim 1998, p. 241; *CHEMTECH* 1992, 498.

[64] (a) Z. Li, Q. Peng, Y. Yuan, *Appl. Catal. A:* 2003, **239**, 79; (b) G. Fremy, E. Monflier, J. -F. Carpentier, Y. Castanet, A. Mortreux, *Angew. Chem.* 1995, **107**, 1608 ; (c) J. N. H. Reek, A. J. Sandee, M. Schreuder Goedheijt, P. C. J. Kamer, P. W. N. M van Leeuwen, *Erdöl,Erdöl,Kohle* 2001, **117**, 134; (d) M. Dessoudeix, U. J. Jáuregui-Haza, M. Heughebaert, A. M. Wilhelm, H. Delmas, A. Lebugle, Ph. Kalck, *Adv. Synth. Catal.* 2002, **344**, 406; (e) J. R. Anderson, E. M. Campi, W. R. Jackson, Z. P. Yang, *J. Mol. Catal. A:* 1997, **116**, 109.

[65] K. Ziegler, H. Gellert, H. Kühlhorn, H. Martin, K. Meyer, K. Nagel, H. Sauer, K. Zosel, *Angew. Chem.* 1952, **64**, 323.

[66] D. Vogt in B. Cornils, W. A. Hermann (Eds.), *Applied Homogeneous Catalysis with Organometallic Compounds*, 2nd ed. , Wiley-VCH, Weinheim 2003, p. 240.

[67] Shell Dev. Corp. (S. R. Baur, H. Chung, P. W. Glockner, W. Keim, H. van Zwet,D. Camel, R. F. Mason), US 3,. 635,937, 3,637,636, 3,644,563, 3,644,564, 3,647,914, 3,647,915, 3,661,803, and 3,686,159 (1972 and 1973); W. Keim, *New J. Chem.* 1987, **11**, 531.

[68] V. Neff, T. E. Müller, J. A. Lercher, *Chem. Commun.* 2002, 906.

[69] W. A. Herrmann, C. W. Kohlpaintner, *Angew. Chem. Int. Ed. Engl.* 1993, **32**, 1524.

[70] (a) W. A. Herrmann, J. A. Kulpe, W. Konkol, H. Bahrmann, *J. Organomet. Chem.* 1990, **389**, 85; (b) W. A. Herrmann, J. A. Kulpe, J. Kellner, H. Riepl, H. Bahrmann, W. Konkol, *Angew. Chem. Int. Ed. Engl.* 1990, **29**, 391; (c) Union Carbide, EP 060,523 (1982); (d) Neftekhim, *Neftekhimija* 1981, **21**, 879.

[71] B. M. Trost , *Science* 1991, **254**, 1471 and *Angew. Chem. Int. Ed. Engl* 1995, **34**, 259.

[72] R. A. Sheldon, *CHEMTECH* 1994, **24**, 38; *Chem. &Ind.* 1992, 903 and 1997, 12; *Green Chem.* 1999, **1**, G3.

[73] (a) M. McCarthy, H. Stemmer, W. Leitner, *Green Chem.* 2002, **4**, 501; (b) S. Kainz, D. Koch, W. Baumann, W. Leitner, *Angew. Chem.* 1997, **109**, 1699.

[74] B. Cornils, E. Wiebus, *Recl. Trav. Chim. Pays-Bas* 1996, **115**, 211.

[75] C. Christ, *Chem. Ing. Tech.* 2000, **72**, 42.

[76] Information given by F. A. Kirk, Synetix, Billingham UK, on the occasion of the DGMK-Conference "Synthesis Gas Chemistry", Sep. 27-29, 2000 at Dresden (Germany), Proceedings, p. 131.

[77] (a) Ruhrchemie AG (G. Dämbkes, H. D. Hahn, J. Hibbel, W. Materne), DE 3,347,406 (1983); (b)

Ruhrchemie AG (G. Dämbkes, H. D. Hahn, J. Hibbel, W. Materne), EP 0,147,824 (1984); (c) Ruhrchemie AG (B. Cornils, W. Konkol, H. Bahrmann, H. W. Bach, E. Wiebus), DE 3,411,034 (1984) and EP 0,156,253 (1985); (d) Ruhrchemie AG (H. Bahrmann, W. Konkol, J. Weber, H. W. Bach, L. Bexten), DE 3,534,314 (1985); (e) Ruhr-chemie AG (L. Bexten, D. Kupies), DE 3,626,636 (1986); (f) Hoechst AG (L. Bexten, D. Kupies), EP 0,255,673 (1987); (g) Hoechst AG (G. Diekhaus, H. Kappeser), EP 0,322,661 (1988); (h) Hoechst AG (P. Lappe, L. Bexten, D. Kupies), DE 3,822,037 (1988), EP 0,348,833 (1989) and US 5,294,415 (1994); (i) Hoechst AG (J. Weber, L. Bexten, D. Kupies, P. Lappe, H. Springer), EP 0,367,957 (1989); (j) Hoechst AG (P. Lappe, H. Springer), DE 3,934,824 (1989) and EP 424,736 (1990); (k) Hoechst AG (P. Lappe, H. Springer), DE 4,025,074 (1990) and EP 0,475,036 (1991); (l) Hoechst AG (P. Lappe, H. Springer), EP 0,510,358 (1992); (m) Hoechst AG (I. Förster, K. Mathieu), EP 0,538,723 (1992); (n) Hoechst AG (W. Konkol, H. Bahrmann, W. A. Herrmann, C. W. Kohlpaintner), EP 0,544,091 (1992); (o) Hoechst AG (G. Diekhaus, H. Kappeser), EP 0,584,720 (1993).

[78] Among a multitude of filings: (a) Union Carbide Corp. (E. Billig, D. B. Stanton), EP 0,049,781 (1981) and 0,083,094 (1982); UCC (S. J. Dougherty), US 4,263,218 (1981); (E. Billig, D. B. Stanton), US 4,605,780 (1986); (b) BASF AG (H. -J. Kneuper, A. Aron, F. -M. Korgitzsch, M. Nilles, W. Harder, M. Roeper, R. Paciello), US 5,919,987 (1999), BASF (R. Paciello, H. -J. Kneuper, □ Geissler, M. Roeper), US 6,107,524 (2000), and BASF (B. Geissler, M. Roeper, R. Paciello, J. Decker, H. Voss, N. Mahr(, US 6,310,261 (2001); (c) Montedison SpA (G. Gregorio, G. Montrasi), US 3,968,134 (1976); (d) Elf (G. S. Silverman, P. Mercando), US 5,773,655 (1998); (e) BP (J. L. Carey), US 5,364,822 (1994); (f) UOP Inc. (E. H. Homeier, T. Imrai, D. E. Mackowiak, C. E. Guzolek), US 4,329,521 (1981); UOP (T. Imrai, E. -H. Homeier), US 4,374,287 (1983); UOP (T. Imrai), US 4,298,499 (1980); UOP (H. M. Quick), US 4,388,279 (1981); (g) Eastman-Kodak (R. L. Barnes), US 4,364,907 (1988), Eastman-Kodak (J. L. Cooper), US 4,388,476 (1983), WP 82-03,856 (1982), UOP (J. L. Dawes, T. J. Devon), US 4,400,547 (1981); (h) Mitsubishi (Y. Tsunoda, S. Tomita, C. Miyazawa, Y. Omori), GB 2,085,747 (1982); JP 81-02,994 (1981) [C. A. 1982, 96, 68297q]; (i) Shell (Z. Ali, J. Radder), FR 2,474,333 (1981); (j) Oxeno (K. -D. Wiese, M. Trocha, D. Röttger, W. Toetsch, A. Kaizek, W. Büschken), US 6,500,991 (2002) and Oxeno (D. Gubisch, K. Armbrust, B. Scholz, R. Nehring), US 6,015,928 (2000); (k) Ets. Kuhlmann (C. Demay), US 4,306,086 (1981); (l) ARCO (T. Chang), US 5,290,743 (1994).

[79] (a) Ruhrchemie AG (R. Gärtner, B. Cornils, L. Bexten, D. Kupies), DE 3,235,029 (1982), EP 0,103,845 (1983), and EP 0,107,006 (1983); (b) Ruhrchemie AG (L. Bexten, B. Cornils, D. Kupies), EP 0,175,919 (1985); (c) Ruhrchemie AG (H. Bahrmann, W. Konkol, J. Weber, H. W. Bach, L. Bexten), EP 0,216,315 (1986); (d) Hoechst AG (W. A. Herrmann et al.), EP 0,352,478 (1989).

[80] K. M. Beirne, *Chem. Engng.* 2002, Jan., 87.

CHAPTER 6 FLUOROUS BIPHASIC CATALYSIS

CLARE R. MATHISON AND DAVID J. COLE-HAMILTON
EaStCHEM, School of Chemistry, University of St Andrews, St Andrews,Fife, KY16 9ST, Scotland

6.1 Introduction

Biphasic systems, in which the catalyst is designed to be dissolved in a liquid phase which is immiscible with the product (either with or without a separate solvent) potentially provide some of the most attractive solutions to the problem of product separation in homogeneous catalysis.

In Chapter 5, aqueous biphasic systems were described. These are very elegant in that a water soluble catalyst is kept completely separate from the lypophilic product, except under conditions of fast stirring. Stopping stirring leads to rapid phase separation and the product can be collected by decanting. Such processes have been commercialised for short chain substrates, which have significant solubility in water, but it has been observed that the rates of reaction when using longer chain, less hydrophilic substrates are too low to be of commercial interest, presumably because mass transport limitations dominate the reaction.

Various other biphasic solutions to the separation problem are considered in other chapters of this book, but an especially attractive alternative was introduced by Horváth and co-workers in 1994.[1] He coined the term "catalysis in the fluorous biphase" and the process uses the temperature dependent miscibility of fluorinated solvents (organic solvents in which most or all of the hydrogen atoms have been replaced by fluorine atoms) with normal organic solvents, to provide a possible answer to the biphasic hydroformylation of long-chain alkenes. At temperatures close to the operating temperature of many catalytic reactions (60-120°C), the fluorous and organic solvents mix, but at temperatures near ambient they phase separate cleanly. Since that time, many other reactions have been demonstrated under fluorous biphasic conditions and these form the basis of this chapter. The subject has been comprehensively reviewed,[2-6] so this chapter gives an overview and finishes with some process considerations.

Fluorous biphasic systems operate on the premise that the catalyst complex is preferentially soluble in the fluorous phase. This is achieved by synthesising fluorinated ligands that have a high weight-percentage of fluorine. It has been reported that for a complex to be preferentially soluble in fluorous solvents it must contain >60

D. J. Cole-Hamilton and R. P. Tooze (eds.), Catalyst Separation, Recovery and Recycling, 145–181.

wt% fluorine.[7] However, work by Hope and co-workers, has shown that this is not necessarily so. In fact, ligands with identical analytical composition, such as *para-*, *meta-* and *ortho-* isomers, may have very different solvent preferences. It is more important that the organic core is enveloped to ensure preferential fluorocarbon solubility rather than containing a specific wt% of fluorine atoms.[8] They also report that preferential solubility of a perfluorinated ligand does not immediately imply preferential solubility of the corresponding catalyst. Rábai and co-workers report that increasing the length of a single fluorous ponytail will be detrimental to the solubilities in both solvents[9] and an increase in the partition coefficients is observed when the number of fluorous ponytails attached to the ligand is increased. Provided that the partition properties of the solvent system are known (solubilities of one within the other), partition coefficients of solutes between fluorous and organic solvents can be calculated[10] with remarkable accuracy using a model which relies only on estimating the molar volume (V_b) and the modified non-specific cohesion parameter (δ_b') of the solute, both of which can be estimated using group contribution incremental methods. The value of the partition coefficient (P) depends on the size and non-specific vaporization energy (E_v') of the organic core of the solute as well as the length and number of attached fluorous ponytails. Generally, log P increases with the chain length and number of chains unless E_v' is very large, when either can lead to a decrease in partition into the fluorous phase. Some systems have been shown to give lower log P on increasing the chain length or number of fluorous chains,[11, 12] but these do not have very high values of E_v' for the organic core. So some modification of this aspect of the modelling is required. Clarke has recently published his findings with phosphino-amine fluorous soluble metal complexes.[13] He demonstrates that the addition of one fluorous ponytail to a $(C_6F_5)_2P$ fragment is sufficient to ensure a high partition co-efficient in perfluoromethylcyclohexane (PFMC)/toluene. Initially the synthesis of fluorous soluble ligands focussed on fluorinating known ligands, for example; fluorous analogues of triphenylphosphine in Vaska's complex[14] and Wilkinson's catalyst[15] or other triaryl phosphines.[16, 17] Further work investigating the influence of the position of these perfluoroalkyl modifying "ponytails" on the aryl ring has been undertaken by Hope and co workers.[18, 19]

With the catalyst immobilised in the fluorous phase, the substrate can be introduced either in solution, e.g. with toluene, or neat.[20] When heated, the two phases form a single homogeneous phase, which allows the substrate to be in intimate contact with the catalyst at all times. With the addition of reacting gases, reaction will occur at this elevated temperature and the catalyst and product are easily separated by cooling the mixture and decanting the product allowing easy reuse of the catalyst phase. The concept is shown schematically in Figure 6.1 for the hydroformylation of alkenes, which was the test reaction first reported by Horváth and co-workers.

Although temperature dependent solubility is generally used to affect the required phase transitions (miscible – immiscible), an alternative involves the use of CO_2 to affect the transition.[21] Being soluble in both organic and fluorous solvents, the CO_2 can render a biphasic system monophasic without raising the temperature. The concept has been demonstrated using hydrogenation and epoxidation of alkenes, and significant rate enhancements (50-70%) were observed at room temperature. Almost similar enhancements can, however also be achieved in the biphasic systems by using more efficient stirring. The CO_2 approach may have particular appeal in systems where one of the substrates or products is thermally sensitive.

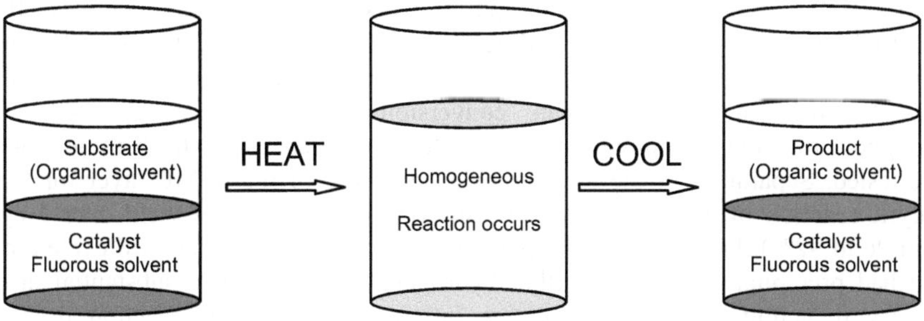

Figure 6.1 The fluorous biphase concept illustrated for the hydroformylation of an alkene (substrate) to an aldehyde (product)

The influence of the fluorous ponytails on the metal complex is an important factor to understand as the strongly electron withdrawing effect of the fluorine atoms may affect the behaviour of the phosphorus atom. A spacer group between the phosphorus and the fluorine tail is usually included in the ligand design. Generally this is an aryl or alkyl group that is effectively acting as a shield to the phosphorus and metal centre from the powerful electron withdrawing effect of the perfluoroalkyl tail. Horváth and co-workers studied the optimum length of an alkyl spacer group and found that the electronic properties of the ligand could be tuned by varying the length of the alkyl spacer group between the phosphorus atom and the perfluoroalkyl tail.[22] The results they reported from theoretical calculations implied that two methylene groups would provide sufficient shielding of the phosphorus atom, but in practice a C_2H_4 spacer does not entirely eliminate the electron withdrawing effect of the perfluoroalkyl tail. An alkyl spacer is expected to be a better insulator of electron effects than an aryl group, which may transmit electronic effects especially to the *o-* and *p-* positions of the ring. Insulation of the effects of the fluorous substituents on aryl groups can be achieved by using a C_2H_4 or $O(CH_2)_n$ (n = 1 or 5) [23] spacer between the aryl group and the fluorous tail, or by attaching the fluorous tail to a $SiCH_2CH_2$ spacer.[11, 24] Ligands such as $P(4\text{-}C_6H_4SiMe_{3\text{-}n}(CH_2CH_2C_6F_{13})_n)_3$ (n = 1-3) and $\{CH_2P(4\text{-}C_6H_4SiMe_{3\text{-}n}(CH_2CH_2C_6F_{13})_n)_2\}_2$ have been shown to be electronically very similar to PPh_3 and $Ph_2PCH_2CH_2PPh_2$ and sterically only marginally larger. Only the ligands with n = 3 are preferentially soluble in fluorous solvents, but the rhodium and nickel complexes tend to show much higher fluorophilicities because the coordination of multiple ligands encapsulates the organic centre of the molecule better. The fluorous content (wt %) actually decreases slightly for the complexes.

There has been great interest since Horváth's original paper, in synthesising new and more fluorous soluble ligands for various reactions. From the wide ranging topics published it would appear that the general consensus is that if the reaction can be carried out under homogeneous conditions, then it should be possible to fluorinate the ligands and perform the reaction under fluorous biphasic conditions. We now discuss the reactions that have been reported using fluorous biphasic systems.

6.2 Alkene Hydrogenation

Having successfully produced a fluorous analogue of Wilkinson's catalyst for the hydroboration of alkenes, Horváth and co-workers applied this complex {ClRh $[P(CH_2CH_2(CF_2)_5CF_3)_3]_3$} to the hydrogenation of a range of alkenes.[25] Using a PFMC/toluene biphasic system at 45°C under a balloon pressure of H_2, the reaction of 2-cyclohexen-1-one showed clean conversion to cyclohexanone, 98% yield. Throughout the study, conjugated alkenes, terminal alkenes and disubstituted alkenes were tested. Separation of the toluene layer and reuse of the PFMC layer for further catalytic reaction of 2-cyclohexen-1-one showed a small drop in yield over 3 cycles (from 96 to 92%). However, for the reaction of 1-dodecene, by the third cycle a black solid had precipitated and any further hydrogenations suffered a substantial drop in rate. By recycling the catalyst solution into new vessels, the reaction proceeded normally. These results do not compare to those for the best homogeneous catalysts, but the ease of recycling the expensive catalyst, does make this a useful reaction system. Hope *et al* have also described a rhodium-catalysed hydrogenation of styrene in the fluorous biphase using a range of fluorinated ligands.[26] The purpose of this study was to investigate the effects of the perfluorocarbon solvent and the fluorous ponytail substituents on the catalytic system. The best rates were achieved with the fluorous ligand $P(4\text{-}C_6H_4OCH_2C_7F_{15})_3$, 201 mmol dm^{-3} h^{-1}, in a toluene/hexane/perfluoro-1,3-dimethylcyclohexane (PFDMCH) solvent system. No free ligand was observed in the organic phase and recharging the reactor with further fractions of substrate resulted in no effect on the catalytic activity, demonstrating a high catalyst stability and recovery.

Hydrogenation of 1-octene was also carried out by van Koten and co-workers using analogues of Wilkinson's catalyst containing $P(4\text{-}C_6H_4SiMe_2(CH_2CH_2R_f))_3$ (R_f = C_6F_{13} or C_8F_{17}).[27, 28] Comparison of this catalyst with $[RhCl(PPh_3)]$ and with $[RhCl(P(4\text{-}C_6H_4SiMe_3)_3]$ under monophasic conditions in $PhCF_3$ showed that the $SiMe_3$ substituent slightly increased the reactivity compared with H, but that the fluorous tail reduced the activity back to the same as that using PPh_3 as the ligand. The ligands with the fluorous ponytails showed good activity under fluorous biphasic conditions (1-octene/PFMC) at 80 °C for the hydrogenation of 1-octene and the fluorous phase could easily be separated from the product by cooling below 25°C. Over 8 cycles using the ligand with $R_f = C_8F_{17}$ and carrying out the separation at 0°C, the rate of reaction increased from 177 catalyst turnovers h^{-1} to 600 h^{-1}, but this was largely attributed to loss of significant amounts of the fluorous solvent into the organic phase. When the lost fluorous solvent was replaced for the 8th run the rate reduced to 155 h^{-1} (87.5% of the initial rate). ICPAAS analysis showed that the overall leaching corresponded to 1% of the Rh after 9 cycles (0.1% after the first cycle, c.f . 0.3% after the first cycle with the ligand with $R_f = C_6F_{13}$). The phosphine leaching to the organic phase was much more significant (1.3% and 8% per cycle for $R_f = C_8F_{17}$ and C_6F_{13} respectively).

Rhodium complexes of triarylphosphines containing an OCH_2 spacer (Figure 6.2) have been used for hydrogenation of methyl *(E)*-cinnamate in Galden D100 (mainly perfluoroctane) or FC-77(mainly C_8 containing fluorocarbons)/ethanol.[23] Generally speaking the reactions were slow (<100 turnovers h^{-1}); they could be recycled twice but the reaction half-lives increased with time. Rhodium and phosphine leaching were of the order of 0.4 and 1.55% respectively for ligands with only one ponytail. Although when the ponytail was in the 2-position, the leaching was much higher (2.8 and 5.6%) because the steric congestion of this ligand inhibited coordination. Leaching was lowest (0.1 and 0.5% in the second and third runs) for the ligand with two ponytails (as was the reaction rate). Higher leaching and lower rate in the first run were attributed to difficulties with coordinating the bulky ligand.

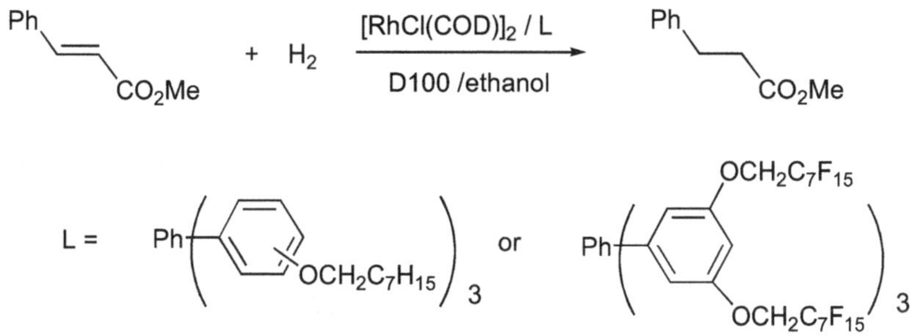

Figure 6.2. Fluorous biphasic hydrogenation of methyl *(E)*-cinnamate catalysed by rhodium complexes.[23]

Fluorous solvents are of very low polarity so it would be expected that ionic compounds would not dissolve readily in them. However, ionic rhodium catalysts have been used successfully in fluorous biphasic systems. $[Rh(COD)(Ar_2PC_2H_4PAr_2)]BF_4$ $(Ar = 4\text{-}C_6H_4SiMe_2(CH_2)_2C_6F_{13})]$ is active for the hydrogenation of 1-octene and partial hydrogenation of 1-octyne.[29] In the case of 1-octene in acetone, the lower rate of alkene isomerisation than for the unsubstituted analogue means that the fluorous compound is a better hydrogenation catalyst. This has been attributed to aggregation, which has also been observed for other analogues in which there is an Me_3Si- or $Me_2Si(C_8H_{17})$- substituent, but to a lesser extent. 1-octyne was hydrogenated (40°C, 1 bar H_2) under fluorous biphasic conditions with phase separation at 0°C.

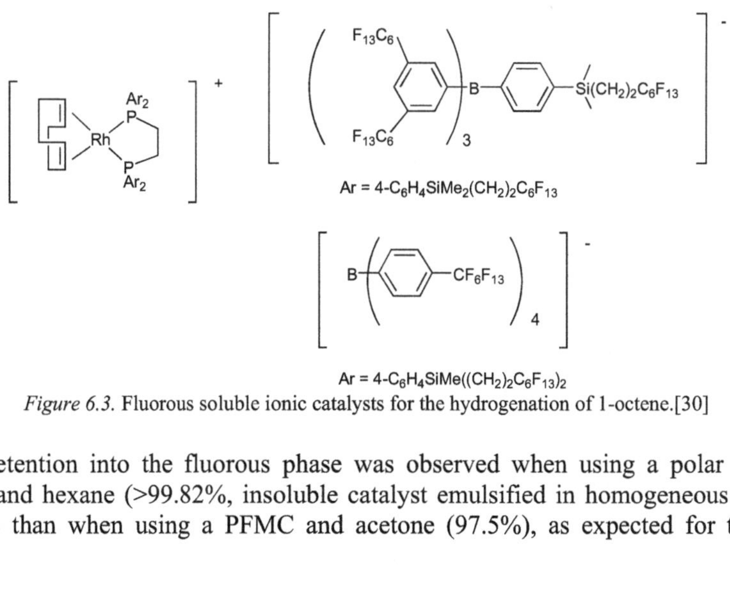

Figure 6.3. Fluorous soluble ionic catalysts for the hydrogenation of 1-octene.[30]

Better retention into the fluorous phase was observed when using a polar fluorous solvent and hexane (>99.82%, insoluble catalyst emulsified in homogeneous reaction mixture) than when using a PFMC and acetone (97.5%), as expected for the ionic catalyst. Phosphine leaching was very similar to that of rhodium in this case suggesting that the diphosphine remains coordinated throughout the reaction. Turnover frequencies for these reactions were low (10-30 h^{-1}) and the selectivities to *cis*-4-octene were 60-70%. The other main product was 4-octyne (15-17%). Using similar cations, but heavily fluorinated anions (Figure 6.3) the catalysts were fully soluble in the fluorous biphasic mixture and could be recycled with only 1% losses after the hydrogenation of 1-octene.[30] The complex containing more fluorous tails on the diphosphine and $[B(4\text{-}C_6H_4C_6F_{13})_4]^-$ was the more active hydrogenation catalyst.

(R)-BINAP ligands bearing fluorinated substituents (Figure 6.4) have been prepared by Hope and co-workers.[31] The ruthenium complexes of the two ligands were compared with Ru-*(R)*-BINAP for the asymmetric hydrogenation of dimethyl itaconate in methanol at room temperature and 20 bar H_2 with a substrate/catalyst ratio of 2000. The enantioselectivity of the compound, to the *(S)*-enantiomer, was unaffected (>95% each) by the addition of the perfluoroalkyl groups. However, they did affect the conversions, only 42% with Ru-(B) and 83% with Ru-(A) compared to 88% conversion with the non-fluorous analogue, the more electron-withdrawing ponytails lowering the activity of the catalyst. Theses reactions were not carried out under fluorous biphasic conditions.

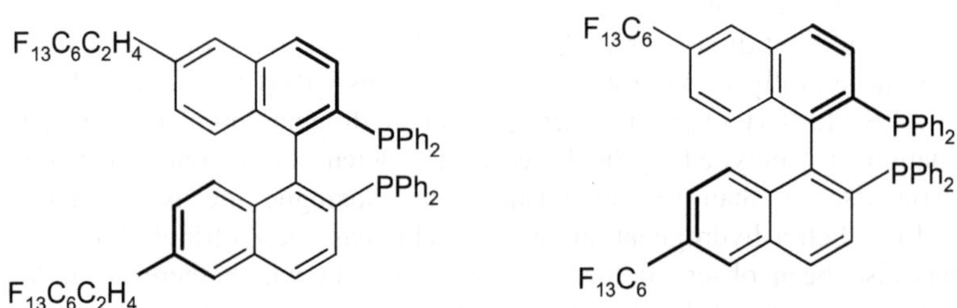

A B

Figure 6.4. (R)-BINAP ligands used in ruthenium catalysed asymmetric hydrogenation of dimethyl itaconate.[31]

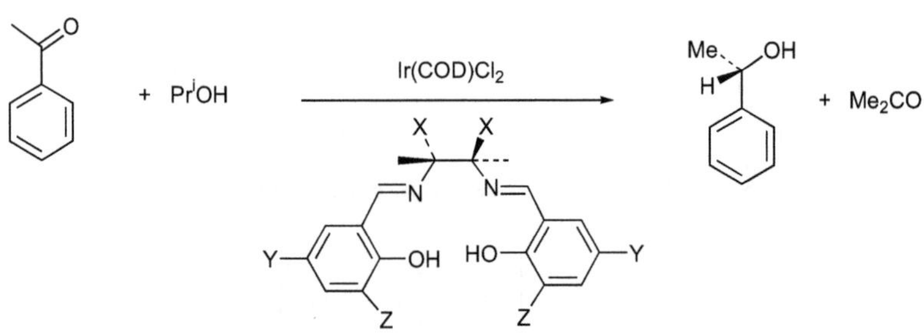

X	Y	Z		e. e.
Ph	C_8H_{17}	H	**1a**	0
Ph	C_8H_{17}	Bu^t	**1b**	56
X_2				
$(CH_2)_4$	C_8H_{17}	Bu^t	**2a**	5
$(CH_2)_4$	$3,5\text{-}C_6H_3(C_8H_{17})_2$	Bu^t	**2b**	10
$(CH_2)_4$	C_8H_{17}	C_8H_{17}	**3**	47

Figure 6.5. Transfer hydrogenation of acetophenone in Galden D-100.[32, 33]

The asymmetric transfer hydrogenation of acetophenone by 2-propanol has been carried out in Galden D-100 (mainly perfluorooctane) using iridium complexes of salicaldehyde diaminoethane Schiff bases as ligands (Figure 6.5).[32, 33] High yields (>90%) with enantiomeric excess up to 57% were observed when the reactions were carried out at 70°C. These e. e.'s are higher than observed with non-fluorous analogues (<22%), but the catalyst was destroyed during the reaction so that the recovered fluorous phase was much less active giving an e. e. of only 6%. The organic phase was also active, but not enantioselective. Other ketones and □–diketoesters were also reduced with e. e. between 18 and 60%.

6.3 Alkene Hydrosilation

Complexes of the form $[RhCl(P(4\text{-}C_6H_4SiMe_{3\text{-}n}((CH_2)_2C_8F_{17})_n)_3]$ have been used for the fluorous biphasic hydrosilation of 1-hexene (Figure 6.6) in PFMC without an additional organic solvent (alkene:silane = 2:1).[34] The reactions were monophasic at the reaction temperature (reflux), but the more fluorinated catalyst was not completely soluble in the monophasic mixture. It was redissolved in the PFMC phase, which separated on cooling to room temperature. Reaction rates were similar to those obtained using $[RhCl(PPh_3)_3]$ in benzene. Reactions were carried out for 15 min and the fluorous phase was recycled twice, with 100% conversion of the silane being observed in each case. However, ICPAAS analysis showed that, for less fluorous catalyst, rhodium and ligand leaching were high (12 and 19% respectively), whilst the retention of the more fluorinated catalyst was better (1.7 and 2.2%). In an interesting variant of the fluorous biphasic principle, fluorinated alkenes were hydrosilated with e.g. $HSiMe_2Cl$ using $[RhCl(PPh_3)_3]$ in benzene or toluene. The fluorous product could be extracted using FC-72 (a mixture of perfluoro hexanes) and the organic phase containing the catalyst recycled. Rhodium and phosphorus were undetectable (<1 ppm) by ICPAAS in the fluorous phase.

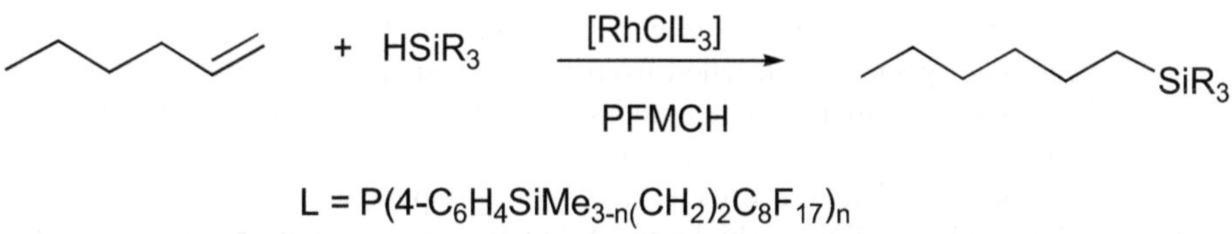

Figure 6.6. Hydrosilation of alkenes under fluorous biphasic conditions.[34]

6.4 Alkene Hydroboration

Fluorous biphasic catalysis (FBC) has been applied to the rhodium-catalysed hydroboration of alkenes and alkynes by Horváth and co-workers.[15, 35] This was achieved by synthesising a fluorous analogue of Wilkinson's catalyst $[RhCl(PPh_3)_3]$, using the perfluorinated phosphine $[P(C_2H_4C_6F_{13})_3]$. The modified catalyst was highly soluble in PFMC and $CF_3C_6H_5$, slightly soluble in THF and acetone, but insoluble in other organic solvents. The catalyst was effective at a loading of 0.01 – 0.25 mol%, affording turnover numbers (TON) as high as 8500. The reaction was carried out in PFMC without an added organic solvent as it was observed that the reactions occurred more rapidly in the absence of a non-fluorous solvent. A high degree of catalyst recovery was observed in repeat reactions of norbornene although some loss of catalyst was indicated by a drop in yield and TON, 90% yield with TON 854 in the first run compared to 83% yield and TON 785 in the second run. In their later paper,[35] this catalyst loss was studied further and quantitatively by atomic absorption analysis, the rhodium loss was found to be 0.4% per cycle, for the Rh complex bearing the above ligand. However, the Rh complex with $[P(C_2H_4C_8F_{17})_3]$ only showed a loss to the organic phase of 0.2% per cycle. It was also noted that after three cycles, the perfluorohexyl complex deposited a black solid and accordingly the activity dropped, suggesting decomposition of the catalyst. Low catalyst loading and Rh loss do make this a favourable method of hydroboration of alkenes and alkynes especially due to the ease of separation of the products from the fluorous phase.

6.5 Alkene Hydroformylation

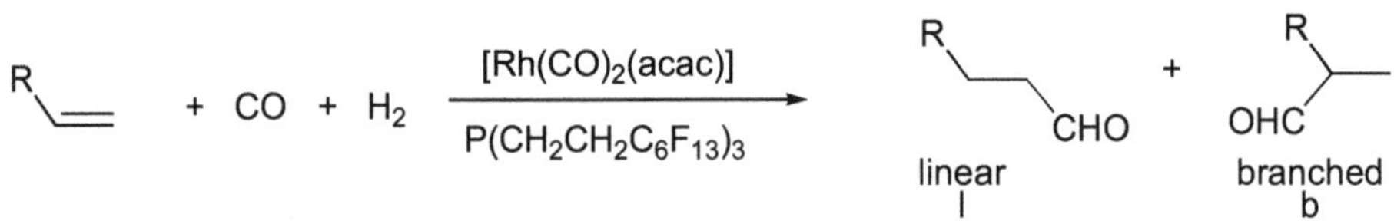

Figure 6.7. Hydroformylation of an alkene using a rhodium complex bearing a fluorous ponytail.[1, 22]

In his original paper introducing FBC,[1] the reaction Horváth chose to demonstrate the principle was the hydroformylation of 1-decene (Figure 6.7 ($R = C_8F_{17}$). Due to the lack of fluorous soluble ligands available, the synthesis of a new fluorinated phosphine was also necessary, $P(CH_2CH_2(CF_2)_5CF_3)_3$ was chosen. PFMC was used as the fluorous phase, toluene as the organic phase and the catalyst was prepared *in situ* from the Rh precursor $[Rh(acac)(CO)_2]$ and the phosphine. After 24 hours, they report that only trace amounts of conversion products were observed by GC analysis indicating that very little catalytic species has leached from the fluorous phase throughout the initial reaction. In a later paper by the same authors,[22] further investigation of the shielding of the phosphine from the electron-withdrawing fluorous tail indicated that, although it does not completely shield the effects, a $-C_2H_4-$ group between the phosphine and fluorous tail provides sufficient shielding. A kinetic investigation carried out on this reaction showed it to be first order in both [Rh] and 1-decene. Increasing $[P(C_2H_4C_6F_{13})_3]$ increased the l:b ratio but decreased the activity of the hydroformylation, which is consistent with the known negative order in [phosphine] for hydroformylation reactions. This paper also reports the semi-continuous hydroformylation of 1-decene, recycling the catalyst phase in nine successive runs. A slight rise in activity and a slight decrease in l:b ratio (4.5 – 3.5) and selectivity to aldehydes, 91.5% to 88.3% were observed, with a corresponding rise in isomerisation from 7.8% to 11.0%. These results are as expected if phosphine is being lost to the organic phase as the reaction has a negative order in [phosphine] and the linear selectivity is lower at lower [phosphine]. The Rh loss from the fluorous phase was measured as 4.2% over 9 runs. This is equivalent to 1.18 ppm Rh loss/mol aldehyde. Being a trialkyl phosphine and not very bulky, very high [phosphine] (0.3 mol dm^{-3}) was at required for high l:b ratios (7.8:1) in the product aldehyde. The catalytic system was also tested for the hydroformylation of ethene using FC-70 (predominantly perfluoro compounds with 15 carbon atoms) as the fluorous solvent under continuous flow operation and gave a selectivity of 99.3% to propanal. This reaction was successfully carried out for two months with no loss in activity, but the product was removed in the gas phase.

Various groups have studied fluorous analogues of triphenylphosphine in attempts to increase the linear selectivity. Masdeu-Bultó and co-workers have reported the successful fluorous biphasic hydroformylation of 1-octene using a Rh complex with the fluorinated phosphine $P(C_6H_4\text{-}4\text{-}OCH_2C_7F_{15})_3$.[36] Rh/P = 5, at 80ºC and 40 bar. The biphasic solvent system used was PFMC/toluene (60/40, v/v). After 1 hour of reaction a conversion of 98% with an aldehyde selectivity of 97% was achieved. The l:b ratio was 2.6:1% . However, on recycling the catalyst solution, a drop in the aldehyde selectivity was observed without a drop in conversion, indicating a loss of ligand to the organic/product phase. On cooling, the toluene phase was observed to be slightly coloured symptomatic of catalyst leaching from the fluorous phase. By increasing the temperature of separation from 10ºC to 50ºC, the separation improved. Reusing this catalyst phase allowed three recycles with little reduction in aldehyde selectivity observed (99%, 97% and 91%). After the first recycle, analysis of the product phase showed Rh and P in the toluene phase and traces of aldehyde are reported as being observed by GC analysis in the fluorous phase. An increase of the substrate/Rh ratio from 500/1 to 2000/1 showed a lower conversion, 20%, after 1 hour, as expected, but the selectivity to aldehydes remained high, 94%. By removing the toluene from the reaction an increase in activity (68% conversion in 1 h) was observed, this effect has also been observed by others.[20]

Hope and co-workers have synthesised a large number of perfluorinated ligands and carried out testing of these ligands in the hydroformylation of 1-hexene[37] using $[Rh(acac)(CO)_2]$ as the precursor and a P/Rh ratio of 3. After I h in a high pressure reactor using a 1:1 organic:fluorous solvent mixture, 70ºC, 20 bar CO/H_2(1:1) the fluorous analogue of triphenyl phosphine, $P(4\text{-}C_6H_4C_6F_{13})_3$, gave 89.2% selectivity to aldehydes, with an l:b ratio of 3.8 compared with values for triphenylphosphine of 98.2% selectivity and l:b ratio of 3.1. Further studies[20, 38] on this ligand with 1-octene as the substrate demonstrated that using $[Rh(acac)(CO)_2]$ (2 mmol dm^{-3}) and $[P(4\text{-}C_6H_4C_6F_{13})_3]$ (20 mmol dm^{-3}) at 60ºC and 20 bar in PFMC, an l:b ratio of 6.3:1 (selectivity to linear aldehyde = 80.9%) could be obtained with a turnover frequency of 4 400 h^{-1} and with 99.95 and 96.7% retention of rhodium and phosphine into the fluorous phase. The rhodium retention is better than that of the phosphine because the form of the catalyst at room temperature, where the separation was carried out, is $[RhH(CO)(P(4\text{-}C_6H_4C_6F_{13})_3)_3]$, anchoring it into the fluorous phase by 9 fluorous ponytails.[39] Higher rates and more Rh leaching were observed at higher reaction temperatures. By omitting the organic solvent, higher rates, linear selectivity and retention into the fluorous phase were observed. Even though the system becomes biphasic under the reaction conditions at 20% conversion,[40] the kinetics were first order up to at least 80% conversion,[20] showing that mass transport does not become rate determining provided that the mixture is stirred efficiently. The rate and selectivity are much better than those obtained using PPh_3 under the same conditions. This is due to the electron withdrawing effect of the fluorous tail, which is transmitted to the rhodium centre *via* the phenyl ring. The higher binding constant of the fluorinated ligand than PPh_3 accounts for the better linear selectivity at low phosphine loadings. This conclusion is supported by the observation that the selectivity under comparable conditions but with less phosphine (80°C, 30 bar, P:Rh = 3) for this system is higher (l:b = 3.5)[20] than for the analogous ligand with an $-OCH_2$- spacer (80°C, 40 bar, P:Rh = 5, l:b = 2.6).[36] although it should be noted that toluene, which may reduce the l:b ratio was present in the latter system. In the presence of toluene under slightly milder conditions and with the lower P:Rh ratio (70°C, 20 bar, P:Rh = 3) the selectivity of the complex without the spacer was still higher (l:b = 3.8).[37] The high rates, selectivities and the omission of the organic solvent meant that this system was chosen for more detailed studies in a continuous flow system (see Section 6.13).[41, 42]

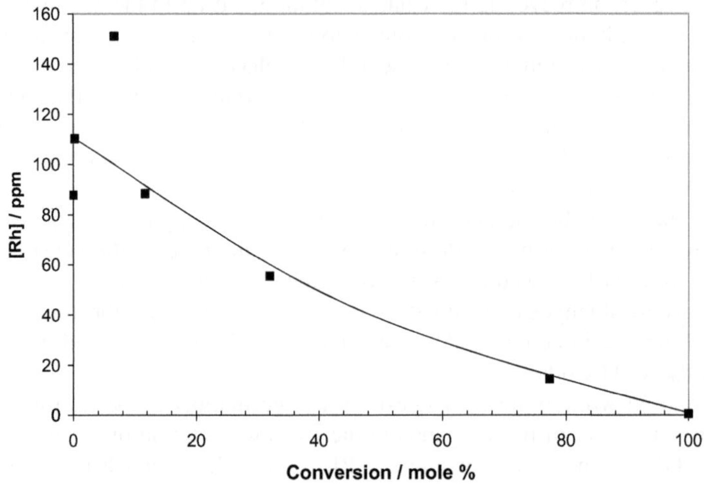

Figure 6.8. Concentration of rhodium in the organic phase as a function of conversion during the hydroformylation of 1-octene catalysed by Rh/ $[P(4-C_6H_4C_6F_{13})_3]$.[41]

A comparison between $[P(4-C_6H_4C_6F_{13})_3]$, $[P(4-C_6H_4C_8F_{17})_3]$ and $[P(3,5-C_6H_4(CF_3)_2)(4-C_6H_4C_6F_{13})_2]$ suggests that a catalyst bearing the ligand containing the C_8 fluorous ponytail undergoes only 10% of the leaching of the other two ligands, whilst their activities (first order rate constant, k = 1.9×10^{-3}, 1.3×10^{-3}, 2.5×10^{-3} s^{-1}) and selectivities (l:b = 6.3, 6.7 and 4.7 respectively) are similar.[8] Other studies have shown that the retention of the catalyst within the fluorous phase is very dependent upon the conversion, being much enhanced at high conversions (see Figure 6.8).[43]

The fluorinated analogue of triphenyl phosphate, $P(O(C_6H_4C_6F_{13})_3$, also performed well in the hydroformylation of 1-hexene when compared with $P(OPh)_3$ with a selectivity of 82.3% and l:b ratio of 6.4 compared to 92.0% and 2.9 respectively.[8] This represented one of the first examples of phosphite ligands being used for hydroformylation in fluorous biphasic catalysis. Further optimisation[20, 38] of this system in perfluorodimethylcyclohexane (PFDMCH) using 1-hexene and 1-octene showed that high rates and linear selectivities (l:b up to 9.9:1) could be achieved, but alkene isomerisation was always *ca.* 13%, whilst leaching of Rh (>2.5%) and phosphine (>4%) were much higher than for the phosphine analogue. This leaching, which was much worse when toluene was present, was attributable to degradation of the phosphite, either through direct reaction with the aldehyde, or through reaction with water formed by aldol condensation of the product. Either would produce complexes in which one or more fluorous tails had been removed, so that the fluorous solubility would be reduced.

Mathivet *et al* have extended this work on phosphites in the hydroformylation of higher alkenes under fluorous biphasic conditions.[44, 45] They have synthesised and studied the affect of a number of ligands with the fluorous ponytails in various positions on the aryl group, Figure 6.9. The two ligands with a single fluorous ponytail, **A** and **B**, partition in a solvent mixture of 1-decene and $C_8F_{17}H$ in a 5/95 ratio and the ligand with two fluorous tails, **C**, partitions with a ratio of 1/99. There is likely to be some loss of ligand through leaching throughout the reaction as all the ligands dissolve to some extent in the substrate.

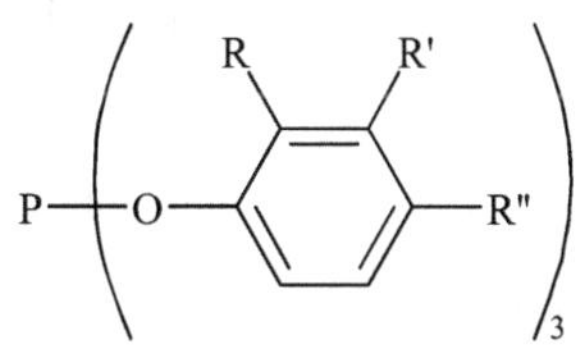

A: R= $C_2H_4C_8F_{17}$, R'=R"= H | **D:** R = C_8F_{17}, R' = R" = H | **G:** R = CH_3, R' = H, R" = C_8F_{17}

B: R=R'= H, R"= $C_2H_4C_8F_{17}$ | **E:** R =H, R' = C_8F_{17}, R" = H | **H:** R = C_8F_{17}, R' = H, R" = C_8F_{17}

C: R= R"= $C_2H_4C_8F_{17}$, R'=H | **F:** R = R' = H, R" = C8F17 | **J:** R = C_8F_{17}, R' = H, R" = COC_7F_{15}

Figure 6.9. Fluorinated ligands used in the rhodium catalysed hydroformylation of alkenes.[44, 45]

The three ligands were tested in the hydroformylation of 1-decene with no organic solvent using only $C_8F_{17}H$ at 80°C under 40 bar CO/H_2 (1:1) and run to 100% conversion. **A** and **C** are both sterically hindered and gave similar results, 100% conversion was achieved quickly after 15 and 12 min respectively with an aldehyde selectivity of 85% in both cases, this results in TOFs of 10 000 and 11 000 h^{-1}. The unhindered ligand **B** achieved 100% conversion in 30 minutes, but showed a higher aldehyde selectivity of 95%, TOF 3900 h^{-1}. Results with **B** and various terminal alkenes demonstrate the high selectivity of this phosphite to aldehyde formation. Catalyst recycling was also considered. Using 1-decene as the substrate allowed direct comparison with Horváth's work. Each reaction was run for just 10 minutes and the first three runs for each ligand show a slight increase in conversion with the fourth run showing a marked drop for ligand **A**, however the selectivities for each ligand dropped slightly in each cycle. In the fourth recycle, the organic phase was noted as being yellow in colour, which indicates that leaching of Rh is occurring. The reason for the differences in results between the various ligands lies in the steric hindrance that exists with ligands **A** and **C**, which will result in only one phosphite coordinating to the Rh making the active species $[HRhL(CO)_3]$ and allowing facile binding to the alkene, causing the fast reaction. The single ligand makes the catalyst less selective, whereas **B** is more similar to the triphenyl phosphite modified catalyst, which forms the bis phosphite. In order to try to overcome some stability issues with these ligands, further syntheses of ligands **D-J** were carried out and reported in a second paper.[44] An investigation of the effect of the position of the fluorous group on the ring **D-F** show that the activity increases as the fluorous group gets closer to the phosphorus (TOF of 10 000 h^{-1}, 6300 h^{-1} and 3500 h^{-1} for **D, E** and **F**). Conversely, aldehyde selectivity increases as the fluorous group moves away from the phosphorus atom; thus the *para*-substituted phosphite shows the highest selectivity towards aldehyde (71%, 80% and 85% for **D, E** and **F**). It is also noted from the comparison of **C** and **H** that the removal of the ethyl spacer group has a marked effect on the aldehyde selectivity (85% and 46% respectively). This is perhaps due to the electronic effect of the fluorous tail on the metal or may simply be due to the steric hindrance of the ligand about the metal centre. Again the hydroformylation of internal and different terminal alkenes was examined using **E.** It was found that decreasing the length of the substrate caused the activity, l:b ratio and aldehyde selectivity to increase (see Table 6.1)

Recycling of the ligands was investigated for ligands **D, E** and **J**. over four recycles the conversions dropped for all three phosphites, however the most marked decrease was with **E** (<10% conversion in the fourth recycle). The selectivity of the catalysts to aldehyde products did not decrease so drastically but a decrease in the selectivity of **E** was most notable in the second recycle. This suggests that some leaching of the three phosphites to the organic phase does occur.

TABLE 6.1. The results of hydroformylation reactions carried out using rhodium complexes ligand **E** (Figure 6.9).[44, 45]

Alkene	TOF / h^{-1}	l:b	Selectvity to aldehyde / %
1-hexene	7800	6.5	94
1-octene	6900	6.3	87
1-decene	6300	5.8	80

Another approach to the fluorous biphasic hydroformylation of alkenes has been investigated by Xiao and co-workers using a fluorous soluble polymer catalyst.[46] The poly(fluoroacrylate-*co*-styryldiphenylphosphine) ligands are of the form shown in Figure 6.10. Neither ligand is soluble in normal organic solvents but both are soluble in fluorinated solvents. The hydroformylation of 1-decene was performed in a batch reactor using a solvent mixture of hexane-toluene-PFMC (40:20:40) (the hexane proved necessary to form a homogeneous phase under the reaction conditions). The catalyst was formed *in situ* from $[Rh(acac)(CO)_2]$ and the ligand (P/Rh = 6) under CO/H_2 (30 bar, 1:1). The results showed a turnover frequency of 136 mol aldehyde (mol Rh h)$^{-1}$ and a selectivity to aldehyde of 99%, the l:b ratios of 4.8 (polymer **a** in Figure 6.10) and 5.9 (polymer **b** in Figure 6.10) are also competitive with those obtained from homogeneous hydroformylation reactions in conventional solvents. The recycling of the catalyst phase was carried out in the hydroformylation of 1-hexene. Three consecutive reactions were carried out and a 1 ppm loss of Rh was observed for each recycle. However, by the end of the third run all PFMC had leached to the product phase, which in turn caused loss of ligand and catalyst to the product phase.

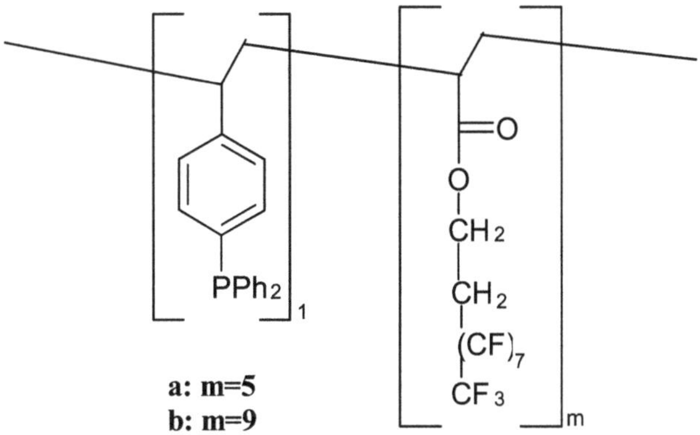

Figure 6.10. Fluorous polymers with bound PPh_3 used in the hydroformylation of 1-hexene.[46]

Ojima and co-workers[47] have reported fluorinated analogues of BINAPHOS, an unsymmetrical bidentate ligand having one phosphine and one phosphite donor, for rhodium-catalysed asymmetric hydroformylation. Unlike the previous papers, this deals with the hydroformylation of styrene, which preferentially forms the branched product. This is desirable because the linear product has no chiral centre and because the branched product is a possible intermediate in the formation of ibuprofen type anti-inflammatories. The ligands were found to be preferentially soluble in toluene over PFMC, possibly due to the aromatic nature of the naphthalene groups and the slightly low fluorine content. All ligands displayed good solubilities in perfluorotoluene (PFT).

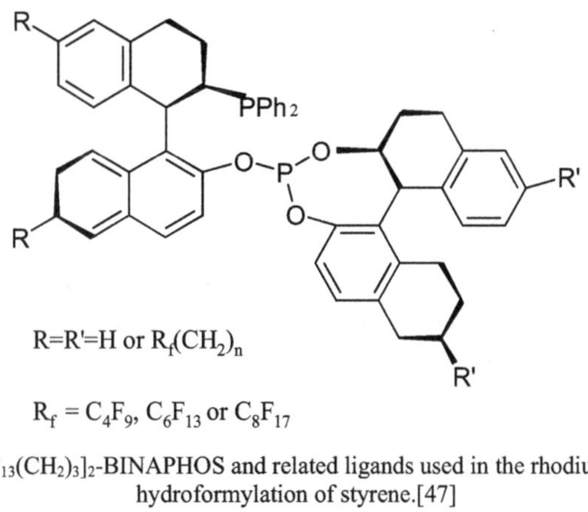

Figure 6.11. (S,R)- $[C_6F_{13}(CH_2)_3]_2$-BINAPHOS and related ligands used in the rhodium catalysed asymmetric hydroformylation of styrene.[47]

The hydroformylation reactions were carried out in a number of different solvent systems, Table 2, at 50-60°C and 40 bar CO/H_2 (1:1) using the (S,R)- $[n\text{-}C_6F_{13}(CH_2)_3]_2$-BINAPHOS ligand, Figure 6.11. After 18 hours in the PFMC/toluene system, 100% conversion was achieved with 100% selectivity to aldehydes of which 92% were branched, with an enantiomeric excess of 85%. With only PFMC as solvent only 88% conversion was observed, with a b:l ratio of 94:6 and an e. e. of 87%. Changing the fluorous solvent to PFT showed 73% conversion after 16 hours with 100% selectivity to aldehydes of which 91% were the branched form with an overall e. e. of 82%.

TABLE 6.2. The Rh/*(S,R)*- [n-$C_6F_{13}(CH_2)_3]_2$-BINAPHOS catalysed hydroformylation of styrene.[47]

Solvent System	Styrene:Rhodium	Time (h)	Conv. (%)	b/l	e. e. (%)
PFMC/Toluene	2235	24	84	91/9	90
PFMC/Toluene	1042	18	100	92/8	85
PFMC	1042	18	88	94/6	87
PFT	2235	2	11	91/9	94
PFT	2235	16	73	91/9	82

6.6 Alkene Epoxidation

The large amounts of molecular oxygen that can be dissolved into fluorous solvents, together with their oxidation resistance makes them potentially very suitable solvents for oxidation reactions. Fish and co-workers reported the synthesis of a novel perfluorinated triazacyclononane (TACN) ligand, shown in Figure 6.12, which is soluble in perfluoroalkanes.[48] Using this ligand along with a fluorinated carboxylate synthon, $[M(O_2C(CH_2)_2C_8F_{17})_2]$ (where M is Co or Mn), they have successfully carried out the epoxidation of a variety of alkenes, in a fluorous biphasic system. The suggested form of the catalyst produced *in situ* is $[C_8F_{17}CH_2CH_2CO_2M\text{-}\{(C_8F_{17}(CH_2)_3)_3TACN\}]^{2+}$. The reactions were carried out in perfluoroheptane with the substrate acting as the organic layer, in an oxygen barosphere in the presence of *tert*-butyl hydroperoxide (TBHP). Alkenes bearing allylic hydrogens gave the highest yields under these conditions. Removal of the upper, organic layer and adding new cyclohexene and TBHP provided similar results to the first run, showing that leaching of the catalyst into the organic phase was negligible, but no quantitative data was provided. $[CuCl\{(C_8F_{17}(CH_2)_3)_3TACN\}]$ is also active for alkene oxidation by Bu^tOOH/O_2 in perfluoroheptane at room temperature (biphasic).[49] The best results were obtained with cyclohexene, which gave cyclohexanone and cyclohexanol (78:22) in 435% conversion based on Bu^tOOH, O_2 being responsible for the majority of the oxidation. The catalyst was recycled twice with the yield dropping to 75% and 50% of that obtained in the first run, suggesting catalyst instability or leaching. Cyclohexane could be oxidised to cyclohexanone and cyclohexanol and toluene to benzaldehyde using the same system, but the yields were very low 97.5 and 17% based on Bu^tOOH respectively. These values represent catalyst turnover numbers of 5 and 13 respectively. A copper complex of bipyridine bearing fluorous ponytails, which also contained fluorinated carboxylate groups, was not successful in these oxidation reactions. The organic phase was blue at the end of the reaction and conversions were very low.

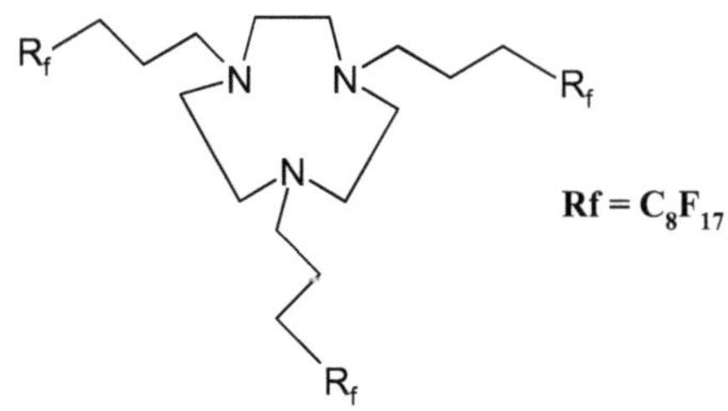

Figure 6.12. Fluourinated TACN ligand used in epoxidation reactions.[48]

Pozzi and co-workers have also reported a fluorous soluble cobalt complex, which is active in the aerobic epoxidation of alkenes in a fluorous biphasic system (FBS).[50] The ligand used in this complex was a fluorinated tetraarylporphyrin, with eight perfluorooctyl chains shown in Figure 6.13. The cobalt complex was dissolved in perfluorohexane and added to a solution of the alkene with 2-methylpropanal (aldehyde: substrate = 2:1) at room temperature.

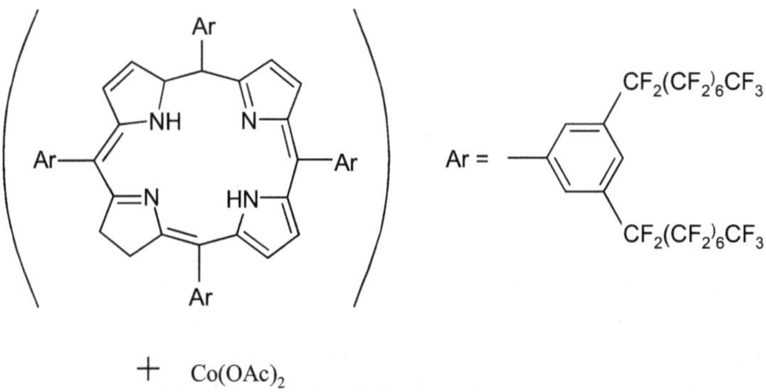

Figure 6.13. Fluorinated porphyrin ligand used in the cobalt catalysed aerobic oxidation of alkenes.[50]

The yields for reactions of unsubstituted terminal alkenes were lower than for substituted alkenes but they were still reasonable and could be increased further by increasing the aldehyde:alkene ratio. Total conversions of substrate were reported with epoxide selectivity as high as 95% in some cases. The FBC system allows for a much higher substrate:catalyst ratio (1000:1) than the non-fluorous epoxidation reported (20:1) previously. Recycling the fluorous layer once showed no reduction in conversion or selectivity.

Further efficient ligands for the epoxidation of alkenes have been reported by Pozzi, but using PhIO as the oxidant and pyridine *N*-oxide as an additive in FBS.[7, 51-53] Chiral (salen)Mn complexes have been synthesised, which are soluble in fluorous solvents and active in the epoxidation of a variety of alkenes. The catalysts were of the form shown in Figure 6.14.

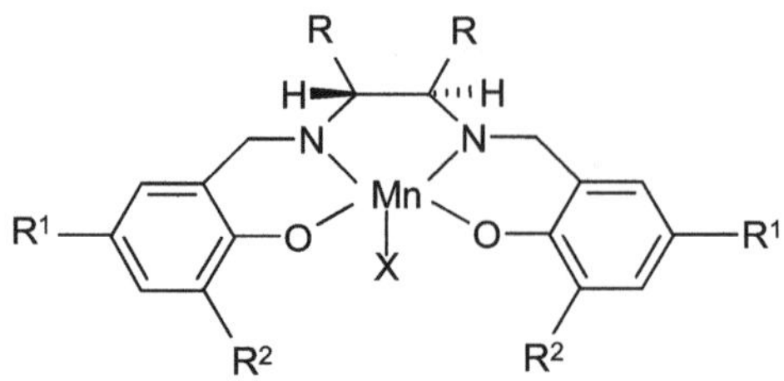

1 R_2 = $-(CH_2)_4-$; $R^1 = R^2 = -C_8F_{17}$; $X = Cl^-$

2 R = Ph; $R^1 = R^2 = -C_8F_{17}$; $X = Cl^-$

3 $R_2 = -(CH_2)_4-$; $R^1 = Bu^t$; $R^2 =$
$X = C_7F_{15}COO^-$

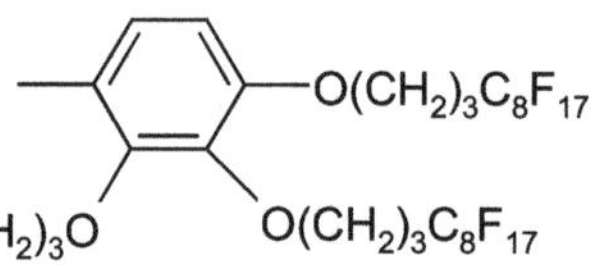

4 $R_2 = -(CH_2)_4-$; $R^1 = Bu^t$; $R^2 =$
$X = C_7F_{15}COO^-$

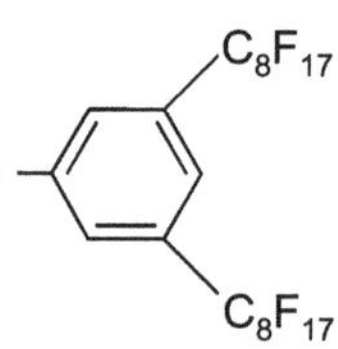

Figure 6.14. Chiral salen complexes used in the asymmetric epoxidation of alkenes.[7]

1 and **2** were initially reported in 1998 although the free ligands were soluble in diethyl ether and ethanol the Mn-complexes were insoluble in organic solvents. The FBS allowed a much smaller catalyst:substrate ratio (0.005:0.33 mol dm^{-3}) compared to other homogeneous systems and provided good yields of epoxides (up to 85%); no significant decrease in activity was observed when the fluorous phase was recycled a second time. However, only the epoxidation of indene showed high enantioselectivity (92%). The low enantioselectivity (<15%) of the ligands **1** and **2** was attributed to the low steric hindrance of the perfluoroalkyl groups, as well as possible inefficient shielding of the strong electron-withdrawing effect of these fluorinated groups. Further work and synthesis produced **3** and **4** as "second generation (salen)Mn^{III} complexes. They were, indeed, more enantioselective than the original complexes. In the case of 1,2-dihydronaphthalene, the e. e. increased from 10% to 50% with the second-generation complexes. Further investigation of the catalyst was carried out and it was found that the fluorous layer could be recycled up to three times after the first reaction, but the catalytic activity dropped significantly in the fourth run. This was attributed to oxidative decomposition of the catalyst and was indicated by the gradual disappearance of the characteristic UV-Vis-absorption bands of the (salen)Mn^{III} from the fluorous layer and their non-appearance in the organic layer.

The same catalysts have been investigated for the asymmetric oxidation of dialkyl sulfides using PhIO as the oxidant in CH_3CN:perfluorooctane.[54] Although the conversions (>80%) and selectivities to sulfoxides (>90%) were generally good, and the more heavily fluorinated catalysts could be recycled 4 times with only small drops in activity, which were attributed to oxidation rather than leaching, the e. e.'s in these reactions were poor (<20%).

6.7 Other Oxidation Reactions

Knochel and co-workers have reported biphasic aerobic oxidation of aldehydes, sulfides and alkenes, using nickel and ruthenium catalysts with a perfluorinated 1,3-diketone ligand, see Figure 6.15.[55] The nickel complex, when reacted with aldehydes in a toluene/perfluorodecalin reaction mixture at 64°C, provided yields of the expected carboxylic acids of 71-87%. The same catalytic system was active for the oxidation of sulphides, but required the presence of isobutyraldehyde and the nickel catalyst did not oxidise alkenes. The ruthenium form of the catalyst in $C_8F_{17}Br$ was active in the epoxidation of disubstituted alkenes. After "several" recycles, the ruthenium complex was recovered in 95% yield. The catalytic leaching was small as yields of 70% were still obtained after 6 runs, but this corresponds to a drop in yield from 87% - 70% over the six recycles. Quantitative data of the metal loss to the organic phase is desirable for comparison with other systems. Using the same fluorinated diketonate ligand, Knochel and co-workers have also reported a fluorous biphasic system for the Wacker oxidation of alkenes to aldehydes, suitable for a variety of alkenes.[56] The catalyst system was a palladium (II) complex of the same fluorinated β-diketonate, as shown in Figure 6.15. Reactions were carried out in $C_8F_{17}Br$ with the substrate in benzene and *tert*-butylhydroperoxide as the oxidant. Although a wide variety of alkenes was tolerated in this system, aliphatic alkenes required longer reaction times than styrene derivatised substrates (8-20 h compared with 2-5 h). The catalyst for the reaction of 4-methoxystyrene was reused 8 times and initially gave yields of 4-methoxyacetophenone of 78% but dropped to 72% yield in the 8th recycle, which indicated low catalyst leaching to the organic solvent. The catalyst was also active for the epoxidation of some activated disubstituted alkenes.

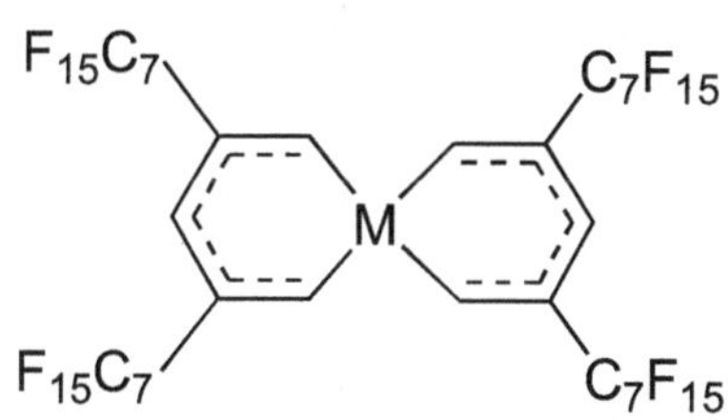

Figure 6.15. Perfluorinated β-diketonate complex used for a variety of oxidation reactions (M = Ni)[55] or Wacker oxidation (M = Pd).[56]

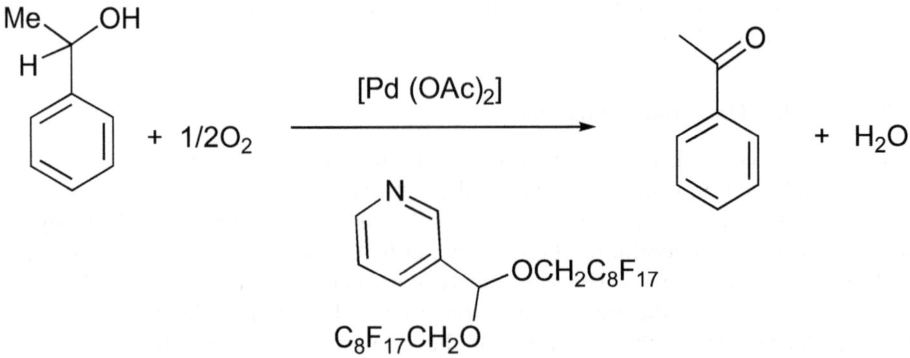

Figure 6.16. Aerobic oxidation of 1-phenylethanol catalysed by palladium complexes of a fluorous pyridine in toluene/perfluorodecalin.[57]

Uemura and co-workers have reported the aerobic oxidation of alcohols to aldehydes and ketones using palladium complexes containing pyridines with fluorous tails in a toluene/perfluorodecalin biphasic system at 80°C, using molecular sieves to remove the generated water.[57] Some catalyst decomposition occurred, but this was minimised by using excess pyridine ligand. Generally the organic phase contained palladium, but using the ligand shown in Figure 6.16, the organic phase was colourless. Inductively coupled plasma optical emission spectroscopy (ICPOES) analysis showed that only 0.8% of the palladium leached to the organic phase during oxidation of 1-phenylethanol to acetophenone. Recycling could be carried out 3 times, but the yield dropped form 90 to 74% even though extra ligand was added before the third cycle. A wide range of primary and secondary alcohols was oxidised in yields usually >75%. 4-Nitrobenzyl alcohol is oxidised to the corresponding aldehyde using air and TEMPO (1-oxy-2,2,6,6-tetramethylpiperidine) in perfluoroheptane in the presence of $[Cu(O_2CCH_2CH_2C_6F_{17})\{R_f(CH_2CH_2)_3TACN]$ at 90°C, when the system is monophasic.[49] Yields >90% can be obtained and cooling to room temperature allows phase separation and reuse of the catalyst. Using 6 mol % TEMPO in each run, the catalyst could be recycled 7 times with yields >90% for the first 5 cycles. Thereafter the yield dropped to 51 and 8% in runs 6 and 7.

Copper complexes synthesised *in situ* from $[CuBr(SMe_2)]$ and bipyridine bearing $-(CH_2)_4C_8F_{17}$ in the 5 and 5' positions are also active for the oxidation of a wide range of primary and secondary alcohols to aldehydes and ketones by air in the presence of TEMPO under fluorous biphasic conditions ($C_8F_{17}Br$/PhCl at 90°C).[58] Yields were generally very high (>80%). For 4-nitrobenzyl alcohol, the yield of aldehyde was 93% after 1.5 h. The fluorous phase was separated and reused 7 times. In the eighth run, the yield of 4-nitrobenzaldehyde was still 86%. Sterically hindered alcohols are rather unreactive in this system, but this allowed the selective oxidation of *cis*-4-methylcyclohexanol to the corresponding ketone in the presence of *trans*-4-methylcyclohexanol, which remained essentially unreacted.

Pozzi has also reported a fluoro-functionalised tetraazacyclotetradecane macrocycle, which is selectively soluble in fluorocarbons and active in the fluorous biphasic oxidation of hydrocarbons.[59] This ligand (Figure 6.17) was produced whilst trying to generate a simple method of synthesising fluorous ligands. Both the copper and cobalt complexes of this ligand, using CuCl and the cobalt salt of pentadecafluorooctanoic acid ($Co(C_7F_{15}COO)_2$) as precursors, catalysed the oxidation of saturated and unsaturated hydrocarbons. The copper complex gave slightly higher yields and higher ketone selectivities in the oxidation of cyclooctane (80% yield, 80% cyclooctanone, 20% cyclooctanol). The Co complex only gave 30% yield, (76% ketone selectivity), but both complexes performed similarly in the oxidation of cyclohexene to give cyclohex-2-en-1-one and cyclohex-2-en-1-ol. Recycling the fluorous layer, in the cyclooctane oxidation and the Co catalysed oxidation of cyclohexene, showed no drop in activity implying negligible catalyst leaching, but the activity of the Cu catalyst for the oxidation of cyclohexene dropped to 50% on the second cycle (c.f. 80% on the first cycle).

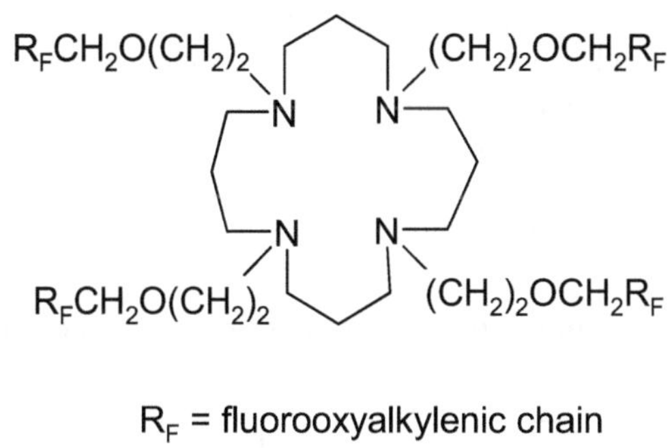

Figure 6.17. Fluorinated macrocycles used for the cobalt or copper catalysed oxidation of saturated or unsaturated hydrocarbons.[59]

6.8 Allylic Alkylation

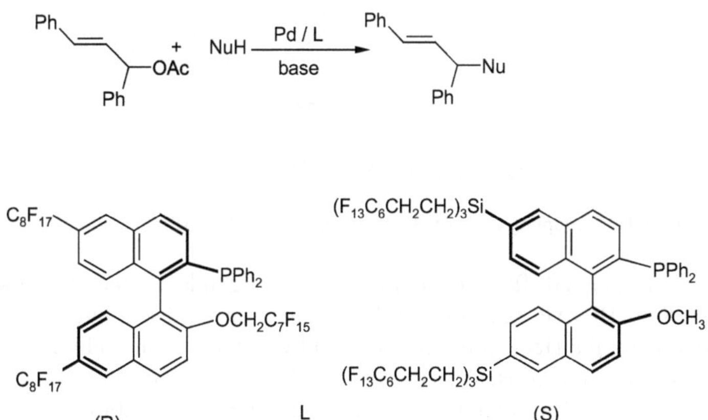

Figure 6.18. Palladium catalysed allylic alkylation using fluorinated ligands.[60, 61]

Maillard *et al* have reported perfluorous analogues of MOP (methoxynaphthyldiphenyl phosphine), see Figure 16.18, and their application in palladium catalysed allylic alkylation.[60, 61] Due to the somewhat low fluorine content in these ligands, (R) 56.0 wt% fluorine, (S) 56.88 wt% fluorine, they are slightly soluble in THF and toluene, which will result in loss of the ligand during reaction and would involve further separation steps to recycle the ligand fully. The ligands were used in Pd-catalysed allylic alkylation but they gave low enantioselectivities (0-37%) for various solvent/base systems. An attempt to recycle the catalyst was made. However, the second run only gave a conversion of 24% as opposed to 100% in the initial run. This drop in activity was attributed to significant leaching of the ligand to the organic phase and is unavoidable unless the fluorine content of the ligands can be increased.

Bayardon and Sinou have reported the synthesis of chiral bisoxazolines, which also proved to be active ligands in the asymmetric allylic alkylation of 1,3-diphenylprop-2-enyl acetate, as well as cyclopropanation, allylic oxidations and Diels-Alder reactions.[62] The ligands do not have a fluorine content greater than 60 wt% and so are not entirely preferentially soluble in fluorous solvents, which may lead to a significant ligand loss in the reaction system and in fact, all recycling attempts were unsuccessful. However, the catalytic results achieved were comparable with those obtained with their non-fluorous analogues.

6.9 Heck, Stille, Suzuki , Sonagashira and Related Coupling Reactions

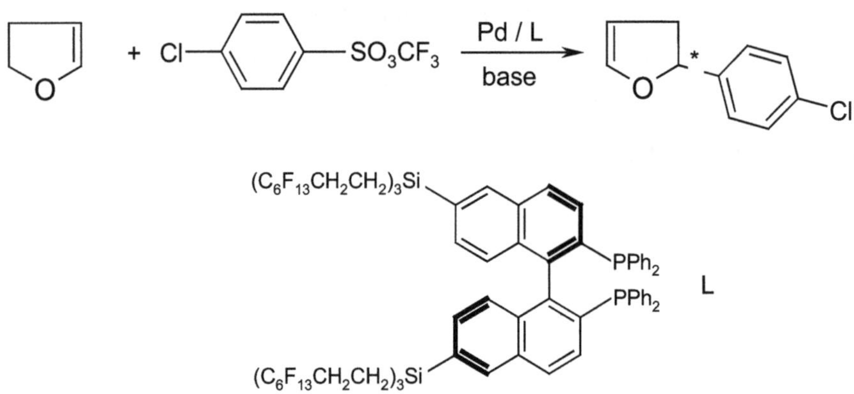

Figure 6.19. Heck coupling catalysed by fluorinated *(R)*-F_{13}BINAP palladium complex.[63]

Nakamura *et al* applied fluorous chiral BINAP ligands to the asymmetric Heck reaction.[63] R-F_{13}BINAP, see Figure 6.19, is easily air oxidised, but in a benzene/FC-72 biphasic system, the Heck coupling of 2,3-dihydrofuran with 4-chlorophenyl triflate occurred with high enantioselectivity (93%), but low yield (39%). Reuse of the fluorous phase gave a 2% yield, most likely because of deactivation of the catalyst by ligand oxidation, TLC monitoring of the reaction showed the ligand to be oxidised in the fluorous phase, but there may also be some catalyst leaching to the organic phase. Compared with the original reaction with non-fluorous BINAP in trifluoromethylbenzene, the e. e. was much higher (90% e. e. F_{13}-BINAP, 76% e. e. BINAP). Therefore, overcoming the air sensitivity of the fluorous BINAP would provide a highly competitive system for Heck couplings.

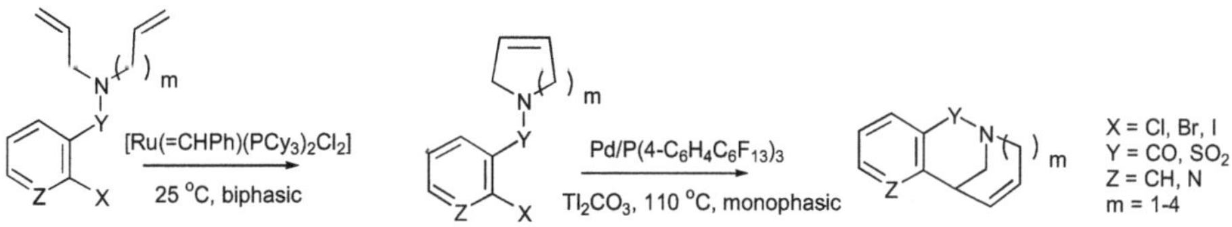

Figure 6.20. Tandem ring-closing metathesis and Heck coupling for the formation of bridged ring systems.[64]

Heck coupling has also been used in tandem with ring closing metathesis to give bridged ring systems (Figure 6.20).[64] In conventional solvents, the Heck catalyst poisoned the ruthenium catalyst used for the ring-closing metathesis. However, by carrying out the reaction in a fluorous biphasic system with a standard metathesis catalyst but a fluorinated palladium catalyst for the Heck reaction, the catalysts could be kept separate during the low temperature ring-closing metathesis, but on heating the system became monophasic, thus allowing the Heck reaction to proceed. Yields were significantly better (37-67%) than in a homogeneous system (0-37%) for reactions where the Heck reaction could be carried out under relatively mild conditions, but more forcing conditions led to decomposition of the fluorous tagged palladium complex. This method for keeping incompatible catalysts separate may have potential for a range of reactions.

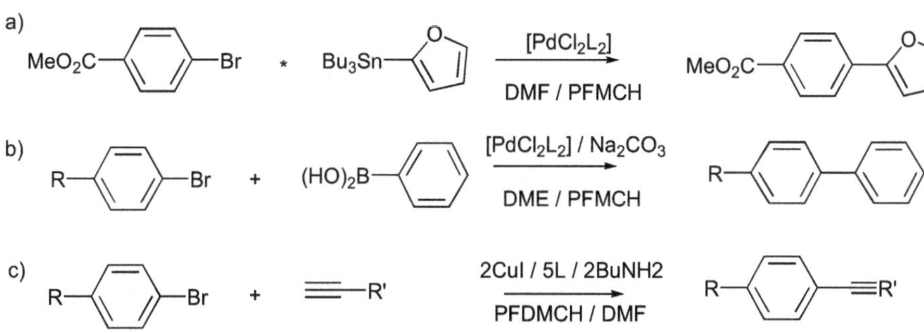

L = P(x-C$_6$H$_4$(CH$_2$)$_n$C$_8$H$_{17}$)$_3$ (x = 3 or 4, n = 0 or 2)

Figure 6.21. a) Stille,[65] Suzuki[66] and Sonigashira[67] coupling catalysed by complexes of fluorous analogues of PPh_3 (R = NO_2, COMe, CO_2Et or OMe; R' = Pr^i_3, Ph or CMe_2OH)

Stille,[65] Suzuki[66] and Sonigashira[67] couplings (Figure 6.21) have also been successfully carried out under fluorous biphasic conditions. Stille couplings of a variety of electron poor aryl bromides and electron rich aryl stannanes catalysed by $[PdCl_2(P(x\text{-}C_6H_4(CH_2)_nC_8F_{17})_3)_2]$ (x = 3, n = 0; m = 4, n = 0 or 2) were carried out in DMF/PFMC at 80°C in the presence of LiCl.[65] The reaction shown in Figure 6.21a gave 90% conversion with all three catalysts and could be repeated at least 6 times with minimal loss of activity. Reactions using 2-tributylstannylanisole were less successful and could not be recycled, probably because the ortho methoxy group inhibited the coordination of the ligand.

The same palladium complexes have been successfully employed for the Suzuki coupling (Figure 6.21b) of a wide range of aromatic bromides (electron rich and electron poor) with phenylboronic acid.[66] Using 1.5 mol% of catalyst in PFMC/1,2-dimethoxyethane, at 75°C, >90% conversion was achieved in 2 h. The catalysts were recycled 5 times, in most cases without appreciable loss of activity. At lower catalyst loadings (0.1 mol%) using an electron rich bromide and an electron rich aryl boronic acid, the conversions were high (100%) in the first run, but diminished substantially on recycling, generally to <10% by the third or fourth run.

The Sonigashira couplings (Figure 6.21c) were carried out using copper (I) iodide in PFDMCH/DMF in the presence of $BuNH_2$ at 100°C over 4 h.[67] The best results (>98% conversion, recycled twice with little loss of activity) were obtained with electron poor substrates (R = NO_2 or CO_2Me in Figure 6.21) and $Pr^i{}_3SiC$ CH. For less electron rich substrates and the other alkynes, yields were between 10 (R = OMe, R' = CMe_2OH) and 80% and for the electron poor substrates yields fell off dramatically on recycling.

Knochel and Betzemeier have described palladium(0) - catalysed cross coupling between arylzinc bromides and aryl iodides as a facile method of forming carbon-carbon bonds and retaining the expensive catalyst for reuse.[68] The perfluorinated phosphine used in this reaction was $P(C_6H_4\text{-}4\text{-}C_6F_{13})_3$ in a toluene/$C_8F_{17}Br$ solvent system. The Pd catalyst was preferentially soluble in fluorous solvents and reuse of this phase did not cause any significant change in the reaction yield. The presence of the electron-withdrawing perfluoroalkyl groups has a positive effect on the reaction. The activity of the catalyst was in fact higher than with the non-fluorous catalyst $[Pd(PPh_3)_4]$. This is explained by the electron-deficient phosphine favouring reductive elimination in the cross-coupling reaction. This is obviously an important effect, which is likely to be utilised further.

6.10 Asymmetric Alkylation of Aldehydes

Chan and co-workers have synthesised perfluoroalkyl-BINOLs and shown them to be active ligands for titanium catalysed asymmetric carbon-carbon bond formation[69, 70] (the best one for fluorous biphasic reactions is shown in Figure 6.22). The reaction was carried out in perfluoro(methyldecalin)/hexane. PhCHO were dissolved in hexane along with the $Ti(O^iPr)_4$ and added to a solution of the ligand in perfluoromethyldecalin. The catalyst complex was formed *in situ* when the homogeneous phase was formed at 45°C and Et_2Zn was then added. After the reaction, the mixture was cooled to 0°C to induce phase separation. Although there was no apparent loss of the *(S)*-R_fBINOL to the organic phase in repeated cycles, addition of $[Ti(O^iPr)_4]$ was required with every cycle. The catalyst containing four C_8F_{17} substituents was recycled nine times without substantial loss of yield or drop in e. e., but it should be noted that although the 2nd – 9th cycles had yields of 80-70%, the first cycle had a yield of only 69%. This may indicate inadequate formation of the catalyst Generally, this system provided yields of 50-80% with e. e.'s as high as 58%. It was observed that the more electron deficient substrate, 4-chlorobenzaldehyde, reacted slightly faster than the electron neutral benzaldehyde and in turn the more electron rich substrate, 4-methoxybenzaldehyde, reacted more slowly.

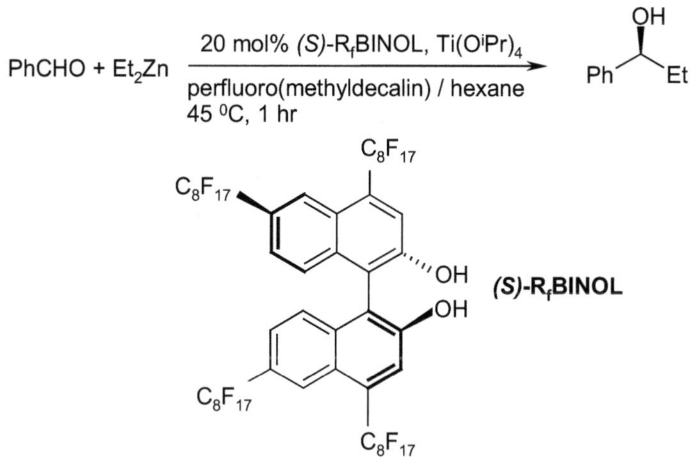

Figure 6.22. Asymmetric alkylation of benzaldehyde catalysed by a titanium complex of *(S)*-R$_f$BINOL.[69, 70]

Simultaneously, Takeuchi and co-workers reported another chiral fluorous BINOL, active in a similar reaction. However, their isomer is the (R)-R$_f$HBINOL[71] as shown in Figure 6.23. The ligand was dissolved in FC-72 (a perfluorous solvent containing primarily compounds with 6 carbons) along with [Ti(O-*i*Pr)$_4$]. Diethyl zinc in hexane was added to this and the solution was cooled to 0°C before adding a benzaldehyde solution in toluene, thus creating an organic phase consisting of toluene and hexane. Under these conditions, the system is biphasic at all times. The organic phase was removed and quenched, before isolating the product. This system produced a higher enantioselectivity than Chan's (>80% compared to 58% e. e. respectively) with comparable yields. Repeat cycles using the fluorous phase showed a slight drop in yield and e. e. over 5 recycles. Again the yield obtained in the first run was lower (81%) than those achieved in the subsequent runs (89-87%). Unusually, the ligand loss to the organic phase has also been reported, 0.2 mmol of the ligand was recovered from the organic layer per cycle. From this data, the authors conclude that the asymmetric catalyst is present in the organic phase and that the asymmetric reaction must occur in this phase. Separating the organic and fluorous layers and performing the reaction in each, showed some catalytic activity in the organic phase, however a lower e. e. was achieved, than when using the fluorous phase or the biphasic system, suggesting a lower concentration of catalyst present in this phase than the fluorous phase.

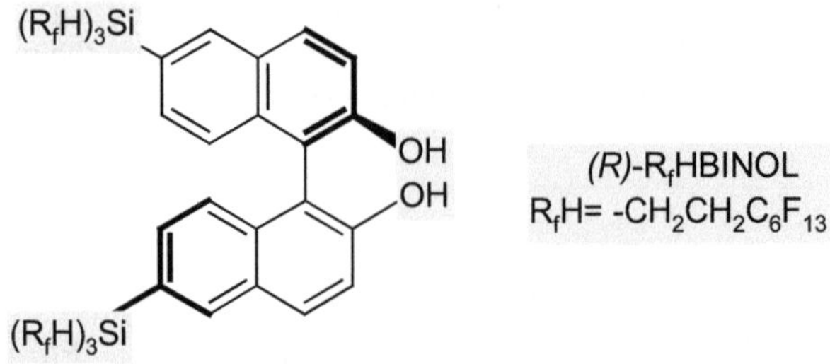

Figure 6.23. (R)-RfHBINOL used in the asymmetric alkylation of benzaldehyde.[71]

Using $AlEt_3$ in place of Et_2Zn in the reaction shown in Figure 6.24, but carried out at 53°C afforded higher e. e. (63% rising to 82% at run 5) and high conversions (59% in the first run, rising to 88% in the third), but it was again confirmed that additional $[Ti(OPr^i)_4]$ was needed for each run.[70] Higher enantioselectivities were obtained using zinc aminothiolates of the kind shown in Figure 6.24 ($R_2 = (CH_2)_4$, n = 10, e. e = 94%).[72] The reactions were carried out in PFMC/hexane at room temperature and the catalysts could be recycled up to 4 times, although the e. e. dropped steadily after the second run to 29% in the fifth run.

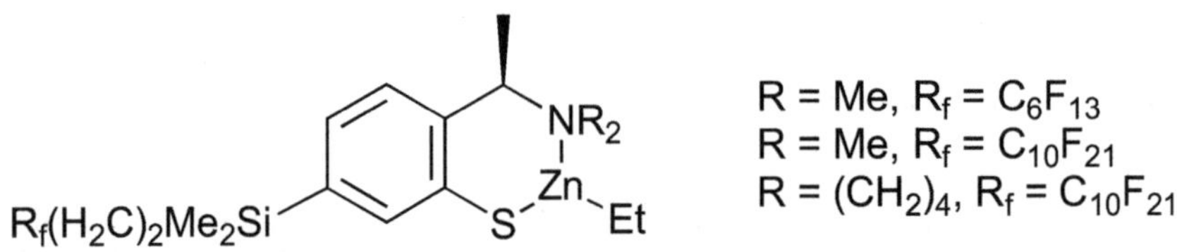

Figure 6.24. Alternative catalysts for the asymmetric alkylation of benzaldehyde.[72]

High enantioselectivities could be obtained using $[Ti(OPr^i)_4]$ and a BINOL ligand modified with $CH_2CH_2C_6F_{13}$ only in the 6 and 6' positions in the allylation of benzaldehyde using 3-tributylstannylpropene (Figure 6.25).[73] This system did not give good results in either the organic or the fluorous solvent alone, but in the biphasic system (hexane/FC72, FC72 mainly contains C_6 fluorinated hydrocarbons) it gave 85% yield with 90.1% e. e. Very similar results were obtained using a C_8F_{17} chain (Yield 83%, e. e. 89.8%), but in a much shorter time (5 h c. f. 10 h for the ligand shown in Figure 6.25). Studies of different aldehyde substrates showed that good yields and e. e.s are only obtained from electron withdrawing aromatic aldehydes.

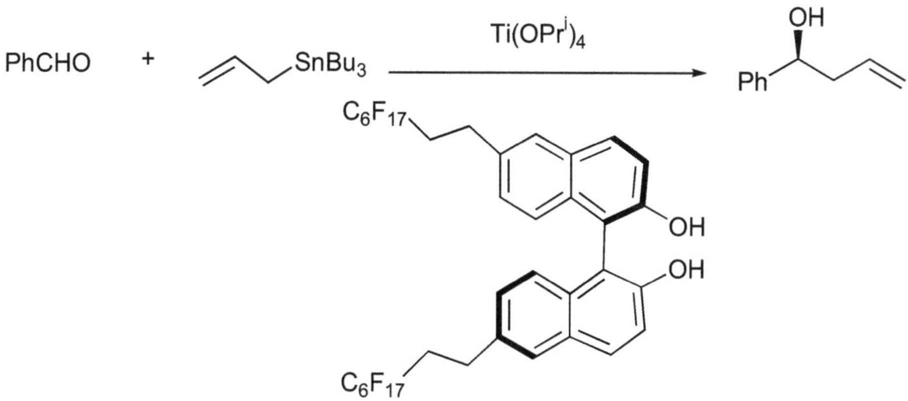

Figure 6.25. Asymmetric allylation of benzaldehyde catalysed by a fluorinated Ti/BINOL complex under fluorous biphasic catalysis.[73]

6.11 Miscellaneous Catalytic Reactions

Another fluorous biphasic system, which showed desirable results was reported by Biffis and co workers[74] who described the preferential silylation of primary alcohols using $[Rh_2(PFTD)_4]$, (PFTD = perfluorotetradecanoate). Ligands of lower carbon number were found to leach to the organic (CH_2Cl_2) phase during the reaction. The best results were obtained with a DCM/FC-77 (mainly C_8 containing fluorocarbons) solvent system, at room temperature, 1 mol% catalyst and a 1:1 alcohol to silane ratio for 6 hours. This gave a yield of 68%, which was not as good as the non-fluorous reaction (96% yield in 3 h). Catalyst decomposition caused a decrease in yield by the second recycle of the fluorous layer. An interesting property of the system was observed when a range of alcohols was tested and the preference for primary alcohols over secondary and hydrophilic alcohols became apparent.

Endres and Maas have also described the use of rhodium(I) perfluoro carboxylates, see Figure 6.26, as active catalysts in the carbenoid reaction of diazoacetate with toluene in a toluene-PFMC solvent system.[75] Both catalysts are insoluble in toluene. The complex containing the aryl carboxylic acid gave a total yield of 71% and the other complex a total yield of 78%. Decomposition of both catalysts was observed. This explains why only partial catalyst recovery could be made. In a more recent paper,[76] further perfluorinated chains have been attached to the dirhodium complex, introducing spacer groups. Only the ligand with $R_f = CH_2C_6F_{13}$ was soluble in PFMC. The spacer group had been introduced to minimise the electronic effect of the electron withdrawing fluorous ponytail on the dirhodium complex, in order that it more resembled its non-fluorous analogue, $[Rh_2(OAc)_4]$. The polar carboxylate groups make these fluorinated catalysts different from others reported, which are generally non-polar. These complexes are also soluble in THF and diethyl ether at room temperature, which causes the catalyst to leach to the organic phase during a reaction. Unfortunately, this study shows the limitations of the fluorous analogue of Rh-catalysed carbene transfer. As the electron-withdrawing effect has to be restricted, two or more methylene spacer groups are required, but this in turn makes the catalyst insoluble in fluorous solvents. Clearly fluorous biphasic catalysis does have its limitations.

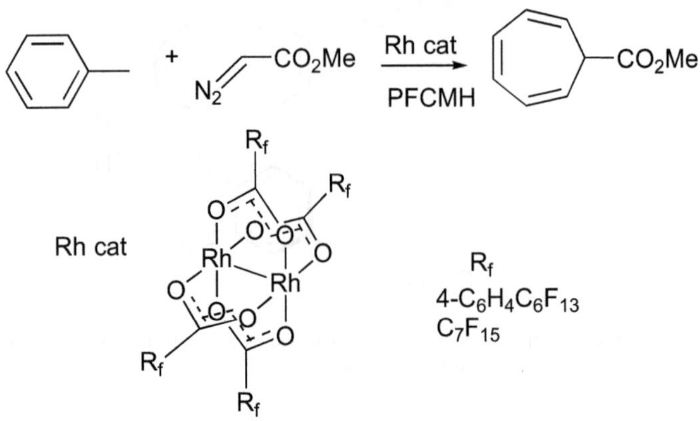

Figure 6.26. Carbene reaction catalysed by rhodium(II) complexes containing fluorinated carboxylates.[75, 76]

6.12 Fluorous Catalysis without Fluorous Solvents

Wende and Gladysz have investigated the reaction of a perfluorinated catalyst system in the absence of any fluorous solvent.[77] This system catalyses the conjugate addition of alcohols to methyl propiolate under homogeneous conditions in *n*-octane at 65°C (Figure 6.27). The fluorinated phosphine, $P(C_2H_4C_8F_{17})_3$, without metal, shows temperature dependent solubility in n-octane such that it is essentially insoluble at -30°C and it is reported that >97% phosphine recovery is made every run. Through synthesis of a large number of fluorous ligands, it was observed that the solubilities of these compounds increased with increasing temperature. The process simply relies upon the same concept as an ordinary FBS, whereby at room temperature there are two phases, in this case solid and liquid; at reaction temperature a single homogeneous solution is formed and upon cooling the solid catalyst is again precipitated and can easily undergo a heterogeneous separation. However, if there is incomplete conversion, unreacted substrate and the organic solvent still have to be removed from the final product.

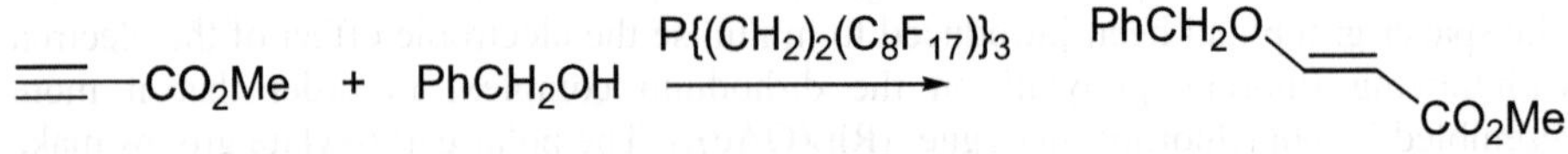

Figure 6.27. Conjugate addition of benzyl alcohol to methyl propiolate catalysed by a fluorous phosphine without a fluorous solvent.[77]

The reaction was carried out by combining the phosphine ligand $\{P(C_2H_4C_8F_{17})_3\}$, the alcohol and methyl propiolate in *n*-octane at room temperature, the authors report no visual dissolution of the phosphine. On heating to 65°C a homogeneous phase was attained. After 8 hours the sample was cooled to –30°C and the phosphine precipitated. The sample was kept at –30°C whilst the supernatant liquid was removed. It is unclear why the phosphine undergoes a colour change during the reaction. Before heating it is a white solid and the authors report that most samples on cooling precipitated an orange solid, which upon recycling often darkened to red. For benzylic alcohol ($PhCH_2OH$) a yield of 82% was achieved in the first cycle. This was maintained for three recycles but in the fifth cycle, dropped to 75%. Interestingly, by omitting the *n*-octane from the reaction and making this a totally solvent free system, the initial yield rose to 99% and was maintained for two further recycles before dropping to 95% in the fourth run. By total omission of any solvent, not only is the cost reduced but also the process becomes much more environmentally friendly. There has been some work published by other groups working on a similar principle, these include catalysed condensations of carboxylic acids and amines to amides,[78] transesterifications[79] and benzoylations of alcohols and esterifications of carboxylic acids.[80] This is an interesting approach from an economic point of view as there is no unnecessary expense on costly fluorinated solvents. However, it would appear that a phase change of the catalyst at or above the collection temperature, to provide a much less soluble form of the catalyst could provide even greater advantages. Such systems do not appear to have been reported.

6.13 Continuous Processing

Yoshida *et al* recently reported the development of a continuous flow reaction system using a fluorous biphasic system.[81] They report the acetylation of cyclohexanol and the Baeyer-Villiger reaction of 2-adamantanone using fluorous biphasic conditions (see Figures 6.28 and 6.29. The acetylation of cyclohexanol employed a fluorinated catalyst, ytterbium(III) bis(perfluorooctanesulfonyl)amide, PFMC at 40°C. The organic phase was made up of a solution of cyclohexanol in toluene to which was added acetic anhydride. This organic mixture was continuously added to the stirred tank reactor and once part filled with reaction mixture the mixture overflowed to a decanter (25°C) where separation occurred. Once this decanter had filled the organic phase began to overflow to the product tank whilst the fluorous phase could be recycled to the reactor (see Figure 6.30). The product was analysed by GC and the results showed conversion was maintained at >90% over 500 hours (nearly 3 weeks) with less than 2 ppm ytterbium in the organic phase.

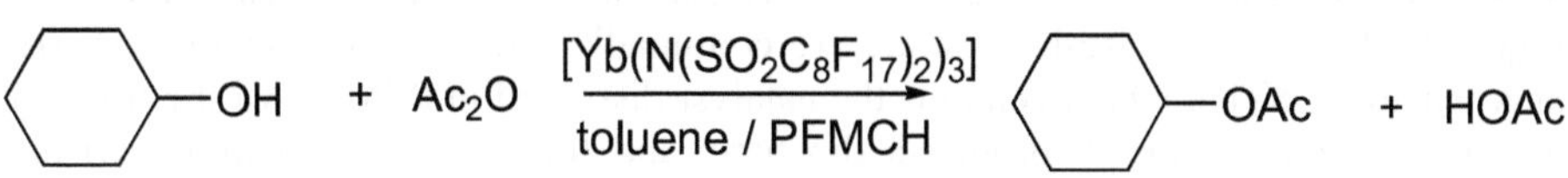

Figure 6.28. Acetylation of cyclohexanol carried out continuously under fluorous biphasic conditions for 500 h.[81]

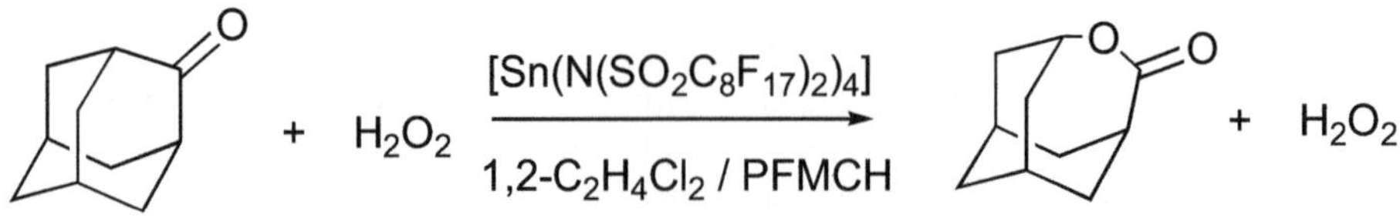

Figure 6.29. Baeyer-Villiger reaction carried out continuously over 200 h under fluorous biphasic conditions.[81]

In order to see if the reaction system was suitable for water forming reactions the Baeyer-Villiger reaction of 2-adamantanone with 35% aqueous solution of H_2O_2 was investigated. Using tin(IV) bis(perfluorooctanesulfonyl)amide in 1,2-dichloroethane/PFMC a conversion in the region of 50-60% was achieved with a high selectivity. This reaction was run for 200 hours. The success of these reactions indicates what may be possible with future continuous flow fluorous biphasic systems, however no high pressure or gas substrates were involved in these reactions.

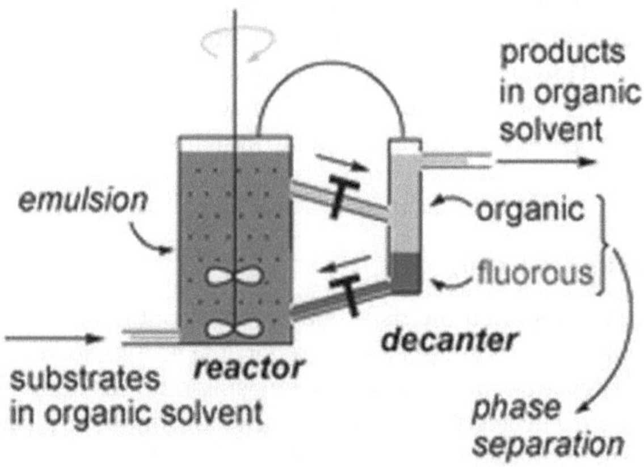

Figure 6.30. Schematic diagram of a reactor used continuously for up to 500 h for fluorous biphasic reactions without gaseous reagents.[81] (A. Yoshida et al, *Development of the continuous-flow reaction system based on the Lewis acid-catalysed reactions in a fluorous biphasic system*, Green Chemisty, 5, (2003), 555) *Reproduced by permission of The Royal Society of Chemistry.*

A reactor which allows the continuous operation of fluorous biphasic reactions involving gases has been described by Manos, Hope, Cole-Hamilton and co-workers and demonstrated in the hydroformylation of 1-octene catalysed by complexes formed *in situ* from $[Rh(CO)_2(acac)]$ and $P(4\text{-}C_6H_4C_6F_{13})$.[41-43] Shown schematically in Figure 6.31, gases and fresh substrate are continually fed to the continuously stirred tank reactor (CSTR), which contains the catalyst dissolved in PFMC. All the time, the mixture is being removed through a dip tube, and flow controller (capillaries) to a gravity separator. The phases separate and the organic phase is allowed to overflow through a tube into a collection vessel, whilst the fluorous phase is continuously fed back into the reactor *via* an HPLC pump.

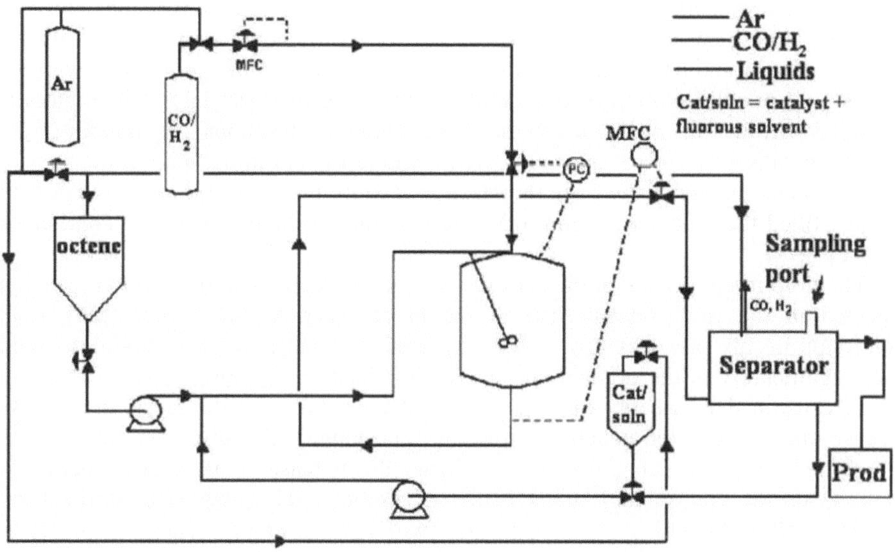

Figure 6.31. Schematic diagram of continuous flow reactor for fluorous biphasic reactions under gas pressure.[42]

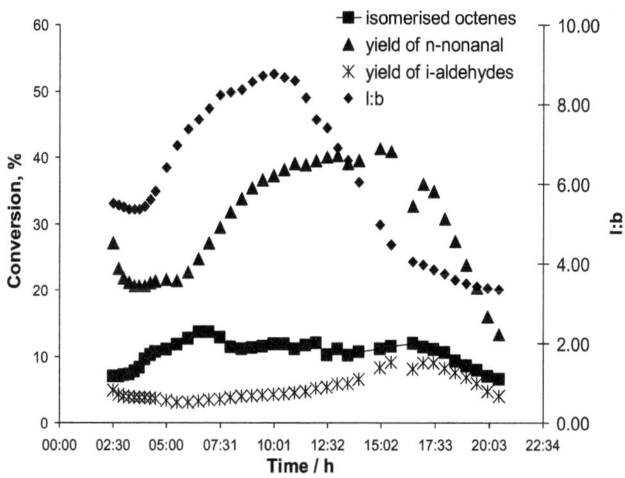

Figure 6.32. Results for the continuous hydroformylation of 1-octene catalysed by Rh/ $P(4\text{-}C_6H_4C_6F_{13})_3$ in fluorocarbon solvents.[42]

Studies of the phase behaviour at ambient temperature within the separator [43] show that there is significant solubility of the product nonanal within the fluorous phase and *vice versa*. Although this does not present a problem for the nonanal (it will simply be recycled to the reactor and create a steady state, it does mean that fluorous solvent is always being lost. The loss of the fluorous solvent (2.8 mol% into pure nonanal), as for the catalyst and the free ligand [41] is much more significant at low conversion, so optimum results should be obtained if the reaction is run at high conversion. This is a fundamental problem with a continuous flow CSTR since some substrate will always enter the reactor and pass straight out of it without having significant contact with the catalyst. A possible solution to this problem, involving two parallel batch reactors is discussed below. In order to work at as high conversions as possible, the hydroformylation reaction was run in batch mode (all flows turned off) at the start of the continuous run and again once the reaction mixture had passed through the system and just filled the separator. Figure 6.32 presents the results obtained in a continuous run.[42]

The drop in conversion at the start of the reaction represents the lower conversion expected in the flow system than in the batch reaction. Since the steady state conversion in this case was low, it was expected that phosphine and rhodium would leach significantly to the organic phase (which indeed was yellow). Loss of rhodium should cause a decrease in rate, whilst loss of phosphine should cause an increase because the reaction is negative order in [phosphine]. As discussed above, the phosphine leaches more than the rhodium, hence the increase in rate over the period 5-10 h. Once the [phosphine] drops below a certain level, complexes of the form $[RhH(CO)_n(P(4\text{-}C_6H_4C_6F_{13})_3)_{3-n}]$ (n = 2-3) will start to form. These have fewer ponytails than $[RhH(CO)(P(4\text{-}C_6H_4C_6F_{13})_3)_3]$ and will leach more heavily, accounting for the reduction in rate at 17-20 h. The l:b ratio is high throughout the main part of the reaction (7:1), but this may in part be attributed to the high levels of isomerised alkene (the branched alkyl intermediate in hydroformylation reactions can lead to the branched aldehyde (reaction with CO) or to isomerised alkene (β-hydrogen abstraction), so that high isomerisation can give high l:b ratios, see Figure 6.33)[82]. The high levels of isomerisation in this continuous reaction has been traced to inefficient stirring in the reactor.[41] Although leaching was a major problem in this reaction, the catalyst performed >15 000 turnovers at an average rate of 750 h^{-1} over the 20 h period of the reaction.[42]

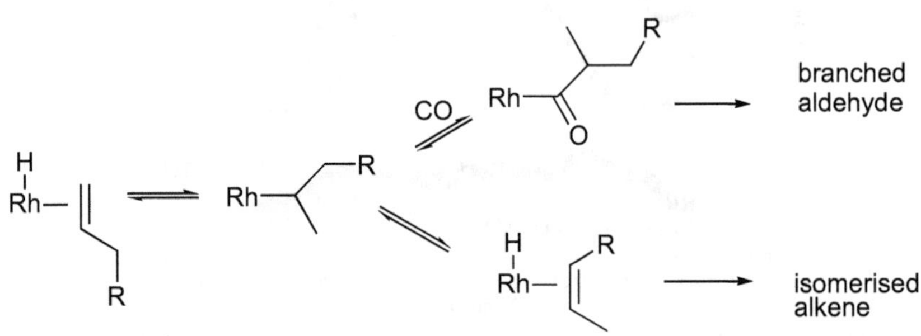

Figure 6.33. Relationship between isomerisation and branched aldehyde formation during hydroformylation reactions

Figure 6.34 shows another continuous reaction.[41] In this case, more catalyst solution was fed when the reaction rate slowed (represented by the dots in the figure) so that the conversion was up to 70% for part of the reaction. The l:b ratio was very high (13:1) for much of the reaction, but interestingly dropped because of a drop in isomerisation activity at the highest conversions. Paradoxically this suggests that the gas mixing is better at high conversion, when the phases are better separated. Perhaps it reflects the higher concentration of rhodium in the active phase of the two-phase system.

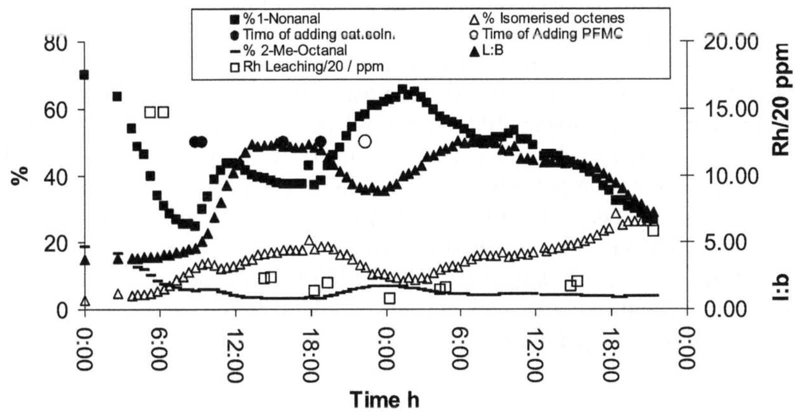

Figure 6.34. Results from the hydroformylation of 1-octene carried out under continuous flow conditions with extra catalyst solution addition.[41]

The drop in conversion at the start of the reaction is much greater than expected just on the basis of transferring from a batch to a continuous reaction. It occurs because there is also substantial leaching of rhodium (300 ppm) at the start of the reaction, either because the catalyst has not preformed properly or because there is oxygen in the system and some of the phosphine is oxidised. Rhodium leaching increases at the end of the reaction (115 ppm), presumably because phosphine is lost to the organic phase and there is insufficient to keep the catalyst as $[RhH(CO)\{P(1\text{-}C_6H_4C_6F_{13})_3\}_3]$, but is about 20-30 ppm for most of the reaction.

The main conclusions to be drawn from this study are that the reactor design works well, and that steady state continuous flow operation requires excellent mixing of the gases and two liquid phases and high conversions. Improvements in the catalyst (ligand) are required to reduce leaching still further, but commercialisation will also require a different reactor design or more than one CSTR in series.

6.14 Process Synthesis for the Fluorous Biphasic Hydroformylation of 1-Octene

The fluorous biphasic concept was first tested on the hydroformylation of 1-decene[1] and has been used more for this type of reaction than for any other. A continuously operational reactor - separator system using 1-octene as the substrate has been reported.[41, 42] The major problem that is encountered is that the fluorous solvent, ligand and catalyst all show some solubility in the organic phase. Significant advantages have been demonstrated when the reaction is carried out without an additional organic solvent.[20] This improves the reaction rate and selectivity as well as the retention of ligand and catalyst in the organic phase. In a commercial plant it would also remove one complete separation step, that of the product from the organic solvent. Certain other advantages compared with triphenylphosphine also accrue if the fluorous pony-tail is attached in the 4-position of the phenyl ring, as in $P(4\text{-}C_6H_4C_6F_{13})_3$. The electron withdrawing nature of the fluorous substituent increases both the rate of reaction and the selectivity to linear aldehyde.[20] This means that lower phosphorus loadings are required (Rh:P = 1:10, compared with 1:150-300 in commercialised PPh_3 systems)[83] to achieve l:b ratios of 6:1 in batch processes (10:1 in some continuous reactions and a linear selectivity of 81% compared with 83-90% for commercialised propene hydroformylation using PPh_3. It should be noted that propene cannot give isomerised alkene and *ca.* 4% of 1-octene is isomerised during the fluorous biphasic hydroformylation. For comparison, commercialised cobalt based hydroformylation processes can give linear selectivities of 10:1, but 10% of alkene is lost to hydrogenation so that the selectivity to linear aldehyde product is again *ca.* 81%.[83]

The fact that less ligand is required when using the fluorous substituent not only saves in terms of ligand cost, but it also means that the reaction can be carried out with very high rates at low temperature. At 70°C, the turnover frequency (TOF) is 4400 h^{-1} compared with 500-700 in commercial propene based systems, which are operated at 100°C.[20] This in turn should mean that ligand degradation, one of the chief repetitive costs of the processes using PPh_3, should be greatly reduced. One possible disadvantage of omitting the organic solvent is that phase separation occurs under the reaction conditions at *ca.* 20% conversion. This might appear to negate the major advantage of the fluorous biphasic system (homogeneous reaction conditions), but kinetic measurements have shown that the reaction remains first order way beyond this conversion and even as far as 80% conversion.[20]

Even when omitting the organic solvent, some leaching of the ligand and the catalyst still occurs. Catalyst retention is improved because the form of the catalyst, which is present in the separator $[RhH(CO)(P(4\text{-}C_6H_4C_6F_{13})_3]$ contains the maximum number of fluorous substituents and hence is most fluorophilic.[41] It is also very much dependent on the degree of alkene conversion to aldehyde because the fluorous ligand and catalyst have very much higher solubility in 1-octene than in nonanal.[41] It is clear from Figure 6.8 that reactions should be conducted to 100% conversion if leaching of the expensive catalyst and ligand are to be minimised. This could be achieved by using stirred tank reactors in series or in parallel. In Figure 6.35, we show the use of two tank reactors (**R1** and **R2**) operating in parallel, both feed from the same supplies and both are attached to the same collecting vessel (**C**) attached to the same separator. This design allows the reactions to be carried out to high conversion in batch mode and allows for the bulk of the catalyst solution to be in the reactor under the conditions for which it has been optimised at all times. By tunrng off the stirrer each reactor also acts as the separator. Since only the product phase is withdrawn from the reactor, the catalyst solution is not itself heated and cooled, so heat losses can be minimised. A heat exchanger (**HE**) is proposed for using the heat from the product leaving the reactor to preheat the fresh substrate. Because the product phase is withdrawn hot from the reactor, a small amount of fluorous solvent (up to 4 mole %) [40] and catalyst will be dissolved in it. On cooling in the heat exchanger and separator, where dissolved gases will also be vented and recycled, the small amount of fluorous solvent and catalyst will separate. It will be recycled to the reactors once it has built up to a significant level. The reactors would be operated in batch mode with a reaction time of 20 min followed by a settling, separation and refilling period also of 20 min. The two reactors would be operated in orthogonal cycles, so that the size of the collecting tank can be reduced to a minimum.The reactors would have feeds for the alkene, the fluorous solvent/catalyst solution and systems for recycling the unreacted gases and the small amount of catalyst solution that separates in the collecting vessel. The product solution will be passed from the collecting vessel to a single fractionating column (**D**), where the isomerised alkene, dissolved fluorous solvent and two aldehyde products will be fractionally separated. The recovered fluorous solvent (expected to be as much as 2.8 mol % [43]) will be returned to the reactor, whilst the isomerised alkenes will be burnt as fuel. The complete system is shown schematically in Figure 6.35.

To produce 100 000 tonnes of nonanal per year (25% down time, 100% conversion of substrate, 80% selectivity to nonanal) requires a production rate from the reactors of 19 tonne h^{-1}, so that each batch must be 6.3 tonnes. Assuming a 1:1 ratio by volume of fluorous solvent:liquid substrate and a 75 % loading, each reactor must have a volume of 20 m^3. If the distillation column were fully integrated into the system it would be required to handle 19 tonnes aldehyde h^{-1}. An increase in selectivity to the linear product, which could be achieved using careful ligand design would reduce the reactor size by up to 25%.

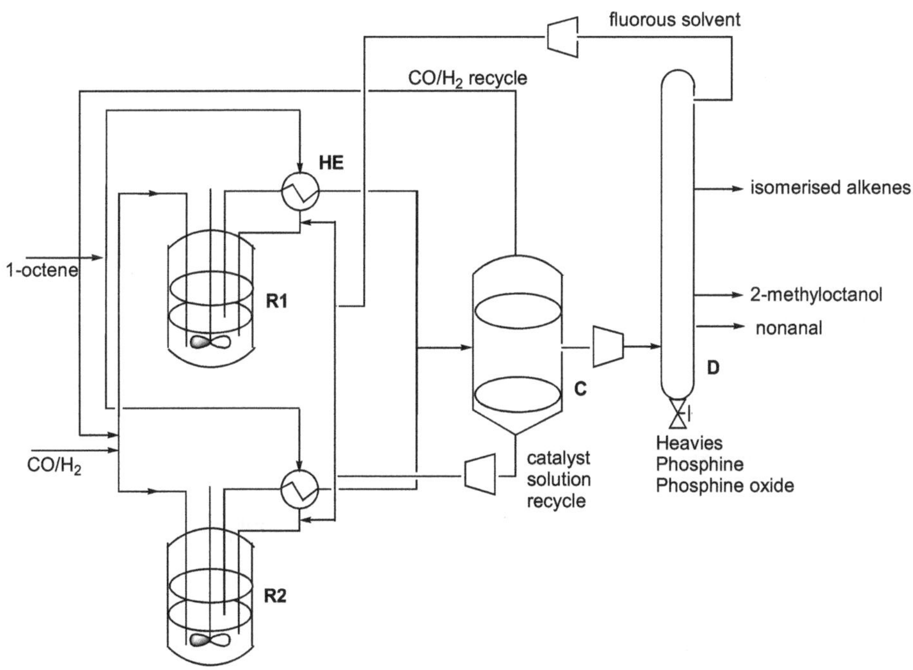

Figure 6.35. Schematic design of a full-scale fluorous biphasic reactor for the hydroformylation of 1-octene

The main costs of the process are in the fluorous solvent and in the ligand. The total estimated cost of the ligand will be of the order of 150 000 € for 30 kg. If we assume that the rhodium losses can be reduced to 1 ppb (this will require improved ligand design) and that ligand losses may be 10 times this (0.12 g tonne^{-1}), the added cost for the ligand replenishment will be 60 000 €, increasing the cost of the nonanal by 0.6 € per tonne. Although some of the fluorous solvent (*ca.* 0.2 mol%) will be dissolved in the product, it is expected that almost all of this will be recovered in the distillation unit and recycled. The total cost of the inventory of the fluorous solvent is 3.5 M€ (based on a current quotation for 25 tonnes of 140 € kg^{-1}), but this could be reduced to <30 € kg^{-1} (0.75 M € for 25 tonnes) if perfluorohexane were used as the solvent. This is a capital investment, and a loss of 1% per annum would increase the cost of the nonanal by 0.35 € tonne^{-1} (0.075 € tonne^{-1} using perfluorohexane). Even though the solvent and the ligand are very expensive, realistic estimates of losses that might be obtainable on optimisation of the reaction suggest that the major additional cost to the final price of the nonanal will still be because of rhodium losses (1 € tonne^{-1}). These costings are collected in Table 6.3.

TABLE 6.3. Process for optimized nonanal production of 100 000 tonnes y^{-1} using perfluoro-methylcyclohexane in a fluorous biphasic system [a]

		Total cost / k€	Cost of replenishment / k€ y^{-1}	Increased cost of nonanal / € tonne^{-1}
T / °C	70			
p / bar	20			
Organic:fluorous (v/v)	1			
Reactor volume / m^3	20			
Collecting vessel volume	20			
Rhodium inventory / kg	2.5	1000	100	1
Ligand inventory / kg	30	150	60	0.6
Fluorous solvent inventory / tonne	25	750	7.5	0.075
Synthesis gas : alkene ratio	1.2			
Selectivity to linear product / %	80			
Isomerised alkene / %	4			
Fluorous solvent in organic phase g kg^{-1}	3			

[a] Based on 25% down time, perfluorohexane as fluorous solvent and $P(4\text{-}C_6H_4C_6F_{13})_3$ as the ligand.

6.15 Conclusions

The fluorous biphasic concept, which was first suggested by Horváth in 1994, has now been applied to a very wide range of catalytic reactions, in many cases with the recycling of the catalyst being demonstrated. In most cases there is a fall off of conversion with reuse and significant leaching of ligand and metal containing catalyst, although in some cases very good catalyst retention does appear to have been demonstrated by ICP analysis and by successive experiments. Two different types of continuous reactor have been developed for laboratory demonstration on a small scale, one with and one without the capacity for handling pressurised gases. Two reactions have been operated, both without gaseous reagents, continuously for up to three weeks with no noticeable loss in activity, suggesting that the technique should definitely be considered for scale-up. Preliminary costings using realistic price estimates suggest that losses of fluorous ligand and of fluorous solvent during the commercial hydroformylation of long chain alkenes would not greatly inflate the costs of the process, despite the high costs of these chemicals. The major renewable cost would still be from loss of rhodium.

6.16 Acknowledgement

We thank Sasol Technology (UK) Ltd and the EPSRC for a studentship (C. R. M.).

6.17 References

[1] I. T. Horvath and J. Rabai, *Science*, 1994, **266**, 72.
[2] C. C. Tzschucke, C. Markert, W. Bannwarth, S. Roller, A. Hebel, and R. Haag, *Angew. Chem.-Int. Edit.*, 2002, **41**, 3964.
[3] G. Pozzi and I. Shepperson, *Coord. Chem. Rev.*, 2003, **242**, 115.
[4] A. P. Dobbs and M. R. Kimberley, *J. Fluor. Chem.*, 2002, **118**, 3.
[5] R. H. Fish, *Chem.-Eur. J.*, 1999, **5**, 1677.
[6] E. G. Hope and A. M. Stuart, *J. Fluor. Chem.*, 1999, **100**, 75.
[7] M. Cavazzini, A. Manfredi, F. Montanari, S. Quici, and G. Pozzi, *Chem. Commun.*, 2000, 2171.
[8] D. J. Adams, D. J. Cole-Hamilton, E. G. Hope, P. J. Pogorzelec, and A. M. Stuart, *J. Organomet. Chem.*, 2004, **689**, 1413.
[9] L. E. Kiss, I. Kovesdi, and J. Rabai, *J. Fluor. Chem.*, 2001, **108**, 95.
[10] E. de Wolf, P. Ruelle, J. van den Broeke, B. J. Deelman, and G. van Koten, *J. Phys. Chem. B*, 2004, **108**, 1458.
[11] B. Richter, E. de Wolf, G. van Koten, and B. J. Deelman, *J. Org. Chem.*, 2000, **65**, 3885.
[12] E. de Wolf, B. Richter, B. J. Deelman, and G. van Koten, *J. Org. Chem.*, 2000, **65**, 5424.
[13] M. L. Clarke, *J. Organomet. Chem.*, 2003, **665**, 65.
[14] M. A. Guillevic, C. Rocaboy, A. M. Arif, I. T. Horvath, and J. A. Gladysz, *Organometallics*, 1998, **17**, 707.
[15] J. J. J. Juliette, I. T. Horvath, and J. A. Gladysz, *Angew. Chem.-Int. Edit. Engl.*, 1997, **36**, 1610.
[16] D. Sinou, G. Pozzi, E. G. Hope, and A. M. Stuart, *Tetrahedron Lett.*, 1999, **40**, 849.
[17] P. Bhattacharyya, D. Gudmunsen, E. G. Hope, R. D. W. Kemmitt, D. R. Paige, and A. M. Stuart, *J. Chem. Soc.-Perkin Trans. 1*, 1997, 3609.
[18] B. Croxtall, J. Fawcett, E. G. Hope, and A. M. Stuart, *J. Chem. Soc.-Dalton Trans.*, 2002, 491.
[19] E. G. Hope, R. D. W. Kemmitt, D. R. Paige, A. M. Stuart, and D. R. W. Wood, *Polyhedron*, 1999, **18**, 2913.
[20] D. F. Foster, D. Gudmunsen, D. J. Adams, A. M. Stuart, E. G. Hope, D. J. Cole-Hamilton, G. P. Schwarz, and P. Pogorzelec, *Tetrahedron*, 2002, **58**, 3901.
[21] K. N. West, J. P. Hallett, R. S. Jones, D. Bush, C. L. Liotta, and C. A. Eckert, *Ind. Eng. Chem. Res.*, 2004, **43**, 4827.
[22] I. T. Horvath, G. Kiss, R. A. Cook, J. E. Bond, P. A. Stevens, J. Rabai, and E. J. Mozeleski, *J. Am. Chem. Soc.*, 1998, **120**, 3133.
[23] D. Sinou, D. Maillard, A. Aghmiz, and A. M. M. i-Bulto, *Adv. Synth. Catal.*, 2003, **345**, 603.
[24] E. de Wolf, A. J. M. Mens, O. L. J. Gijzeman, J. H. van Lenthe, L. W. Jenneskens, B. J. Deelman, and G. van Koten, *Inorg. Chem.*, 2003, **42**, 2115.
[25] D. Rutherford, J. J. J. Juliette, C. Rocaboy, I. T. Horvath, and J. A. Gladysz, *Catal. Today*, 1998, **42**, 381.
[26] E. G. Hope, R. D. W. Kemmitt, D. R. Paige, and A. M. Stuart, *J. Fluor. Chem.*, 1999, **99**, 197.
[27] B. Richter, B. J. Deelman, and G. van Koten, *J. Mol. Catal. A-Chem.*, 1999, **145**, 317.
[28] B. Richter, A. L. Spek, G. van Koten, and B. J. Deelman, *J. Am. Chem. Soc.*, 2000, **122**, 3945.

[29] E. de Wolf, A. L. Spek, B. W. M. Kuipers, A. P. Philipse, J. D. Meeldijk, P. H. H. Bomans, P. M. Frederik, B. J. Deelman, and G. van Koten, *Tetrahedron*, 2002, **58**, 3911.
[30] J. van den Broeke, E. de Wolf, B. J. Deelman, and G. van Koten, *Adv. Synth. Catal.*, 2003, **345**, 625.
[31] D. J. Birdsall, E. G. Hope, A. M. Stuart, W. P. Chen, Y. L. Hu, and J. L. Xiao, *Tetrahedron Lett.*, 2001, **42**, 8551.
[32] D. Maillard, C. Nguefack, G. Pozzi, S. Quici, B. Valade, and D. Sinou, *Tetrahedron: Asymmetry*, 2000, **11**, 2881.
[33] D. Maillard, G. Pozzi, S. Quici, and D. Sinou, *Tetrahedron*, 2002, **58**, 3971.
[34] E. de Wolf, E. A. Speets, B. J. Deelman, and G. van Koten, *Organometallics*, 2001, **20**, 3686.
[35] J. J. J. Juliette, D. Rutherford, I. T. Horvath, and J. A. Gladysz, *J. Am. Chem. Soc.*, 1999, **121**, 2696.
[36] A. Aghmiz, C. Claver, A. M. Masdeu-Bulto, D. Maillard, and D. Sinou, *J. Mol. Catal. A-Chem.*, 2004, **208**, 97.
[37] P. Bhattacharyya, B. Croxtall, J. Fawcett, D. Gudmunsen, E. G. Hope, R. D. W. Kemmitt, D. R. Paige, D. R. Russell, A. M. Stuart, and D. R. W. Wood, *J. Fluor. Chem.*, 2000, **101**, 247.
[38] D. F. Foster, D. J. Adams, D. Gudmunsen, A. M. Stuart, E. G. Hope, and D. J. Cole-Hamilton, *Chem. Commun.*, 2002, 916.
[39] C. R. Mathison and D. J. Cole-Hamilton, Unpublished observations.
[40] Y. L. Huang, E. Perperi, G. Manos, and D. J. Cole-Hamilton, *J. Mol. Catal. A-Chem.*, 2004, **210**, 17.
[41] E. Perperi, Y. L. Huang, P. Angeli, G. Manos, C. R. Mathison, D. J. Cole-Hamilton, D. J. Adams, and E. G. Hope, *Chem. Eng. Sci*, 2004, In the press.
[42] E. Perperi, Y. L. Huang, P. Angeli, G. Manos, C. R. Mathison, D. J. Cole-Hamilton, D. J. Adams, and E. G. Hope, *Dalton Trans.*, 2004, 2062.
[43] E. Perperi, Y. L. Huang, P. Angeli, G. Manos, and D. J. Cole-Hamilton, *J. Mol. Catal. A-Chem.*, 2004, **221**, 19.
[44] T. Mathivet, E. Monflier, Y. Castanet, A. Mortreux, and J. L. Couturier, *C. R. Chim.*, 2002, **5**, 417.
[45] T. Mathivet, E. Monflier, Y. Castanet, A. Mortreux, and J. L. Couturier, *Tetrahedron*, 2002, **58**, 3877.
[46] W. P. Chen, L. J. Xu, and J. L. Xiao, *Chem. Commun.*, 2000, 839.
[47] D. Bonafoux, Z. H. Hua, B. H. Wang, and I. Ojima, *J. Fluor. Chem.*, 2001, **112**, 101.
[48] J. M. Vincent, A. Rabion, V. K. Yachandra, and R. H. Fish, *Angew. Chem.-Int. Edit. Engl.*, 1997, **36**, 2346.
[49] M. Contel, C. Izuel, M. Laguna, P. R. Villuendas, P. J. Alonso, and R. H. Fish, *Chem.-Eur. J.*, 2003, **9**, 4168.
[50] G. Pozzi, F. Montanari, and S. Quici, *Chem. Commun.*, 1997, 69.
[51] G. Pozzi, F. Cinato, F. Montanari, and S. Quici, *Chem. Commun.*, 1998, 877.
[52] G. Pozzi, M. Cavazzini, F. Cinato, F. Montanari, and S. Quici, *Eur. J. Org. Chem.*, 1999, 1947.
[53] M. Cavazzini, A. Manfredi, F. Montanari, S. Quici, and G. Pozzi, *Eur. J. Org. Chem.*, 2001, 4639.
[54] M. Cavazzini, G. Pozzi, S. Quici, and I. Shepperson, *J. Mol. Catal. A-Chem.*, 2003, **204**, 433.
[55] I. Klement, H. Lutjens, and P. Knochel, *Angew. Chem.-Int. Edit. Engl.*, 1997, **36**, 1454.
[56] B. Betzemeier, F. Lhermitte, and P. Knochel, *Tetrahedron Lett.*, 1998, **39**, 6667.
[57] T. Nishimura, Y. Maeda, N. Kakiuchi, and S. Uemura, *J. Chem. Soc.-Perkin Trans. 1*, 2000, 4301.
[58] G. Ragagnin, B. Betzemeier, S. Quici, and P. Knochel, *Tetrahedron*, 2002, **58**, 3985.
[59] G. Pozzi, M. Cavazzini, S. Quici, and S. Fontana, *Tetrahedron Lett.*, 1997, **38**, 7605.
[60] D. Maillard, J. Bayardon, J. D. Kurichiparambil, C. Nguefack-Fournier, and D. Sinou, *Tetrahedron: Asymmetry*, 2002, **13**, 1449.
[61] J. Bayardon, M. Cavazzini, D. Maillard, G. Pozzi, S. Quici, and D. Sinou, *Tetrahedron: Asymmetry*, 2003, **14**, 2215.
[62] J. Bayardon and D. Sinou, *Tetrahedron Lett.*, 2003, **44**, 1449.
[63] Y. Nakamura, S. Takeuchi, S. L. Zhang, K. Okumura, and Y. Ohgo, *Tetrahedron Lett.*, 2002, **43**, 3053.
[64] R. Grigg and M. York, *Tetrahedron Lett.*, 2000, **41**, 7255.
[65] S. Schneider and W. Bannwarth, *Angew. Chem.-Int. Edit.*, 2000, **39**, 4142.
[66] S. Schneider and W. Bannwarth, *Helv. Chim. Acta*, 2001, **84**, 735.
[67] C. Markert and W. Bannwarth, *Helv. Chim. Acta*, 2002, **85**, 1877.
[68] B. Betzemeier and P. Knochel, *Angew. Chem.-Int. Edit.*, 1997, **36**, 2623.
[69] Y. Tian and K. S. Chan, *Tetrahedron Lett.*, 2000, **41**, 8813.
[70] Y. Tian, Q. C. Yang, T. C. W. Mak, and K. S. Chan, *Tetrahedron*, 2002, **58**, 3951.
[71] Y. Nakamura, S. Takeuchi, Y. Ohgo, and D. P. Curran, *Tetrahedron Lett.*, 2000, **41**, 57.

[72] H. Kleijn, E. Rijnberg, J. Jastrzebski, and G. van Koten, *Org. Lett.*, 1999, **1**, 853.
[73] Y. Y. Yin, G. Zhao, Z. S. Qian, and W. X. Yin, *J. Fluor. Chem.*, 2003, **120**, 117.
[74] A. Biffis, E. Castello, M. Zecca, and M. Basato, *Tetrahedron*, 2001, **57**, 10391.
[75] A. Endres and G. Maas, *Tetrahedron Lett.*, 1999, **40**, 6365.
[76] A. Endres and G. Maas, *J. Organomet. Chem.*, 2002, **643**, 174.
[77] M. Wende and J. A. Gladysz, *J. Am. Chem. Soc.*, 2003, **125**, 5861.
[78] K. Ishihara, S. Kondo, and H. Yamamoto, *Synlett*, 2001, 1371.
[79] J. N. Xiang, A. Orita, and J. Otera, *Adv. Synth. Catal.*, 2002, **344**, 84.
[80] K. Ishihara, A. Hasegawa, and H. Yamamoto, *Synlett*, 2002, 1299.
[81] A. Yoshida, X. H. Hao, and J. Nishikido, *Green Chem.*, 2003, **5**, 554.
[82] 'Rhodium catalysed hydroformylation', ed. P. N. W. M. Van Leeuwen and C. Claver, Kluwer, Dordrecht, 2000.
[83] C. D. Frohning and C. W. Kohlpaintner, in 'Applied Homogeneous Catalysis with Organometallic Compounds', ed. W. A. Herrmann, VCH, Weinheim, 1996.

CHAPTER 7 CATALYST RECYCLING USING IONIC LIQUIDS

P. WASSERSCHEID, M. HAUMANN
Lehrstuhl für Chemische Reaktionstechnik, Universität Erlangen
Egerlandstraße 3, D-91058 Erlangen, Germany
e-mail: wasserscheid@crt.cbi.uni-erlangen.de

7.1. Introduction

7.1.1 INTRODUCTION TO IONIC LIQUIDS

Ionic liquids are characterised by the following three definition criteria. They consist entirely out of ions, they have melting points below 100 °C and they exhibit no detectable vapour pressure below the temperature of their thermal decomposition. As a consequence of these properties most ions forming ionic liquids display low charge densities resulting in low intermolecular interaction. Figure 7.1 displays some of the most common ions used so far for the formation of ionic liquids.

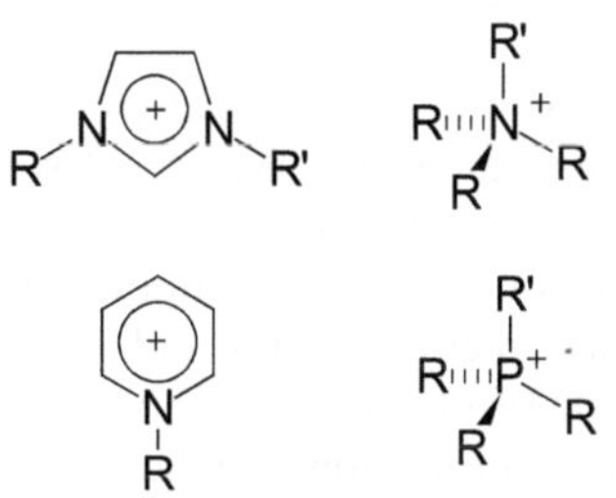

Cl^-, Br^-
$[AlCl_4]^-$, $[Al_2Cl_7]^-$
$[PF_6]^-$, $[BF_4]^-$
$[CF_3COO]^-$, $[CF_3SO_3]^-$
$[(CF_3SO_2)_2N]^-$
$[HSO_4]^-$
$[RSO_4]^-$, $[RSO_3]^-$

Figure 7.1. Typical cations and anions used for the formation of ionic liquids

The historical development of ionic liquids can be structured according to the different classes of anions that were found to form low melting salts in combination with imidazolium, pyridinium, ammonium and phosphonium cations. Low melting chloroaluminate salts can be regarded as the first ionic liquids to be studied in detail and in the modern literature they are often referred to as "first generation ionic liquids". They were described as early as in 1948 by Hurley and Wier at the Rice Institute in Texas as bath solutions for the electroplating of aluminium [1]. Later in the seventies and eighties, these systems were intensively studied by the groups of Osteryoung [2], Wilkes [3], Hussey [4] and Seddon [5].

D. J. Cole-Hamilton and R. P. Tooze (eds.), Catalyst Separation, Recovery and Recycling, 183–213.

In 1992, the ionic liquid methodology received a substantial boost when Wilkes and Zaworotko described the synthesis of non-chloroaluminate, room temperature liquid melts (e. g. low melting tetrafluoroborate melts) which may be regarded as "second generation ionic liquids" [6]. Nowadays, tetrafluoroborate and (the slightly later published [7]) hexafluorophosphate ionic liquids are still widely used in ionic liquid research. However, their use in many technical applications will be clearly limited by their relatively high sensitivity towards hydrolysis. Of course, the tendency of their anions to hydrolyse is much less pronounced than for the chloroaluminate melts but it still clearly exists. Consequently, the technical application of tetrafluoroborate and hexafluorophosphate ionic liquids will be effectively restricted to those applications where water-free conditions can be realised at acceptable costs.

In 1996, Grätzel, Bonhôte and co-workers published the synthesis and properties of ionic liquids with anions containing CF_3- and other fluorinated alkyl groups [8]. These do not show the same sensitivity towards hydrolysis as $[BF_4]^-$ and $[PF_6]^-$ based systems. In fact, heating $[BMIM][(CF_3SO_2)_2N]$ with excess of water to 100°C for 24 h does not reveal any hint of anion hydrolysis [9]. However, despite the very high stability of imidazolium bis(trifluoromethanesulfonyl)imides against hydrolysis and a number of other very suitable properties (e. g. low viscosity, high conductivity, high thermal stability, easy preparation in halogen-free form due to the miscibility-gap with water etc.) the high price of the Li $[(CF_3SO_2)_2N]$ starting material may be a major problem for their practical application on a larger scale. Moreover, the presence of fluorine in the anion may still be problematic even if hydrolysis is not an issue. The obvious disposal option for technical amounts of spent ionic liquid is thermal treatment and this would require great additional efforts in the case of fluorinated ionic liquids in order to avoid the liberation of toxic and highly corrosive HF.

Nowadays, the use of ionic liquids with halogen-free anions is becoming more and more popular. In 1998, Andersen et al. published a paper describing the use of some phosphonium tosylates (all with melting points >70°C) in the rhodium catalysed hydroformylation of 1-hexene [10]. More recently, we found that ionic liquids with halogen-free anions can be synthesised with much lower melting points and used as solvents in transition metal catalysis. For example, $[BMIM][n\text{-}C_8H_{17}SO_4]$ (mp = 35°C) could be used as solvents in the rhodium catalysed hydroformylation of 1-octene [11] (more details are given in Section 7.2).

Two groups of newly developed ionic liquids can be distinguished: The first group can be considered as "bulk ionic liquids". These ionic liquids are designed to be produced, used and somehow consumed in larger quantities. Applications for these ionic liquids are expected to be solvents for organic reactions, homogeneous catalysis, biocatalysis and other synthetic applications with some ionic liquid consumption as well as some non synthetic applications (e. g. lubricants, performance additives, surfactants, phase transfer catalysts, extraction solvents, solvents for extractive distillation, antistatics etc.). The cations and anions of these "bulk ionic liquids" are chosen to make a relatively cheap (expected price on a multi-ton scale: ca 30€/litre), halogen-free (e. g. for easy disposal of spent ionic liquid by thermal treatment) and toxologically well-characterized liquid (preliminary studies concerning the toxicity of a non-chloroaluminate ionic liquid has been recently published [12]). From all ionic liquids meeting these requirements only a limited number of candidates will be produced on a large scale. However, these candidates will become well characterised and – as a result of their larger production quantities - readily available.

Promising examples for this type of "bulk ionic liquids" include benzenesulfonate [13], toluenesulfonate [10], octylsulfate [11] and hydrogensulfate [14] ionic liquids. Some examples are given in Figure 7.2.

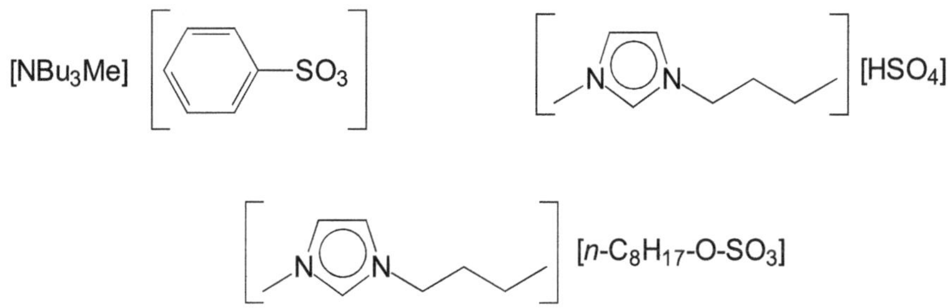

Figure 7.2. Examples for bulk ionic liquids

The second group of recently developed ionic liquids is often referred to as "task specific ionic liquids" in the literature [15]. These ionic liquids are designed and optimised for the best performance in high-value-added applications. Functionalised [16], fluorinated [17], deuterated [18] and chiral ionic liquids [19] are expected to play a future role as special solvents for sophisticated synthetic applications, analytical tools (stationary or mobile phases for chromatography, matrixes for MS etc.), sensors and special electrolytes.

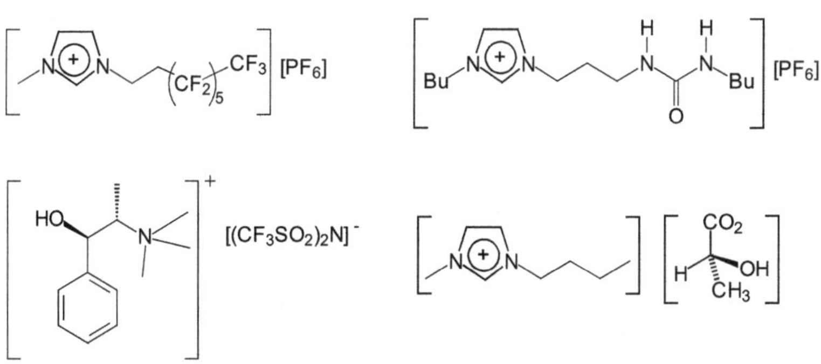

Figure 7.3. Examples of task specific ionic liquids

Historically, the know-how to synthesise and handle ionic liquids has been treated in some ways as a "holy grail". Indeed, only a small number of specialised industrial and academic research groups were able to prepare and handle the highly hygroscopic chloroaluminate ionic liquids which were the only ionic liquid systems available in larger amounts before the mid-nineties.

Nowadays, a number of commercial suppliers [20] offer ionic liquids, some of them in larger quantities, [21] and the quality of commercial ionic liquid samples has clearly improved in recent years. The fact that small amounts of impurities significantly influence the properties of the ionic liquid and especially its usefulness for catalytic reactions [22] makes the quality of an ionic liquid an important consideration [23]. Without any doubt the improved commercial availability of ionic liquids is a key factor for the strongly increasing interest in this new class of liquid materials.

From the historic examples above it is quite clear that there are many different ionic liquids. More important than the impressive number of possible ion combinations is the fact, that the physical and chemical properties of these ionic liquid materials can cover a wide range of density, viscosity, surface tension, heat capacity, polarity, acidity, coordination etc. Consequently, it is a key interest in ionic liquid development to understand structure-property relationships in a way that allows a specific ionic liquid to be designed for a given technical application. In this respect, we are still far from a conclusive general understanding. However, in the last three years, the available and reliable data has increased exponentially, providing a much better base for predictive modelling. Table 7.1 displays some key properties of well-investigated ionic liquids that may also be relevant for applications in liquid-liquid biphasic catalysis.

TABLE 7.1. Comparison of some properties of well-established ionic liquid systems with the 1-ethyl-3-methyimidazolium ([EMIM]) and 1-butyl-3-methylimidazolium ([BMIM]) ions

ionic liquid	mp./gp. [°C]	viscosity [cP]	density[a)] [g/ml]	tendency for hydrolysis[c)]	ref.
[EMIM][$AlCl_4$]	7 (mp)	18[a]	1.240	very high	24
[EMIM][BF_4]	6 (gp)	34 [a]	1.240	existent	25
[EMIM][CF_3SO_3]	-3 (mp)	45 [a]	1.390	very low	8
[EMIM][$(CF_3SO_2)_2N$]	-9 (mp)	31 [a]	1.518	very low	26
[BMIM][PF_6]	6 (mp)	207[a]	1.363	existent	27
[BMIM][CF_3SO_3]	16 (mp)	90[b]	1.290	very low	8
[BMIM][$(CF_3SO_2)_2N$]	-4 (mp)	52[b]	1.429	very low	8
[BMIM][n-$C_8H_{17}OSO_4$]	35 (mp)	875[b]	1.060	low	11

gp = glass point; mp = melting point; a) at 25°C; b) at 20°C; c) very high = immediate hydrolysis even with traces of water; existent = slow hydrolysis with water accelerated at higher temperature; low = slow hydrolysis only in acidic solutions at elevated temperature; very low = no hydrolysis even at higher temperature in acidic solutions; n.d. = not determined.

Since the focus of this contribution is clearly on catalysis and catalyst recycle using the ionic liquid methodology it is not possible to go into more detail on the materials science aspects of ionic liquids. However, it should be clearly stated that at least some understanding of the ionic liquid material is a prerequisite for its successful use as a liquid catalyst support in catalysis. Therefore, the interested reader is strongly encouraged to explore the more specialized literature [28].

Many transition metal complexes dissolve readily in ionic liquids, thus enabling their use as solvents for transition metal catalysis. Sufficient solubility for a wide range of catalyst complexes is an obvious, but not trivial, prerequisite for a versatile solvent for homogenous catalysis.

The first example of homogeneous transition metal catalysis in an ionic liquid dates back to 1972 when Parshall described the platinum catalysed hydroformylation of ethene in tetraethylammonium trichlorostannate (mp. 78°C) [29]. This work was followed by the pioneering studies of Knifton who reported the ruthenium- and cobalt-catalysed hydroformylation of internal and terminal alkenes in molten $[Bu_4P]Br$ in 1987 [30]. The first biphasic, liquid-liquid catalysis with room temperature ionic liquids was carried out by Chauvin and Olivier-Bourbigou in the early nineties [31]. However, only the development and broader availability of non-chloroaluminate ionic liquids led to the much larger number of publications in this research field starting from the late nineties. Today, a significant part of the ballooning number of publications on ionic liquid chemistry (more than 1700 in 2004 according to SciFinder) deals with transition metal catalysis in these unusual liquid materials. This intense publication activity has been documented in a number of comprehensive review articles and book chapters [32]. Therefore, there is no need to add another list of reactions and catalyst systems that have been successfully applied in ionic liquids.

In contrast, we intend to demonstrate the principle aspects of catalyst recycling and regeneration using the ionic liquid methodology. These aspects will be explored in more detail for the example of Rh-catalysed hydroformylation (see Section 7.2). First, however, we will briefly introduce important general facts concerning transition metal catalysis in ionic liquids (see Section 7.1.2). This will be followed by a consideration of liquid-liquid biphasic reactions in these media from an engineering point of view (see Section 7.1.3).

7.1.2 INTRODUCTION TO TRANSITION METAL CATALYSIS IN IONIC LIQUIDS

There are many good reasons for applying ionic liquids as alternative solvents in transition metal catalysed reactions. Besides their very low vapour pressure and their good thermal stability [33], an important advantage is the possibility of tuning their solubility [34] and acidity/coordination properties [35] by varying the nature of the anions and cations systematically.

The possibility of adjusting solubility properties is of particular importance for liquid-liquid biphasic catalysis. Liquid-liquid catalysis can be realised when the ionic liquid is able to dissolve the catalyst, especially if it displays partial solubility of the substrates and poor solubility of the reaction products. Under these conditions, the product phase, which also contains the unconverted reactants, is removed by simple phase decantation. The ionic liquid containing the catalyst can then be recycled. In such a scenario the ionic catalyst solution may be seen as part of the capital investment for a potential technical process (in an ideal case) or at least as a "working solution" (only a small amount has to be replaced after a certain time of application). A crucial aspect of this concept is the immobilisation of the transition metal catalyst in the ionic liquid. While most transition metal catalysts easily dissolve in an ionic liquid without any special ligand design, ionic ligand systems have been applied with great success to prevent catalyst leaching under the conditions of intense mixing in continuous liquid-liquid biphasic operation. Since many of the important ligand developments have been made for the biphasic hydroformylation reaction more details on specific ligand systems developed for the use in ionic liquids will be given in Section 7.2.

Apart from these recycling aspects, liquid-liquid biphasic catalysis can also help to improve the selectivity of a given reaction. Attractive options arise from the preferential solubility of only one reactant in the catalyst solvent or from the *in-situ* extraction of desired reaction products out of the catalyst layer in order to avoid unfavourable consecutive reactions (see Figure 7.4) [36].

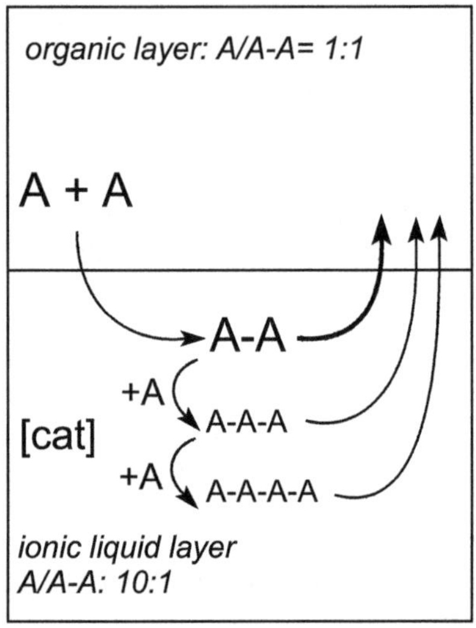

Figure 7.4. Enhanced dimer selectivity in the oligomerisation of compound A due to the biphasic reaction mode with an ionic liquid of high preferential solubility for A

The possibility of adjusting acidity/coordination properties opens up a wide range of possible interactions between the ionic liquid solvent and the dissolved transition metal complex. Depending on the acidity/coordination properties of the anion and on the reactivity of the cation (the possibility of carbene ligand formation from 1,3-dialkylimidazolium salts is of particular importance here [37]), the ionic liquid can be regarded as an "innocent" solvent, as a ligand precursor, as a co-catalyst or as the catalyst itself.

Ionic liquids with weakly coordinating, inert anions (e. g., $[(CF_3SO_2)_2N]^-$, $[BF_4]^-$ or $[PF_6]^-$ under anhydrous conditions) and inert cations (cations that do not coordinate to the catalyst themselves and that do not form species under the reaction conditions that coordinate to the catalyst) can be considered as "innocent" solvents in transition metal catalysis. In these cases, the role of the ionic liquid is solely to provide a more or less polar, more or less weakly coordinating medium for the transition metal catalyst that additionally offers special solubility for the feedstock and products.

However, unlike most conventional solvents, many ionic liquids combine high solvating power for polar catalyst complexes (polarity) with weak coordination (nucleophilicity) [38]. It is this combination that enables a biphasic reaction mode with these ionic liquids even for catalyst systems which are deactivated by water or polar organic solvents.

In many other reactions where ionic or polar transition metal catalysts are used it has been demonstrated that the use of polar and weekly coordinating ionic liquids can result in a clear enhancement of catalytic activity [32a].

A truly co-catalytic effect of ionic liquids is observed with those ionic liquids displaying a certain latent or real Lewis-acid character. These ionic liquids are usually formed by the reaction of a halide salt with a Lewis acid (e. g. chloroaluminate or chlorostannate melts). In many examples, the Lewis acidity of an ionic liquid has been used to convert the neutral catalyst precursor into the active form of the catalyst by halide abstraction (see Scheme 7.1) [39].

$$Cp_2TiCl_2 + [cation][Al_2Cl_7] \rightleftharpoons [Cp_2TiCl][AlCl_4] + [cation][AlCl_4]$$

$$(ligand)_2NiCl_2 + [cation][Al_2Cl_7] + [cation][Al_2EtCl_6] \rightleftharpoons$$

$$[(ligand)Ni\text{-}CH_2\text{-}CH_3][AlCl_4] + 2\,[cation][AlCl_4] + AlCl_3\text{-}ligand$$

Scheme 7.1. Activation of a neutral catalyst precursor by chloroaluminate ionic liquids

Even if the ionic liquid is not directly involved in creating the active catalytic species, a co-catalytic interaction between the ionic liquid solvent and the dissolved transition metal complex often takes place and can result in significant catalyst activation. When a catalyst complex is, for example, dissolved in a slightly acidic ionic liquid some electron-rich parts of the complex (e.g., lone pairs of electrons in the ligand) may interact with the solvent, providing increased activity to the resulting catalytic centre. Acidic ionic liquids can be considered as liquid acid supports for transition metal catalysts dissolved therein.

7.1.3 MULTIPHASIC CATALYSIS WITH IONIC LIQUIDS – ENGINEERING ASPECTS

Liquid-liquid multiphasic catalysis with the catalyst present in the ionic liquid phase relies on the transfer of organic substrates into the ionic liquid or reactions must occur at the phase boundary. One important parameter for the development of kinetic models (which are crucial for up-scaling and proper economic evaluation) is the location of the reaction. Does the reaction take place in the bulk of the liquid, in the diffusion layer or immediately at the surface of the ionic liquid droplets?

The answer to this question depends mainly on the relative speed of the chemical reaction vs. mass transfer of the substrate into the ionic liquid layer. If the chemical reaction is fast vs. the mass transfer rate, a significant part of the reaction will take place at the surface or in the diffusion layer. If the chemical reaction is slow relative to mass transfer, the feedstock concentration will be high in the bulk ionic liquid phase and the major part of the reaction will take place there. Figure 7.5 shows both cases for a reaction where substrate 2 is present in large excess in the ionic liquid layer while substrate 1 has to be transferred from the gas phase into the ionic liquid. For an organic solvent-ionic liquid biphasic reaction the picture becomes slightly more complex, since mass transfer resistance on the organic side of the organic-ionic liquid phase boundary has to be taken into account as well.

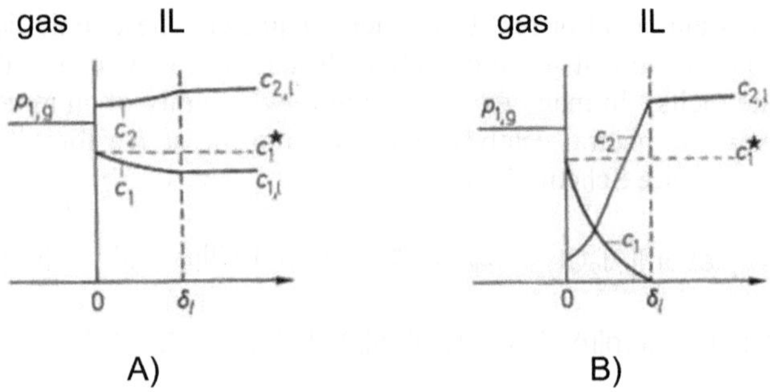

Figure 7.5. Concentration of gaseous feedstock 1 in a gas-ionic liquid biphasic reaction for slow (A) and fast (B) chemical reaction - mass transfer resistance on the gas side is neglected

TABLE 7.2. Viscosities of selected ionic liquids with 1-ethyl-3-methylimidazolium ([EMIM]) as the cation

Anion	T/K	Viscosity/cP	ref.
$[CH_3CO_2]$	293	162	[40]
$[CH_3SO_3]$	298	160	[40]
$[C_3F_7CO_2]$	293	105	[40]
$[CF_3SO_3]$	293	45	[40]
$[CF_3CO_2]$	293	35	[40]
$[BF_4]$	298	34	[40]
$[(CF_3SO_2)_2N]$	293	34	[40]
$Br/AlBr_3$ (34.0/66.0 mol%)	298	32	[40]
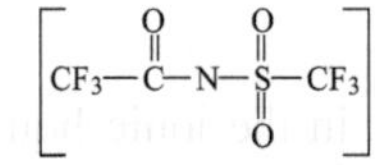	298	25	[41]
$[N(CN)_2]$	298	21	[42]
$Cl/AlCl_3$ (50.0/50.0 mol%)	298	18	[40]
$Cl/AlCl_3$ (34.0/66.0 mol%)	298	14	[40]
for comparison:			
ethylene glycol	*298*	*16*	[40]

It is important to be able to identify mass transfer limitations that occur when the reaction rate is high compared with the rate of mass transfer. For a heavily mass transfer limited reaction, preliminary experiments in a non optimised laboratory reactor

may indicate slow kinetics while the observed overall reaction rate may become much higher by scaling-up with an optimized reactor design. If the reaction is exothermic such a misinterpretation can cause great problems for the heat management in the reactor.
In the case of ionic liquids, these general aspects for all fluid-fluid reactions are of particular importance, since mass transfer into an ionic liquid layer is generally slower than into an organic or aqueous medium. This is because ionic liquids usually have much higher viscosities than organic solvents. The least viscous ionic liquids are somewhat similar to ethylene glycol as demonstrated in Table 7.2. However, many ionic liquids used in liquid-liquid biphasic catalysis are significantly more viscous.

The diffusional mass transfer into an ionic liquid can be expressed by the following equation:

$$dn/dt = D_{1,2}^{IL}\ A\ dc_1/dz$$

with $D_{1,2}^{IL}$ being the diffusion coefficient, A being the liquid-liquid interfacial area and dc_1/dz being the concentration gradient along the z-axis.

As can be seen easily from this equation, relatively high viscosity of the ionic liquid reduces the mass transport into the catalytically active layer in two ways: a) the higher the viscosity, the lower the diffusion coefficient $D_{1,2}^{IL}$; b) the higher the viscosity the larger the droplets formed with a given stirring effort in a given device and – consequently – the smaller the organic - ionic liquid interfacial area A.

The aspects discussed above are not only relevant for the transfer of a component 1 from one liquid organic phase into the ionic liquid but also for the transfer of reactive gases into the ionic liquids. If the chemical kinetics are relatively fast a special stirrer design (as shown in Figure 7.6) can help to reduce problems related to mass transfer of the reactive gas into the ionic liquid.

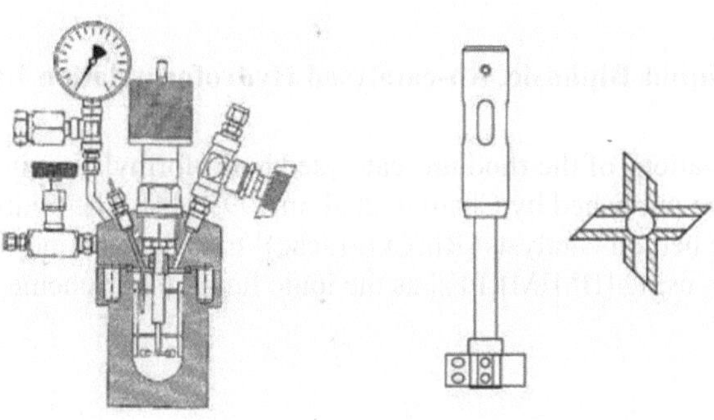

Figure 7.6. A 150 ml autoclave with a special stirrer design to maximise the intake of gaseous reactants into an ionic liquid (for more details see [43])

However, it should be noted that in some cases the existence of mass transfer limitations can be advantageously used to control the exothermicity of reactions. For example, a decrease of stirring can be a simple and efficient way to decrease the reaction rate and hence the heating effect of the reaction.

In many respects liquid-liquid, biphasic transition metal catalysis in ionic liquids is better regarded as heterogeneous catalysis on a liquid support than as conventional homogeneous catalysis in an organic solvent. As mentioned above, support-catalyst interactions are known in ionic liquids and can lead to catalyst activation (or deactivation). Product separation from an ionic catalyst layer is often easy (at least if the products are not too polar and have low boiling points) as in classical heterogeneous catalysis. However, mass transfer limitation problems (when the chemical kinetics are fast) and some uncertainty concerning the exact micro-environment around the catalytically active centre represent common limitations for transition metal catalysis in ionic liquids and on heterogeneous supports.

Of course, there are some still important differences between the use of a liquid catalyst immobilisation phase and a classical heterogeneous support. Obviously, by using a liquid, ionic catalyst support it is possible to integrate some classical features of traditional homogenous catalysis into this type of "heterogeneous" catalysis. For example, a defined transition metal complex can be introduced and immobilised in an ionic liquid providing the opportunity of optimising the selectivity of a reaction by ligand variation, which is a typical approach in homogeneous catalysis. Reaction conditions in ionic liquid catalysis are still mild and comparable to those in homogenous catalysis. Analysis of the active catalyst in an ionic liquid immobilisation phase is, in principle, possible using the same methods developed for homogeneous catalysis, which should enable more rational catalyst design in the future.

In comparison to traditional biphasic catalysis using water, fluorous phases or polar organic solvents, transition metal catalysis in ionic liquids represents a new and advanced way of combining the specific advantages of homogeneous and heterogeneous catalysis.

7.2. Liquid-liquid Biphasic, Rh-catalysed Hydroformylation Using Ionic Liquids

The first investigations of the rhodium-catalysed hydroformylation in room temperature ionic liquids were published by Chauvin et al. in 1995 [44]. The hydroformylation of 1-pentene with the neutral catalysts $[Rh(CO)_2(acac)]$/triarylphosphine was carried out in a biphasic reaction using $[BMIM][PF_6]$ as the ionic liquid (see Scheme 7.2).

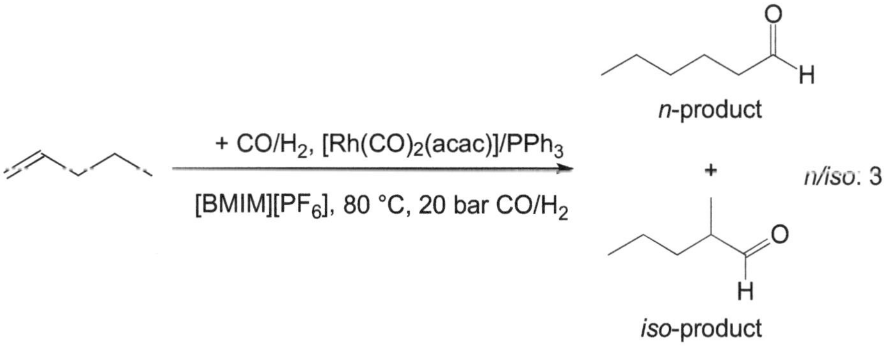

Scheme 7.2. First example of Rh-catalysed hydroformylation with room temperature ionic liquid as published by Chauvin et al. in 1995

However, with none of the well-established ligands tested in this study was it possible to combine high activity, complete retention of the catalyst in the ionic liquid and high selectivity for the desired linear hydroformylation product. The use of PPh_3 resulted in significant leaching of the Rh-catalyst into the product phase, thus the catalyst was active in both phases. The catalyst leaching could be suppressed by the application of the same sulfonated triaryl phosphine ligands that are used with great success in aqueous biphasic catalysis. Surprisingly these ligands were found to cause major deactivation when applied in an ionic liquid medium (TOF = 59 h^{-1} with tppms compared to 333 h^{-1} with PPh_3). Moreover, all of the ligands used in Chauvin's work showed poor selectivity to the desired linear hydroformylation product (n/iso-ratio between 2 and 4).

The first ligand system especially designed for the use in ionic liquids was described in 2000 [45]. Cationic ligands with a cobaltocenium backbone were successfully applied in the biphasic, Rh-catalysed hydroformylation of 1-octene. 1,1'-Bis(diphenylphosphino)cobaltocenium hexafluorophosphate (cdpp) proved to be an especially promising candidate. The compound was synthesised according to Scheme 7.3 by mild oxidation of 1,1'-bis(diphenylphosphino)cobaltocene with C_2Cl_6 followed by anion exchange with $[NH_4][PF_6]$ in acetone. In a typical example, this ionic catalyst system allowed the hydroformylation of 1-octene at a rate of 800 catalyst turnovers per hour (ligand/Rh: 2:1, CO/H_2 =1:1, t = 1h, T = 100°C, p = 10 bar, 1-octene/Rh = 1000, 5 mL $[BMIM][PF_6]$). Under these conditions, the n/iso-selectivity was found to be as high as 16.2 corresponding to a selectivity of 94% to the desired linear aldehyde.

It is important to note that with the phosphinocobaltocenium ligand cdpp the reaction took place almost exclusively in the ionic liquid phase (ca. 0.5% of the Rh was found in the product layer after reaction). The catalyst phase was separated from the product by decantation. Moreover, the recovered ionic catalyst solution could be reused at least once more with similar reactivity to that in the original run.

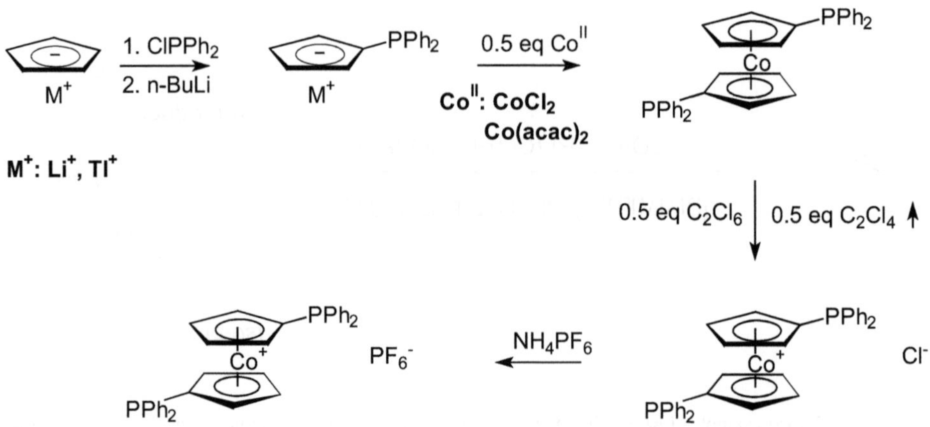

Scheme 7.3. Synthesis of 1,1'-Bis(diphenylphosphino)cobaltocenium hexafluoro-phosphate

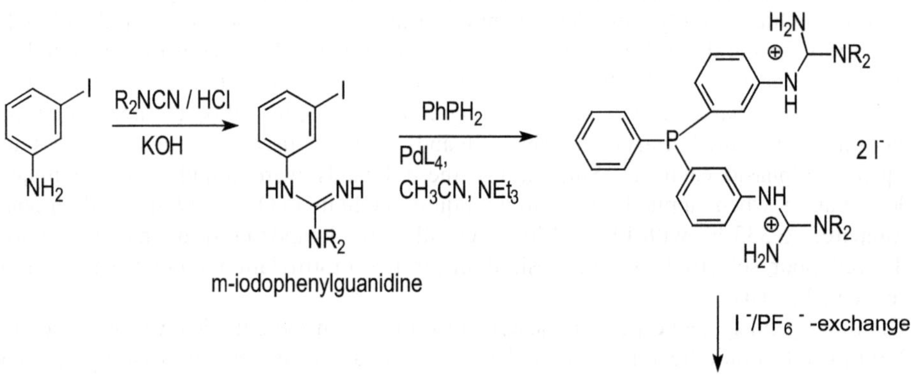

Scheme 7.4. Synthesis of a guanidinium modified triphenylphosphine ligand

Subsequently, it was found that modification of neutral phosphine ligands with cationic phenylguanidinium groups represents a very powerful tool for immobilising Rh-complexes in ionic liquids [46]. Cationic phosphine ligands containing guanidiniumphenyl moieties were originally developed to make use of their pronounced solubility in water [47], [48]. They were shown to form active catalytic systems in Pd mediated C-C coupling reactions between aryl iodides and alkynes (Castro-Stephens-Sonogashira reaction) [49] and Rh catalysed hydroformylation of olefins in aqueous two-phase systems [50]. The guanidinium modified triphenylphosphine ligand was prepared according to Scheme 7.4 by anion exchange with $[NH_4][PF_6]$ in aqueous solution from the corresponding iodide salt [46]. In the biphasic Rh-catalysed hydroformylation experiment using $[BMIM][PF_6]$ as the catalyst solvent, this ligand reduced Rh-leaching to about 0.07% of the Rh used in the experiment. The ionic catalyst solution could be recycled 10 times without significant loss in activity. The n/iso ratio of 2.5-2.8 was in the expected range for a PPh_3 derived ligand system.

Alternative methods of immobilising phosphine ligands by attaching them to ionic groups similar to the ionic liquid cation have been reported. Both pyridinium-modified phosphine ligands [51] and imidazolium modified phosphine ligands [52], [53] have been synthesised and applied in Rh-catalysed hydroformylation (see Figure 7.7). While the presence of the ionic group led to better immobilisation of the Rh-catalyst in the ionic liquid in all cases no outstanding reactivity or selectivity were observed with these ligands. This is not really surprising since all these ligands are electronically and sterically closely related to PPh_3.

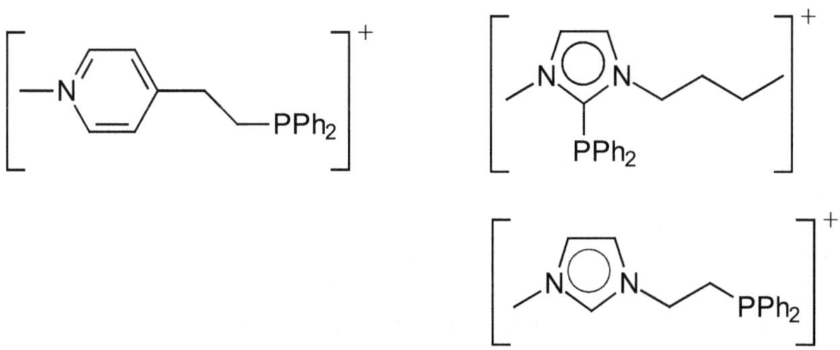

Figure 7. 7. Cationic diphenylphosphine ligands as used in the biphasic, Rh-catalysed hydroformylation of 1-octene in e. g. [BMIM][PF_6]

Further developments were directed towards ligand systems which are ionic (for highly efficient immobilisation in the ionic liquid) but have a ligand backbone that allows very high selectivity towards the linear aldehyde product.

It is well-known that diphosphine ligands with large natural P-metal-P bite angles form highly regioselective hydroformylation catalysts [54]. Here, xanthene type ligands (P-metal-P~ 110 °) developed by van Leeuwen's group proved to be especially suitable allowing an overall selectivity of 98% towards the desired linear aldehyde in 1-octene hydroformylation in traditional solvents [55], [56].

While the unmodified xanthene ligand (compound (a) in Figure 7.8) shows highly preferential solubility in the organic phase in the biphasic 1-octene/[BMIM][PF_6] mixture, both the sulfoxantphos ligand (compound (b) in Figure 7.8) and the guanidinium-modified xanthene ligand (compound (c) in Figure 7.8) [57] have been successfully applied using biphasic catalysis with ionic liquids.

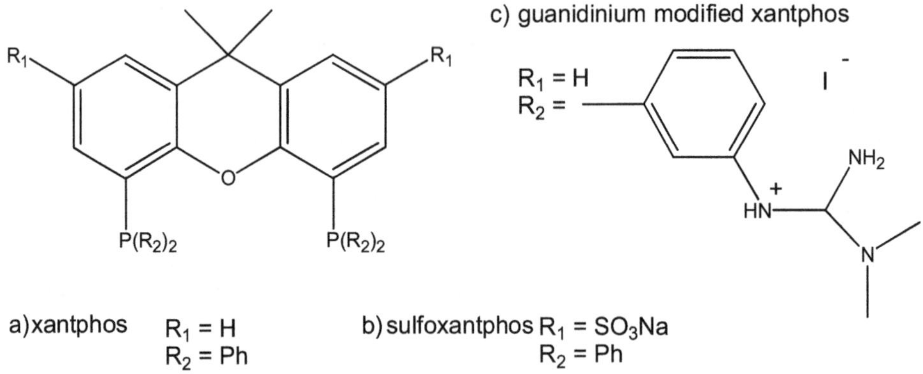

Figure 7.8. Unmodified xantphos (a), sulfoxantphos (b) and guanidinium-modified xantphos ligand (c) as used in the biphasic, Rh-catalysed hydroformylation of 1-octene

The use of the sulfoxantphos ligand (compound (b) in Figure 7.8) in the biphasic hydroformylation of 1-octene with [BMIM][PF_6] has been studied by Dupont and co-workers [58]. The ligand allowed recycling of the catalyst solution up to four times with no loss in activity or selectivity. Highly regioselective hydroformylation (n/iso = 13) was reported for a Rh/phosphine-ratio of 4 (100°C, 15 bar syngas pressure).

The guanidinium modified ligand (compound (c) in Figure 7.8) was synthesised by reacting the corresponding xanthenediphosphine (prepared after [57]) with iodophenylguanidine in a Pd(0)-catalysed coupling reaction (see Scheme 7.5) [46]. The ligand was tested in the Rh-catalysed hydroformylation in consecutive recycling runs both at 10 bar and at 30 bar syngas pressure using [BMIM][PF_6] as the catalyst layer. The recycling was performed by simple decantation of the organic phase after reaction and a replenishment of all feedstock components. Despite this very simple procedure (e.g. no inert gas atmosphere could be guaranteed during the handling) the catalytic performance remained very stable over ten repeated runs. At 30 bar, the overall turnover number reached 3500 mol 1-octene (mol Rh)$^{-1}$ and the n/iso ratio of the product was 19 (95% n-product). At 10 bar, the overall turnover number was 2180 mol 1-octene (mol of Rh)$^{-1}$ and the n/iso ratio of the product was 24 (96% n-product). In agreement with the excellent recylability of the ionic catalyst solution, no Rh could be detected in the product layer by AAS or ICP, indicating a leaching of less then 0.07 %. In all experiments very good selectivities for the linear aldehyde were obtained, thus proving that the attachment of the guanidinium moiety to the xanthene backbone does not influence its known positive effect on the regioselectivity of the reaction. However, it is noteworthy that the catalytic activity increased during the first runs, achieving a stable level only after the forth recycling run. This behaviour has been attributed to a certain catalyst pre-forming in the ionic liquid.

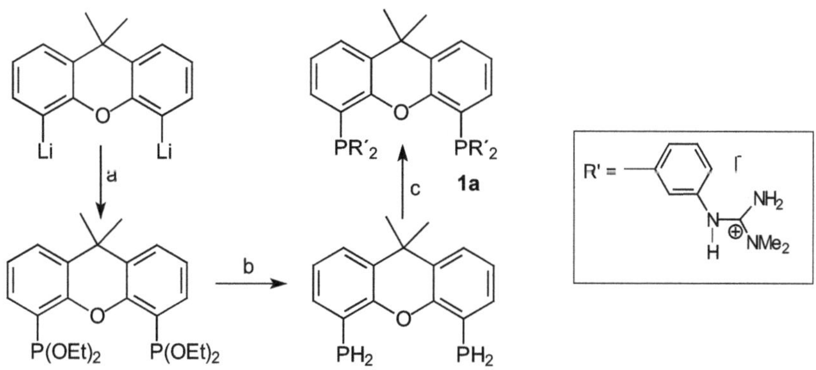

Scheme 7.5. Synthesis of the guanidinium modified xanthene ligand; a) diethylchlorophosphite; b) $LiAlH_4$/chlorotrimethylsilane; c) 4 eq. 3-iodophenylguanidine, 2 mol% Pd_2dba_3 * $CHCl_3$, 80 °C, 24h in DMF

Recently, a novel ionic, phenoxaphosphino-modified xanthene ligand has been prepared by van Leeuwen and co-workers [59]. The ligand, containing two 1-methyl-3-pentylimidazolium hexafluorophosphate moieties, was synthesised via a six-step route from 9,9-dimethylxanthene according to Scheme 7.6 [60].

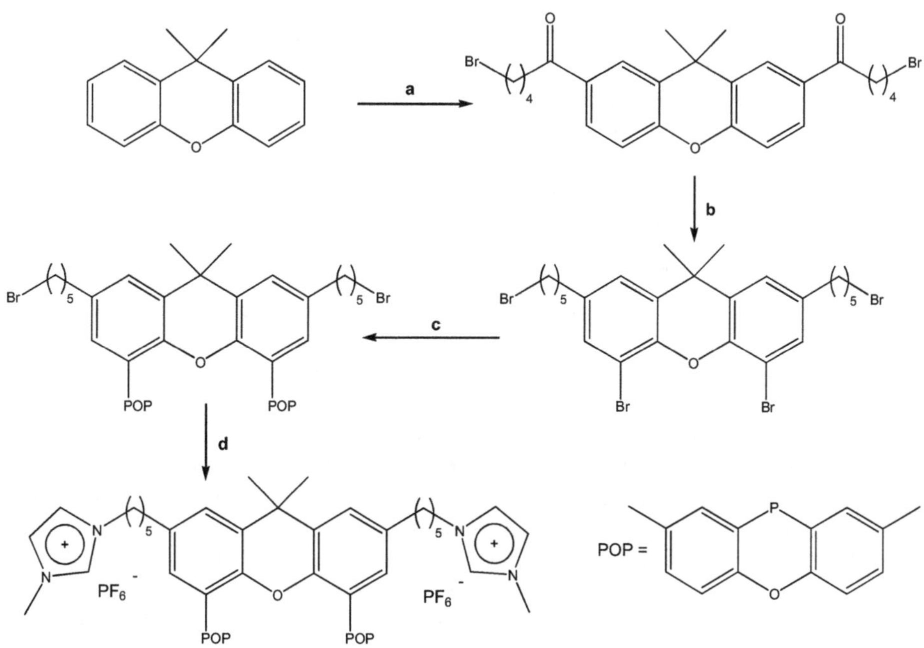

Scheme 7.6. Synthesis of the novel dicationic POP ligand

The ionic catalyst solution was prepared by stirring [$Rh(acac)(CO)_2$] with four equivalents of this ligand in an acetonitrile/[BMIM][PF_6] mixture for 1 h followed by removal of the volatiles. The reaction conditions for the hydroformylation of 1-octene were similar to those used by Dupont et al. [58] (see earlier). The results of seven consecutive recycling experiments are shown in Table 7.3.

TABLE 7.3. Results for the hydroformylation of 1-octene with the Rh-POP catalyst

Cycle	Isomerisation[a)] [%]	TOF[a,b)] [h^{-1}]	n/iso[a)]	Selectivity[a)] [%]
1	11.8	65	44	86.2
2	8.3	88	49	89.8
3	7.8	93	44	90.1
4	7.7	112	44	90.3
5	9.6	107	38	88.1
6[c)]	13.3	318	49	85.0
7[c)]	17.7	305	55	80.8

Reaction conditions: T = 100 °C, p = 17 bar, CO:H_2 = 1:1, ligand/Rh = 4, substrate/Rh = 988; a) percent isomerisation to 2-octene, turnover frequency, n/iso ratio and percent linear aldehyde were determined at ca. 30 % alkene conversion; b) turnover frequency = mol aldehyde per mol rhodium per hour; c) pH_2 = 40 bar and pCO = 6 bar.

Again, the catalytic activity of the system increased within the first four cycles. The authors assume that the conversion of the dimeric species into the active rhodium-hydride species was incomplete at the beginning. The red colour of the rhodium complex in the ionic liquid also suggested the presence of the dimeric complex [61]. An increase in hydrogen partial pressure shifted the equilibrium towards the active hydride species, enhancing the activity to turnover frequencies of higher than 300 h^{-1} (entries 6 and 7). Rh and phosphorous leaching was not detected at all after each recycle and the catalyst proved to be stable in air for more than 14 days without significant loss of activity.

So far, research in the field of biphasic, Rh-catalysed hydroformylation using ionic liquids has been largely dominated by attempts to improve the immobilisation of the phosphine ligands in the ionic medium. Even if the development of such ionic ligands is undoubtedly an important prerequisite for all future research in biphasic hydroformylation using ionic liquids, one should not forget other research activities with slightly different goals.

The group of Olivier-Bourbigou has shown, for example, that phosphite ligands can be used in the Rh-catalysed hydroformylation in ionic liquids as well as the well-known phosphine systems [51]. Since phosphite ligands are usually unstable in aqueous media this adds, apart from the much better solubility of higher olefins in ionic liquids, another important advantage to biphasic hydroformylation using ionic liquids in comparison to the established biphasic reactions in water.

Finally, research efforts to replace hexafluorophosphate (and other halogen-containing) ionic liquids by some cheap and halogen-free ionic liquids in the Rh-catalysed hydroformylation should be mentioned. The first attempts in this direction were made by Andersen et al. [10] These authors investigated the hydroformylation of

1-hexene in high melting phosphonium salts, such as butyltriphenylphosphonium tosylate (mp = 116-117 °C). Obviously, the high melting point of the salts used makes the processing of the reaction difficult, even if the authors describe easy product isolation by pouring the product off the solid catalyst medium at room temperature. Recently, in our group, much lower melting benzenesulfonate, tosylate or octylsulfate ionic liquids have been obtained in combination with imidazolium ions.

These systems have been successfully applied as catalyst media for the biphasic, Rh-catalysed hydroformylation of 1-octene [11], [62]. The catalyst activities obtained with these systems were in all cases equal or even higher than those with the commonly used [BMIM][PF_6]. Table 7.4 compares some typical results using Rh-sulfoxantphos in alkylsulfate and sulfonate ionic liquids with analogous experiments in a hexafluorophosphate system.

TABLE 7.4. Biphasic hydroformylation using halogen-free octylsulfate and tosylate ionic liquids compared with the same reaction in [BMIM][PF_6]

Ionic liquid	TON[a)]	Hydr./Isom. (%)	n/iso
[BMIM] [PF_6]	169	3.0	32.5
[BMIM] [Octylsulfate]	646	9.9	49.5
[BEIM] [Tosylate]	541	8.4	32.8

Reaction conditions: 2 h reaction time, T = 120 °C, p = 11 bar, 5 ml IL, 1-Octene:Rh = 1000, Ligand:Rh = 2.5 ; ligand = sulfoxantphos, a) turnover number = mol aldehyde per mol rhodium

The results clearly prove that higher activity was obtained with the halogen-free ionic liquids. This may be attributed to their higher solubilising power for the feedstock 1-octene. The higher hydrogenation/isomerisation activity of the halogen-free system was related to the higher conversion that was obtained under standard condition with the more active systems. Moreover, the regioselectivity was found to be at least comparable to the halogen-free systems.

Taking into account the much lower costs of the ionic medium, the better stability against hydrolysis and the wider disposal options related to, for example, an octylsulfate or a tosylate ionic liquid in comparison to [BMIM][PF_6], there is no real reason to centre future hydroformylation research around hexafluorophosphate ionic liquids.

Another interesting recent development is the continuous, Rh-catalysed hydroformylation of 1-octene in the unconventional biphasic system [BMIM][PF_6]/$scCO_2$ that has been developed by Cole-Hamilton and co-workers [63].

The concept makes use of the complimentary strengths and weaknesses of the two unconventional media. While ionic liquids are known to be excellent solvents for many transition metal catalysts, the solubility of most transition metal complexes in $scCO_2$ is poor (if not modified with e. g. phosphine ligands with fluorous "ponytails" [64]). However, product isolation from $scCO_2$ is always very simple, while from an ionic catalyst solution it may become more and more complicated depending on the solubility of the product in the ionic liquid and on the product's boiling point.

Beckman, Brennecke and their research groups first realised that $scCO_2$ and ionic liquid exhibit an interesting mixing behaviour that can be used to design continuous catalytic processes. They observed that, although $scCO_2$ is surprisingly soluble in some ionic liquids, the reverse is not the case, with no detectable ionic liquid solubilisation in the CO_2 phase [65].

In addition to Rh-catalysed hydroformylation, this special phase behaviour has been successfully applied to other continuous catalytic reactions – such as Ni-catalysed, enantioselective hydrovinylation [66] and the lipase-catalysed kinetic resolution and enantiomer separation of chiral alcohols [67].

A basic scheme of the apparatus that has been used by Cole-Hamilton et al. for the continuous Rh-catalysed hydroformylation in the system $[BMIM][PF_6]/scCO_2$ is shown in Figure 7.9.

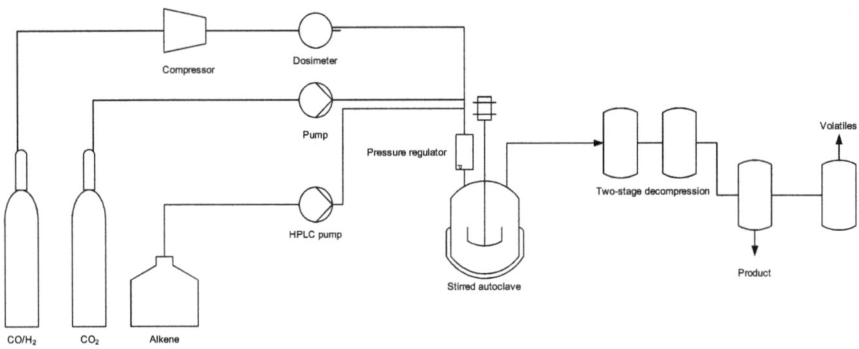

Figure 7.9. Basic scheme of the apparatus used for biphasic Rh-catalysed hydroformylation in an $IL/scCO_2$ mixture

During a 33 h continuous hydroformylation run using this set-up, no catalyst decomposition was observed and Rh leaching into the $scCO_2$/product stream was less than 1 ppm. The selectivity for the linear nonanal was found to be stable over the reaction time with n/iso = 3.1. During the continuous reaction, alkene, CO, H_2 and CO_2 were separately fed into the reactor containing the ionic liquid catalyst solution. Products and unconverted feedstock dissolved in $scCO_2$ were removed from the ionic liquid. After decompression the liquid product was collected and analysed.

In contrast to the reaction in repetitive batch mode, the continuous operation avoids any contact with air which is often unavoidable during the phase separation and refilling procedures. In the repetitive batch mode, Cole-Hamilton et al. observed a continuous drop of the product's n/iso ratio from 3.7 to 2.5, an increased isomerisation activity of the system and significant Rh leaching after the ninth run. The authors concluded from ^{31}P NMR experiments that ligand oxidation occurred. This resulted in the formation of $[RhH(CO)_4]$ which is known to show more isomerisation activity and a lower n/iso ratio than the modified catalyst system. Unmodified $[RhH(CO)_4]$ is also known to display some solubility in $scCO_2$, which explains the observed leaching of Rh into the organic layer.

All these problems could be convincingly overcome by application of the continuous operation mode [63 d)]. Most interestingly, this unusual continuous biphasic reaction mode enabled the quantitative separation of relatively high boiling products from the ionic catalyst solution under mild temperature conditions and without use of an additional organic extraction solvent. More details of this process are included in Chapter 8, Section 8.2.2.3.

7.3 Rhodium Catalysed Hydroformylation Using Supported Ionic Liquid Phase SILP) Catalysis

The term "Supported Ionic Liquid Phase (SILP) catalysis" has recently been introduced into the literature to describe the heterogenisation of a homogeneous catalyst system by confining an ionic liquid solution of catalytically active complexes on a solid support [68]. In comparison to the conventional liquid-liquid biphasic catalysis in organic-ionic liquid mixtures, the concept of SILP-catalysis offers very efficient use of the ionic liquid. Figure 7.10 exemplifies the concept for the Rh-catalysed hydroformylation.

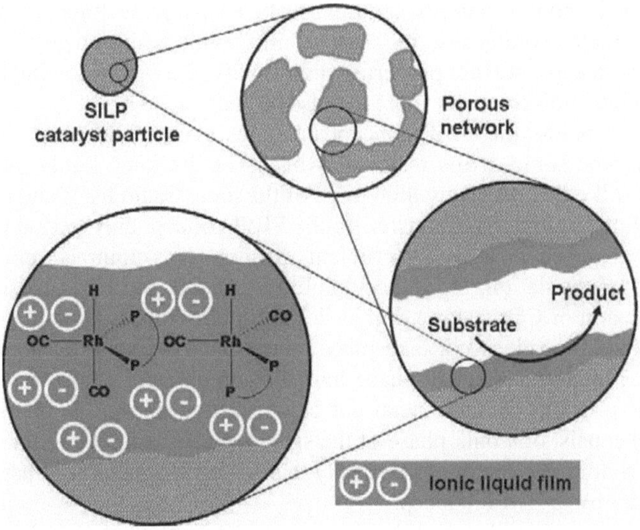

Figure 7.10. Supported ionic liquid phase (SILP) catalysis exemplified for the Rh-catalysed hydroformylation reaction

The basic idea of immobilising a catalytically active solution on a solid support to obtain a solid catalyst with homogeneous type selectivity and efficiency is not new. In the late seventies, Scholten [69] and Hjortkjær [70] described supported liquid phase

(SLP) systems where a high boiling organic solvent was used as the liquid film. Later, supported aqueous phase (SAP) catalysts where intensively studied [71].

To our knowledge, none of the developed SLP and SAP catalysts made their way into a technical process. Obviously, the possibility of using a supported liquid catalyst in a continuous liquid phase reaction is generally very restricted. The reason is that a very low solubility of the liquid in the feedstock/product mixture is enough to remove the catalyst from the surface over time (due to the very small amounts of liquid on the support). Even worse, the immobilised liquid film can be removed from the support physically by the mechanical forces of the continuous flow even in the case of complete immiscibility.

Therefore, the most promising area of application for a supported liquid catalyst is a continuous gas phase process. In this context, the SILP concept offers very important advantages that make a reinvestigation of supported liquid catalysis with these unique liquids highly promising.

Among these advantages, the non-volatile character of the ionic liquid is, of course, of great importance. While SLP and SAP concepts were known to suffer from the fact that the liquid immobilisation phase is more or less rapidly removed from the carrier by evaporation (leading to short catalyst lifetimes) evaporation of the liquid support is negligible in case of an ionic liquid adsorbed on the surface. Moreover, the ionic liquid adds all its advantages to the concept depending on the cation/anion combination chosen, such as its good wetability and high surface tension, its high solubilising power for transition metal complexes, its compatibility with hydrolytically labile catalyst complexes, its tuneable solvent properties and, finally, the option for catalyst activation by liquid support interaction (for more details see chapter 7.1).

Besides these points the SILP concept becomes even more interesting because it helps to overcome some of the major drawbacks of the ionic liquid methodology in catalysis that still exist: relatively high price of the ionic liquid itself and relatively high viscosity leading to slow mass transfer. In the SILP concept only very thin films of the ionic liquids are used in a highly efficient manner. This point is valid for the Rh-catalysed hydroformylation, too. Since solubility for CO and H_2 in the ionic liquid is low (and diffusion of CO is slow), liquid-liquid biphasic catalysis with a highly active ionic catalyst solution does not take place in the bulk of the ionic liquid, but in the diffusion layer between the gas phase and the ionic liquid. This results in a large amount of the precious catalyst metal not being utilised during the reaction. An ideal system would consist of a bulk phase of the size of the diffusion layer thus allowing for the complete rhodium inventory to be catalytically active; a condition that is fulfilled in an ideal manner by the SILP concept.

Two ways of ionic liquid immobilisation have been applied so far: a) the approach in which part of the ionic liquid phase is fixed to a support by chemical bonds between either the cation or the anion of the ionic liquid and the support and b) the impregnation approach in which physisorption fixes the ionic liquid onto the support.

7.3.1 SUPPORTED IONIC LIQUIDS BY CHEMICAL BONDS

The first example of SILP-catalysis was the fixation of an acidic chloroaluminate ionic liquid on an inorganic support. The acidic anions of the ionic liquid, $[Al_2Cl_7]^-$ and $[Al_3Cl_{10}]^-$, react with free OH-groups of the surface to create an anionic solid surface with the ionic liquid cations attached [72]. The catalyst obtained was applied in the Friedel-Crafts acylation of aromatic compounds. Later, the immobilisation of acidic ionic liquids by covalent bonding of the ionic liquid cation to the surface was developed and applied again in Friedel-Crafts chemistry [73].

In 2002 Mehnert and co-workers were the first to apply SILP-catalysis to Rh-catalysed hydroformylation [74]. They described in detail the preparation of a surface modified silica gel with a covalently anchored ionic liquid fragment (Scheme 7.7). The complex N-3-(3-triethoxysilylpropyl)-4,5-dihydroimidazole was reacted with 1-chlorobutane to give the complex 1-butyl-3-(3-triethoxysilylpropyl)- 4,5-dihydroimidazolium chloride. The latter was further treated with either sodium tetrafluoroborate or sodium hexafluorophosphate in acetonitrile to introduce the desired anion. In the immobilisation step, pre-treated silica gel was refluxed with a chloroform solution of the functionalised ionic liquid to undergo a condensation reaction giving the modified support material. Treatment of the obtained monolayer of ionic liquid with additional ionic liquid resulted in a multiple layer of free ionic liquid on the support.

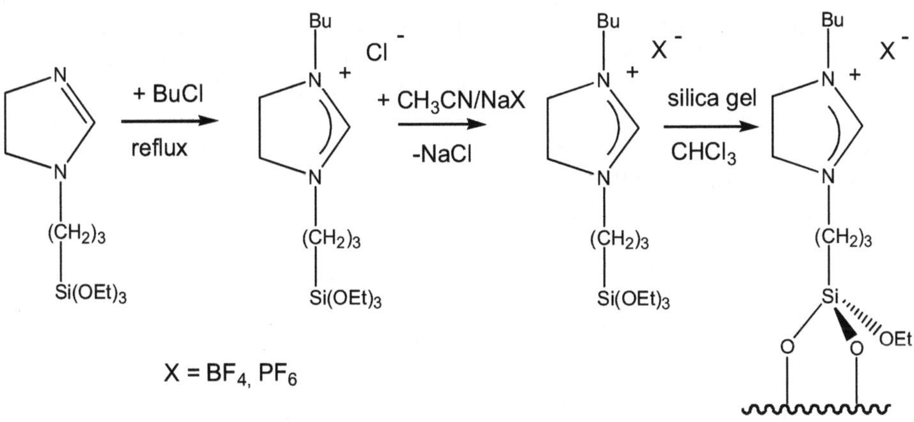

Scheme 7.7. Preparation of surface anchored ionic liquid phases

Analysis of the surface coverage revealed an average of 0.4 ionic liquid fragments nm^{-2}, corresponding to the involvement of approximately 35% of all the hydroxyl groups of the pre-treated silica gel.

To introduce the Rh-centre in the supported ionic liquid, a solution of $[Rh(CO)_2(acac)]$ in acetonitrile was treated with either the ligand tri(m-sulfonyl)triphenyl phosphine trisodium salt (TPPTS) or the ligand tri(m-sulfonyl)triphenyl phosphine tris(1- butyl-3-methyl-imidazolium) salt (TPPTI) (Rh/P ratio of 1:10). The ligand TPPTI was found to dissolve in $[BMIM][BF_4]$ and

[BMIM][PF_6], while the corresponding sodium salt only dissolved in the [BMIM][BF_4] The resulting acetonitrile solutions were combined with the ionic liquid phase (25 wt % loading) and added to the support. After the solvent was removed under reduced pressure, a slightly yellow-coloured powder was obtained.

The obtained SILP-catalyst was tested for the hydroformylation of 1-hexene in a batch-wise, liquid phase reaction. All results obtained where compared to the identical reaction in a liquid-liquid biphasic reaction. Mehnert et al. found that the SILP-catalyst exhibited a higher TOF (63 min^{-1}) compared with the biphasic system (23 min^{-1}) due to the higher concentration of Rh at the interface and the generally larger interface of the supported system. However, at high aldehyde concentrations the ionic liquid [BMIM][BF_4] was found partially to dissolve in the organic phase, thus leading to a rhodium loss of up to 2.1 mol%. Rhodium leaching could be suppressed at lower aldehyde concentrations or by using a higher ligand excess.

7.3.2 SUPPORTED IONIC LIQUIDS BY IMPREGNATION

Recently, the SILP concept has been applied by Riisager, Fehrmann and us to the continuous, Rh-catalysed hydroformylation of propene in a gas-phase reaction [68]. The supported liquid phase catalyst systems were prepared by impregnation. [Rh(acac)$(CO)_2$] and the ligand under investigation were dissolved in dry methanol together with either [BMIM][PF_6] or [BMIM][n-$C_8H_{17}OSO_3$]. Appropriate amounts of silica gel 100 (BET surface area 297.5 m^2, pore volume 1.015 cm^3, mean pore diameter 137 A) were added, resulting in a Rh loading of 0.2 wt%. Three different ligands (guanidinium modified phenylphosphine (Rh-guanidinium), NORBOS-Cs_3 (Rh-norbos [75]) and sulfonated xantphos (Rh-sulfoxantphos)) were tested for the gas-phase propene hydroformylation in a continuous fixed-bed reactor as shown in Figure 7.11 (for detailed description of the set-up see [76]).

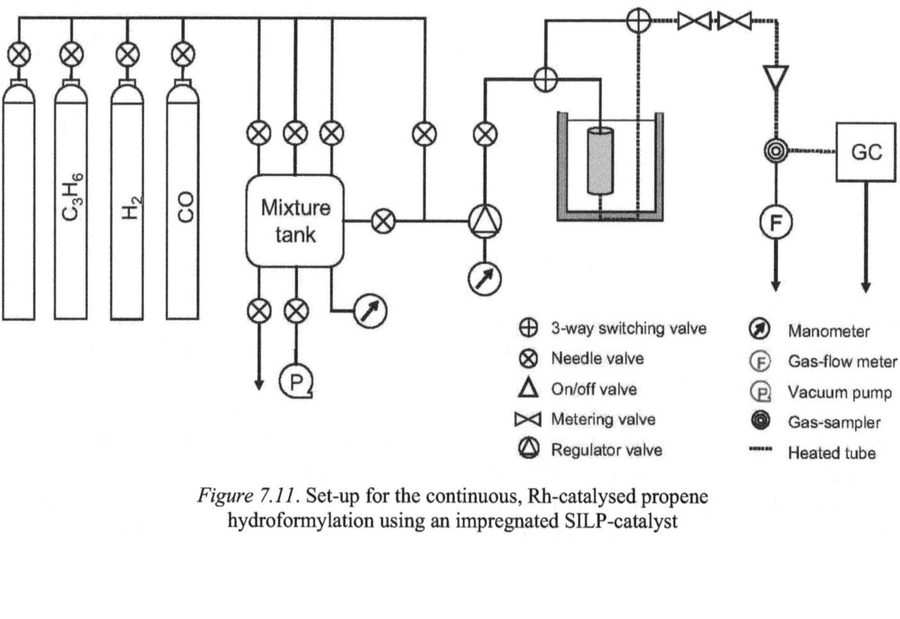

Figure 7.11. Set-up for the continuous, Rh-catalysed propene hydroformylation using an impregnated SILP-catalyst

TABLE 7.5. Continuous, Rh-catalysed hydroformylation of propene using impregnated SILP-catalysts with different ligands – results at 300 min time on stream

No.	ligand [a]	L/Rh	Liquid loading Wt%	∟ [b]	TOF [c] $[h^{-1}]$	n/iso [d]	Linearity [%]
1	guanidinium	2.9	-	0.00	55.5	0.9	47.4
2	guanidinium	2.9	108.4	0.78	20.6	0.9	47.4
3	guanidinium	2.9	138.9	1.00	16.8	1.0	50.0
4	norbos-Cs	11.3	6.9	0.05	88.4	2.0	66.7
5	norbos-Cs	11.3	20.8	0.15	79.4	1.3	56.5
6	norbos-Cs	21.4	-	0.00	45.8	2.8	73.8
7	norbos-Cs	21.3	6.9	0.05	28.2	2.6	72.2
8	sulfoxantphos	2.5	-	0.00	37.4	1.7	63.0
9	sulfoxantphos	2.4	23.6	0.17	1.5	1.8	64.0
10	sulfoxantphos	2.5	68.1	0.49	5.1	2.0	66.4
11	sulfoxantphos	10.2	-	0.00	40.8	16.9	94.4
12	sulfoxantphos	10.0	25.0	0.18	34.9	22.6	95.8
13	sulfoxantphos	10.0	72.2	0.52	25.4	22.0	95.8
14	sulfoxantphos	20.0	27.8	0.20	16.7	23.7	96.0
15	[e]	-	105.6	0.76	-	-	-

Reaction conditions: alkene:CO:H_2 = 1:1:1, T = 100 °C, 0.2 wt% Rh metal loading, silica 100; p = 10 bar for guanidinium and norbos-Cs, p = 5 bar for sulfoxantphos; ionic liquid used for impregnation: [BMIM][PF_6]; a) molar ligand to metal ratio; b) ratio of ionic liquid volume to support pore volume; c) mol aldehyde per mol rhodium per hour; d) ratio between linear and branched aldehyde; e) support loaded with ionic liquid only.

The reaction was not influenced by the type of ionic liquid, since no significant differences were observed when the reaction was carried out in [BMIM][PF_6] or [BMIM][n-$C_8H_{17}OSO_3$]. In Table 7.5 the most relevant results for the hydroformylation of propene with the different rhodium catalyst systems in the ionic liquid [BMIM][PF_6] are compiled.

Turnover frequencies of up to 56 h^{-1} were obtained for Rh-guanidinium catalysts while the linearities were poor at 50 % linear aldehyde being formed. The pore filling degree ∟ was found to influence the activity of the catalyst significantly (entry 1 to 3), indicating that the ionic liquid induced severe mass transfer limitations. The performance of the Rh-norbos system was studied at higher L/Rh ratios and lower ∟ values. TOFs up to 88 h^{-1} were obtained when using an ionic liquid loading of 0.05 and a L/Rh ratio of 11 (entry 4 and 5). Further increase of the L/Rh ratio to 21 resulted in linearities around 73 % (entry 6 and 7). Using the bidentate phosphine ligand sulfoxantphos, n/iso ratios of 23 and linearities up to 96 % were obtained (entries 12 to 14), clearly indicating that in SILP hydroformylation the general principles of homogeneous catalysis hold true, where chelating ligands also were found to be more selective towards the linear products. At low L/Rh ratios around 2, the selectivity of the Rh-sulfoxantphos system was significantly lower and comparable to the best selectivity of the monodentate phosphines. In comparison to experiments in liquid-liquid biphasic mode, this result indicates that there might be some reaction between the ligand and the solid surface that causes some ligand loss. Similar to the monophosphine system the absence of ionic liquid resulted in higher activity (entry 11).

From these short time investigations (300 min time on stream) the question may arise whether the ionic liquid can add anything positive to the reaction system. The adsorbed catalyst without ionic liquid appears to be active and stable. A convincing answer to this question was found in 60 h runs using the conditions of entry 11 (without ionic liquid) and entry 13 (SILP-catalyst) [68, 77]. Without the ionic liquid film the activity and selectivity started to decrease sharply after 10 h on stream (reaching TOF = 14 h^{-1} vs. 41 h^{-1} after 300 min and n/iso = 6 vs. n/iso = 17 after 300 min), whereas the SILP system reached its maximum activity only after 30 h (TOF = 45 h^{-1} vs. 25 h^{-1}) and maintained this level up to 60 h. This indicates that the negative ionic liquid effect on the catalyst activity as observed in the 300 min runs is in fact based only on a delayed pre-formation of the catalyst. Moreover, the SILP system showed very stable selectivity with n/iso still being as high as 21 after 60 h.

The concept of continuous flow SILP hydroformylation was also tested in the biphasic, liquid-liquid hydroformylation of 1-octene using the Rh-norbos catalyst system [78]. TOFs of 44 h^{-1} were achieved after 3-4 h with no sign of deactivation at prolonged reaction times. At steady-state conditions an n/iso ratio of 2.6 was obtained. No leaching of rhodium metal could be detected by ICP-AES analysis of product samples at least after these short reaction times.

The few examples where SILP catalysis has been tested so far showed highly encouraging results. It is very likely that other applications where ionic catalyst solutions were tested in liquid-liquid biphasic reactions could be reinvestigated under SILP conditions. If very high catalyst stability over time can be realised or simple catalyst regeneration protocols can be developed than SILP catalysis can be expected to make its way into industrial processes.

7.4 Costs and Economics

What can drive the switch from existing homogeneous catalytic processes to novel ionic liquids technology? One major point is surely if a higher cost-effectiveness can be obtained in some cases. This can be attributed to an improved reaction rate and selectivity, associated with more efficient catalyst recovery and better environmental compatibility.

The cost of the ionic liquid is still a limiting factor. However, the commercial availability of these liquids has improved considerably over the last few years and prices have already gone down significantly [20]. This development can be expected to continue as ionic liquids continue to make their transition “from curiosities to commodities” [79]. In any case the cost of the ionic liquid has to be weighed against that of current chemicals or catalysts. If the ionic liquid can be recycled and if its lifetime proved to be long enough, then its initial price is probably not the critical point.

To estimate costs for the liquid-liquid biphasic hydroformylation using ionic liquids, a process was designed for the production of 100,000 tons per year of nonanal. The use of ionic liquids in hydroformylation catalysis is a fairly new technology and exact kinetic data are scarce, thus the TOFs reported for the Rh-sulfoxantphos system [80] have been used to determine catalyst inventory and reactor dimensions. In a similar way the plant design for the SILP process for a production capacity of 100,000 tons per year of butanal has been derived based on preliminary literature results [68]. The process flow sheets for both process variations are shown in Figures 7.12 and 7.13.

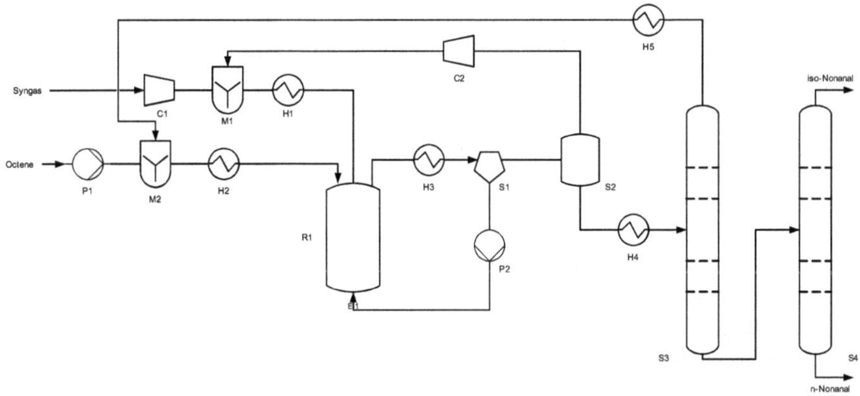

Figure 7.12. Process flow sheet for the production of nonanal using a biphasic ionic liquid system

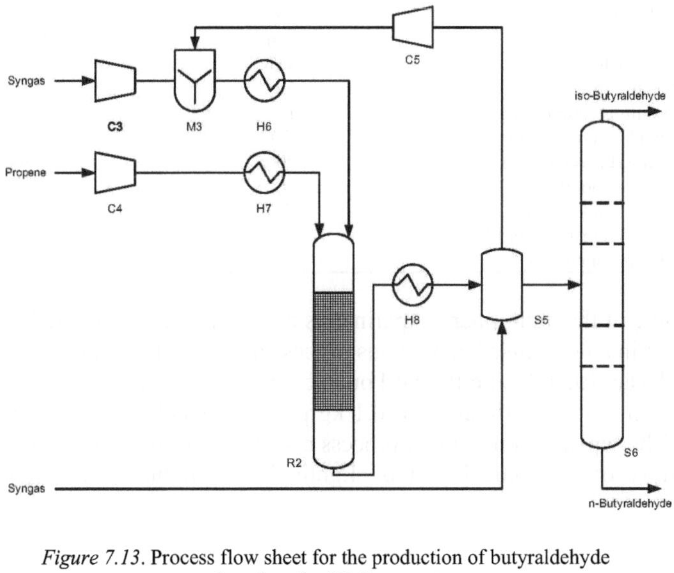

Figure 7.13. Process flow sheet for the production of butyraldehyde based on the SILP concept

In the biphasic system utilising ionic liquids, syngas is compressed to 50 bar in device C1, heated to 100 °C in the preheater H1 and enters reactor R1. Substrate octene is pumped via P1 and heated to 100 °C before entering the continuously stirred tank reactor R1. Reactor dimensions are given in Table 7.6. A constant product stream is taken from the reactor and is cooled under reaction pressure to 40 °C in H3. The ionic liquid is separated from the organic product phase containing unreacted octene in a separation unit S1. In a flash column S2 the syngas is removed from the product phase, compressed to 50 bar and recycled via mixer M1. The raw product is heated to 60 °C and fed into the distillation column S3 in which the octene is removed over top, reheated to 100 °C and fed into the mixer M2 for reuse. In the distillation column S4 the crude product is separated into the two major products, the linear and the branched aldehyde with the latter one being removed over top.

The flow sheet for the SILP process is simplified due to the fixation of the catalyst on a heterogeneous support that allows for fixed bed technology. Since the advantages of the SILP technique are best when using gaseous substrates, propene was used as feedstock. Syngas and propene are compressed to 50 bar in devices C3 and C4, respectively and heated to 100 °C in H6 and H7, before entering the reactor R2. The product stream is cooled to 30 °C in H8 and unreacted propene is removed by stripping with fresh syngas in S5. The raw product is separated into the major product n-butyraldehyde and the by-product iso-butyraldehyde in the distillation column S6. Further process characteristics are summarised in Table 7.6.

TABLE 7.6. Process characteristics for optimised nonanal production (using liquid-liquid biphasic catalysis with ionic liquids) and butanal production (using SILP catalysis) on a 100.000 tons/year scale

	Biphasic	SILP
Temperature / °C	100	100
Pressure / bar	50	50
Alkene	Octene	Propene
Reactor type	CSTR	Fixed bed
Reactor volume / m^3	60	25
Rhodium inventory / kg	15	60
Ligand inventory / kg	450	1800
Ionic liquid inventory / kg	5200	7000
Silica support / kg	-	10000
Syngas to alkene ratio	2	2
Selectivity to linear product / %	>95	>95
By-products / %	< 1	< 1

Since in some of the preliminary experiments no rhodium losses could be detected, it is assumed that in an optimised continuous process the metal leaching will be in the range of the biphasic Ruhrchemie/Rhône-Poulenc process operating around 1 ppb. This would result in a loss of rhodium of 0.1 kg per year which is approximately 0.7 % in the case of the biphasic ionic liquid process and 0.2 % in the case of the SILP process. The inventory of the ionic liquid is slightly lower in the case of the liquid-liquid biphasic reaction than in the case of the SILP catalysis (on the base of a 70% ionic liquid loading on the support). Assuming a bulk price for imidazolium octylsulfate ionic liquid of 50 €/kg the capital investment in ionic liquid for the liquid-liquid biphasic process would be about 260.,000 €. An operational annual loss of 1% of the ionic liquid would increase the costs per ton of product by 0.026 € (replacement of purc ionic liquid).

The ionic liquid investment could be further reduced if future research enables the application of ammonium based alkylsulfate or arylsulfonate ionic liquids. For these systems bulk prices around 15 €/kg are expected. Ammonium based alkylsulfate or arylsulfonate ionic liquids usually show melting points slightly above room temperature but clearly below the operating temperature of the hydroformylation reaction. Therefore these systems may be less suitable for the liquid-liquid biphasic process in which the ionic liquid may be involved in process steps at ambient temperature (e.g. phase separation or liquid storage). In contrast, for the SILP catalyst a room temperature ionic liquid is not necessarily required as long as the film becomes a liquid under the reaction conditions. Assuming an ammonium based SILP catalyst, the capital investment for the ionic liquid for the industrial SILP catalyst would add up to 105,000 €.

On the base of a Rh-price of about 20,000 €/kg and a ligand price of about 1000 €/kg it becomes quite obvious that the loss of the ionic liquid would only be a minor part of the overall cost arising from the case of complete SILP-catalyst deactivation. It should be noted that a deactivated SILP catalyst may still offer some options for regeneration (e.g. extraction with $scCO_2$ to remove heavies). However, these options are not yet developed and their efficiency is unclear at this point.

7.5 Conclusions

Ionic liquids offer a highly attractive alternative for the immobilisation of homogeneous catalysts. Their non volatile nature, tuneable solubilising power and coordination/acidity properties as well as their large liquid ranges make them suitable media for both liquid-liquid biphasic catalysis and supported ionic liquid phase (SILP) catalysis. In all these application the ionic liquid can play its role as either "innocent" solvent, ligand precursor, co-catalyst or the catalyst itself depending on the specific cation/anion combination and the reaction under investigation.

For Rh-catalysed hydroformylation the role of the ionic liquid as an innocent solvent is by far the most important. To our knowledge, none of the published research in this area claims special chemistry. The selectivity found with the different Rh-ligand complexes corresponds in most cases to the values obtained in traditional organic solvent or water (with the surprisingly low selectivity of TPPTS ligands in ionic liquids being a remarkable exception). Overall activities were found to be very comparable if mass transfer effects between the gas phase and the two immiscible liquid phases were overcome by proper stirring.

Recently, van Leeuwen and co-workers provided support for the similarity between the active catalytic species in ionic liquids and in organic solvents by spectroscopic investigations [80]. These authors compared the complex formation of $[Rh(acac)(CO)_2]$ in the presence of 4 equivalents of sulfoxantphos ligand dissolved in $[BMIM][PF_6]$ and organic solvents under different pressures and syngas compositions using high-pressure IR and NMR techniques. A dynamic equilibrium between the complex isomers was found (see Scheme 7.8).

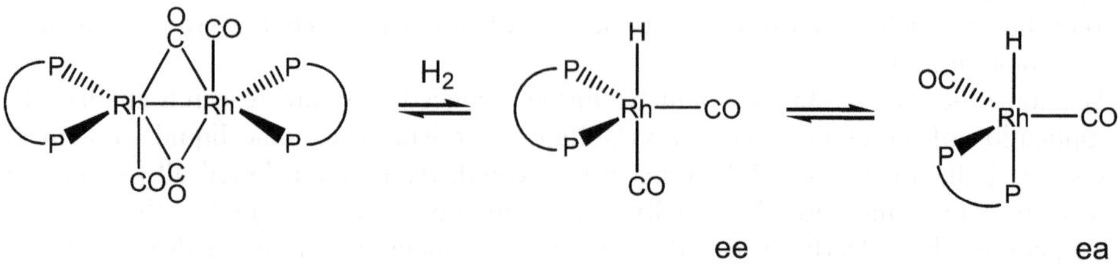

Scheme 7.8. Equilibria between different rhodium species present during hydroformylation in an ionic liquid

The absorbances of the different species were almost identical to those observed when xantphos was used as ligand in organic solvents. The syngas pressure was found to have a direct influence on the equilibrium. The ea isomer giving branched aldehydes, was formed to a larger extent when the syngas pressure was increased thus leading to a lower selectivity. The equilibrium between the two isomers was not affected by other Rh-carbonyl complexes present in the reaction media. All observations in ionic liquids were analogous to homogeneous systems using the same catalyst [81].

Nevertheless, the application of ionic liquids in the liquid-liquid, biphasic Rh-catalysed hydroformylation offers technically interesting advantages vs. the traditional aqueous biphasic catalysis e.g. much higher solubility for longer chain olefins and the compatibility of the ionic liquid with phosphite ligands [51].

A summary of the research activities of the last four years reveals three different important trends: (a) The design of new ionic ligands for excellent catalyst immobilisation in ionic liquids and high regioselectivity; (b) the successful application of cheap, halogen-free ionic liquids in the biphasic Rh-catalysed hydroformylation; (c) the successful development of unusual multiphasic reaction concepts for Rh-catalysed hydroformylation, namely catalysis in ionic liquid/supercritical CO_2 and SILP-catalysts.

Of course, there is still a large amount of research to be done to develop further the very preliminary character of the above described economic evaluation of an ionic liquid hydroformylation process. Only on the basis of more detailed data it will be possible to decide whether we will see an industrial hydroformylation plant using ionic liquids in the future.

7.6 References

[1] a) F.H. Hurley, *U.S. Patent* 1948, 2,446,331, [*Chem. Abstr.* 43, 1949, P7645b]; b) F.H. Hurley, T.P. Wier Jr., *J. Electrochem.* Soc. 98, 1951, 207-212.

[2] a) H.L. Chum, V.R. Koch, L.L. Miller, R.A. Osteryoung, *J. Am. Chem. Soc.* 97, 1975, 3264-3265; b) J. Robinson, R.A. Osteryoung, *J. Am. Chem. Soc.* 101, 1979, 323-327.

[3] J.S. Wilkes, J.A. Levisky, R.A. Wilson, C.L. Hussey, *Inorg. Chem.* 21, 1982, 1263-1264.

[4] a) T.M. Laher, C.L. Hussey, *Inorg. Chem.* 22, 1983, 3247-3251; b) T.B. Scheffler, C.L. Hussey, *Inorg. Chem.* 23, 1984, 1926-1932; c) D. Appleby, C.L. Hussey, K.R. Seddon, J.E. Turp, *Nature* 323, 1986, 614-616.

[5] P.B. Hitchcock, T.J. Mohammed, K.R. Seddon, J.A. Zora, C.L. Hussey, E.H. Ward, *Inorg. Chim. Acta* 113, 1986, L25-L26.

[6] J.S. Wilkes, M.J.J. Zaworotko, *J. Chem. Soc. Chem. Comm.* 1992, 965 967.

[7] J. Fuller, R.T. Carlin, H.C. de Long, D.J. Haworth, *J. Chem. Soc. Chem. Comm.* 1994, 299-300.

[8] P. Bonhôte, A.-P. Dias, N. Papageorgiou, K. Kalyanasundaram, M. Grätzel, *Inorg. Chem.* 1996, 35, 1168-1178.

[9] P. Wasserscheid, A. Bösmann, *unpublished results*.

[10] N. Karodia, S. Guise, C. Newlands, J.-A. Andersen, *Chem. Comm.* 1998, 2341-2342.

[11] P. Wasserscheid, R. van Hal, A. Bösmann, *Green Chem.* 4, 2002, 400-404.

[12] a) J. Pernak, A. Czepukowicz, R. Pozniak, *Ind. Eng. Chem. Res.* 40, 2001, 2379-2383; b) B. Jastorff, R. Stoermann, J. Ranke, K. Moelter, F. Stock, B. Oberheitmann, W. Hoffmann, J. Hoffmann, M. Nüchter, B. Ondruschka, J. Filser, *Green Chem.* 5, 2003, 136-142.

[13] H. Waffenschmidt, PhD-thesis, RWTH Aachen, 2000.

[14] a) P. Wasserscheid, M. Sesing, W. Korth, *Green Chem.* 4, 2002, 134-138; b) W. Keim, W. Korth, P. Wasserscheid, *World Patent* 2000 0016902 (to BP Chemicals Limited, UK; Akzo Nobel NV; Elementis UK Limited) 2000 [*Chem. Abstr.* 132, 2000, 238691].

[15] a) A.E. Visser, R.P. Swatlowski, M.W. Reichert, J.H. Davis, R.D. Rogers, R. Mayton, S. Sheff, A. Wiezbicki, *Chem. Comm.* 2001, 135-136; b) A.E. Visser, R.P. Swatloski, W.M. Reichert, R. Mayton, S. Sheff, A. Wierzbicki, J.H. Davis Jr., R.D. Rogers, *Env. Sci. Techn.* 36, 2002, 2523-2529; c) S. Anjaiah, S. Chandrasekhar, R. Gree, *Tetrahedron Lett.* 45, 2004, 569-571.

[16] P. Wasserscheid, B. Drießen-Hölscher, R. van Hal, C.H. Steffens, J. Zimmermann *Chem. Comm.* 2003, 2038-2039.

[17] T.L. Merrigan, E.D. Bates, S.C. Dorman, J.H. Davis Jr., *Chem. Comm.* 2000, 2051-2052.

[18] C. Hardacre, J.D. Holbrey, S.E.J. McMath, *Chem. Comm.* 2001, 367-368.

[19] a) P. Wasserscheid, A. Bösmann, C. Bolm, *Chem. Comm.* 2002, 200-2001; b) M.J. Earle, P.B. McCormac, K.R. Seddon, *Green Chem.* 1, 1999, 23-25.

[20] a) Solvent Innovation (www.solvent-innovation.com); b) Fluka (www.fluka.com); c) Merck (www.merck.de); d) Acros Organics (www.acros.com); e) Wako (www.wako-chem.co.jp).

[21] e.g. Solvent Innovation offers its ionic liquids routinely in 5 kg batches but some representatives are even commercialised on a ton-scale (for more details see www.solvent-innovation.com)

[22] K.R. Seddon, A. Stark, M.J. Torres, *Pure Appl. Chem.* 72, 2000, 2275-2287.

[23] Among the potential impurities in ionic liquids water, halide ions and organic starting material are of great importance for transition metal chemistry while the colour of an ionic liquid is not a critical parameter in most applications.

[24] a) J.S. Wilkes, J.A. Levisky, R.A. Wilson, C.L. Hussey, Inorg. Chem. 21, 1982, 1263-1264; b) A.A. Fannin Jr., D.A. Floreani, L.A. King, J.S. Landers, B.J. Piersma, D.J. Stech, R.J. Vaughn, J.S. Wilkes, J.L. Williams, *J. Phys. Chem.* 88, 1984, 2614-2621.

[25] J. Fuller, R.T. Carlin, R.A. Osteryoung, *J. Electrochem. Soc.*, 144, 1997, 3881-3886.

[26] A. Noda, K. Hayamizu, M. Watanabe, *J. Phys. Chem.* B, 105, 2001, 4603-4610.

[27] a) melting point: J.D. Holbrey, R.D. Rogers, *Melting Points and Phase Diagrams* in P. Wasserscheid, T. Welton (Eds.) *Ionic Liquids in Synthesis*, Wiley-VCH, Weinheim 2003, pp. 41-55; b) viscosity: S.N. Baker, G.A. Baker, G.A. Kane, F.V. Bright, *J. Phys. Chem.* B, 105, 2001, 9663-9668; c) density: S. Chun, S.V. Dzyuba, R.A. Bartsch, *Anal. Chem.* 73, 2001, 3737-3741.

[28] P. Wasserscheid, T. Welton (Eds.) *Ionic Liquids in Synthesis*, Wiley-VCH, Weinheim, 2002.

[29] G.W. Parshall, *J. Am. Chem. Soc.* 94, 1972, 8716-8719.

[30] J.F. Knifton, *J. Mol. Catal.* 43, 1987, 65-78.

[31] Y. Chauvin, B. Gilbert, I. Guibard, *J. Chem. Soc. Chem. Comm.* 1990, 1715-1716.

[32] (a) P. Wasserscheid *Transition Metal Catalysis in Ionic Liquids* in P. Wasserscheid, T. Welton (Eds.) *Ionic Liquids in Synthesis*, Wiley-VCH, Weinheim 2003, pp. 213-257; (b) H. Olivier-Bourbigou, L. Magna, *J. Mol. Catal. A (Chemical)* 182-183, 2002, 419-437; (c) R. Sheldon, *Chem. Comm.* 2001, 2399-2407; (d) C.M. Gordon, *Appl. Catal. A (General)* 222, 2001, 101-117; (e) P. Wasserscheid, W. Keim, *Angew. Chem. Int. Ed.* 39, 2000, 3772-3789.

[33] a) K.J. Baranyai, G.B. Deacon, D.R. MacFarlane, J.M. Pringle, J.L. Scott, Aust. *J. Chem.* 57, 2004, 145-147; b) M. Kosmulski, M; Gustafsson, J; Rosenholm, J.B., *Thermochim. Acta* 412, 2004, 47-53.

[34] J.D. Holbrey, A.E. Visser, R.D. Rogers, *Solubility and Solvation in Ionic Liquids* in P. Wasserscheid, T. Welton (Eds.*) Ionic Liquids in Synthesis*, Wiley-VCH, Weinheim 2003, pp. 68-81.

[35] T. Welton, Polarity in P. Wasserscheid, T. Welton (Eds.) *Ionic Liquids in Synthesis*, Wiley-VCH, Weinheim 2003, pp. 94-103.

[36] B. Drießen-Hölscher, P. Wasserscheid, W. Keim, *CATTECH*, 3, 1998, 47-52.

[37] a)A.J. Arduengo, R.L. Harlow, M. Kline, *J. Am. Chem. Soc.* 113, 1991, 361-363; b) A.J. Arduengo, H.V.R. Dias, R.L. Harlow, *J. Am. Chem. Soc.* 114, 1992, 5530-5534; c) W.A. Herrmann, M. Elison, J. Fischer, C. Koecher, G.R.J. Artus, *Angew. Chem. Int. Ed.* 34, 1995, 2371-2374; d) D. Bourissou, O. Guerret, F.P. Gabbaï, G. Bertrand, *Chem. Rev.* 100, 2000, 39-91; e) L. Xu, W. Chen, J. Xiao, *Organometallics* 19, 2000, 1123-1127; f) C.J. Mathews, P.J. Smith; T. Welton, A.J.P. White, *Organometallics*, 20, 2001, 3848-3850; g) D.S. McGuinness, K.J. Cavell, B.F. Yates, *Chem. Comm.* 2001, 355-356; h) M. Hasan, I.V. Kozhevnikow, MR.H. Siddiqui, C. Fermoni, A. Steiner, N. Winterton, *Inorg. Chem.* 40, 2001, 795-800.

[38] P. Wasserscheid, C.M. Gordon, C. Hilgers, M.J. Maldoon, I.R. Dunkin, *Chem. Comm.* 2001, 1186-1187.

[39] a) R.T. Carlin, R.A. Osteryoung, *J. Mol. Catal.* 63, 1990, 125-129; b) Y. Chauvin, S. Einloft, H. Olivier, *Ind. Eng. Chem. Res.* 34, 1995, 1149-1155; c) H. Waffenschmidt, P. Wasserscheid, *J. Mol. Catal.* 164, 2001, 61-67.

[40] R.A. Mantz, P.C. Trulove, Viscosity and Density *of Ionic Liquids* in P. Wasserscheid, T. Welton (Eds.*) Ionic Liquids in Synthesis*, Wiley-VCH, Weinheim 2003, pp. 56-68.

[41] H. Matsumoto, H. Kageyama, Y. Miyazaki, *Chem. Comm.* 16, 2002, 1726-1727.

[42] D.R. McFarlane, J. Golding, S. Forsyth, M. Forsyth, G.B. Deacon, *Chem. Comm.* 16, 2001, 1430-1431.

[43] P. Wasserscheid, C. Hilgers, W. Keim, J. Mol. Catal. A 214(1), 2004, 83-90.

[44] a) Y. Chauvin, L. Mussmann, H. Olivier, *Angew. Chem. .Int. Ed.* 34, 1995, 1149-1153; b) Y. Chauvin, H. Olivier, L. Mußmann, Y. Chauvin, *European Patent* 1997 776880 (to Institut Francais Du Petrole, Fr.), [Chem. Abstr. 127, 1997, 65507].

[45] C.C. Brasse, U. Englert, A. Salzer, H. Waffenschmidt, P. Wasserscheid, *Organometallics* 19, 2000, 3818-3823.

[46] P. Wasserscheid, H. Waffenschmidt, P. Machnitzki, K. Kottsieper, O. Stelzer, *Chem. Comm.* 5, 2001, 451-452.

[47] A. Heßler, O. Stelzer, H. Dibowski, K. Worm, F.P. Schmidtchen, J. Org. Chem. 1997, 62, 2362-2369.

[48] P. Machnitzki, M. Teppner, K. Wenz, O. Stelzer, E.J. Herdtweck, *Organomet. Chem.* 602, 2000, 158-169.

[49] H. Dibowski, F. P. Schmidtchen, *Angew. Chem. Int. Ed.* 37, 1998, 476-478.

[50] O. Stelzer, F.P. Schmidtchen, A. Heßler, M. Tepper, H. Dibowski, H. Bahrmann, M. Riedel, *German Patent* 1998 19701245 (to Celanese G.m.b.H., Germany), [Chem. Abstr. 129, 1998, 149094].

[51] F. Favre, H. Olivier-Bourbigou, D. Commereuc, L. Saussine, *Chem. Comm.* 15, 2001, 1360-1361.

[52] J. Sirieix, M. Ossberger, B. Betzemeier, P. Knochel, *Synlett.* 11, 2000, 1613-1615.

[53] a) K.W. Kottsieper, O. Stelzer, P. Wasserscheid, *J. Mol. Catal. A: Chem.* 175, 2001, 285-288; b) D. J. Brauer, K.W. Kottsieper, C. Liek, O. Stelzer, H. Waffenschmidt, P. Wasserscheid, *J. Organomet. Chem.* 630, 2001, 177-184.

[54] C.P. Casey, G.T. Whiteker, M.G. Melville, L.M. Petrovich, L.J.A. Gavey, D.R.J. Powell, *J. Am. Chem. Soc.* 114, 1992, 5535-5543.

[55] M. Kranenburg, Y.E.M. van der Burgt, P.C.J. Kamer, P.W.N.M van Leeuwen, K. Goubitz, J. Fraanje, *Organometallics* 14, 1995, 3081-3089.

[56] P.W.N.M. van Leeuwen, P.C.J. Kamer, J.N.H. Reek, P. Dierkes, *Chem. Rev.* 100, 2000, 2741-2770.

[57] P. Dierkes, S. Ramdeehul, L. Barloy, A. De Cian, J. Fischer, P.C.J. Kamer, P.W.N.M. van Leeuwen, *Angew. Chem. Int. Ed.* 37, 1998, 3116-3118.

[58] J. Dupont, S.M. Silva, R.F. de Souza, *Catal. Lett.* 77, 2001, 131-133.

[59] R.P.J. Bronger, S.M. Silva, P.C.J. Kamer, P.W.N.M. van Leeuwen, *Chem. Comm.* 24, 2002, 3044-3045.
[60] Experimental details available, see http://www.rsc.org/suppdata/cc/b2/b209486j/
[61] A.Castellanos-Páez, S. Castillon, C. Claver, P.W.N.M. van Leeuwen, W.G.J. de Lange, *Organometallics* 17, 1998, 2543-2552.
[62] P.Wasserscheid, R. van Hal in R.D. Rogers, K.R. Seddon (Eds) "Ionic liquids as Green Solvents", ACS Symposium Series 856, 2003, 57-70.
[63] a) M.F. Sellin, P.B. Webb, D.J. Cole-Hamilton, *Chem. Comm.* 8, 2001, 781-782; b) D.J. Cole-Hamilton, M.F. Sellin, P.B. Webb *World Patent* 2002 0202218 (to the University of St. Andrews) [*Chem. Abstr.* 136, 2002, 104215]; c) D.J. Cole-Hamilton, *Science* 299, 2003, 1702-1706., d) P.B. Webb, M.F. Sellin, T.E. Kunene, S. Williamson, A.M.Z. Slawin, D.J. Cole-Hamilton, *J. Am. Chem. Soc.* 125, 2003, 15577-15588.
[64] P.G. Jessop, W. Leitner,(eds.), "Chemical Synthesis Using Supercritical Fluids", Wiley-VCH, Weinheim, 1999.
[65] L.A. Blanchard, D. Hancu, E.J. Beckman, J.F. Brennecke, *Nature* 299, 1999, 28-29.
[66] A. Bösmann, G. Franciò, E. Janssen, M. Solinas, W. Leitner, P. Wasserscheid, *Angew. Chem.* 113, 2001, 2769-2771.
[67] M.T. Reetz, W. Wiesenhoefer, G. Francio, W. Leitner, *Adv. Synth & Catal.* 345, 2003, 1221-1228.
[68] A. Riisager, P. Wasserscheid, R. van Hal, R. Fehrmann, *J. Catal.* 219, 2003, 452-455.
[69] a) L.A. Gerritsen, A. van Meerkerk, M.H. Vreugdenhil, J.J.F. Scholten, *J. Mol. Catal.* 9, 1980, 139-155 ; b) L.A. Gerritsen, J.M. Herman, W. Klut, J.J.F. Scholten, *J. Mol. Catal.* 9, 1980, 157-168 ; c) L.A. Gerritsen, J.M. Herman, J.J.F. Scholten, *J. Mol. Catal.* 9, 1980, 241-256 ; d) L.A. Gerritsen, W. Klut, M.H. Vreugdenhil, J.J.F. Scholten, J. Mol. Catal. 9, 1980, 257-264 ; e) L.A. Gerritsen, W. Klut, M.H. Vreugdenhil, J.J.F. Scholten, J. Mol. Catal. 9, 1980, 265-274.
[70] J. Hjortkjær, M.S. Scurell, P. Simonsen, J. Mol. Catal. 6, 1979, 405-420.
[71] a) J.P. Arhancet, M.E. Davis, J.S. Merola, B.E. Hanson, Nature (London) 339, 1989 454-455; b) J.P. Arhancet, M.E. Davis, J.S. Merola, B.E. Hanson, Journal of Catalysis 121, 1990, 327-339 ; c) M.E. Davis CHEMTECH, 22, 1992, 498-502 ; d) I. Toth, I. Guo, B.E. Hanson, J. Mol. Catal. A 116, 1997, 217-229 ; e) H. Ding, J. Kang, B.E. Hanson, C.W. Kohlpaintner, J. Mol. Catal. A 124, 1997, 21-28.
[72] M.H. Valkenberg, C. deCastro, W.F. Hölderich, Appl. Catal. A General 215, 2001, 185-190.
[73] a) M.H. Valkenberg, C. deCastro, W.F. Hölderich, Topics in Catalysis 14, 2001, 139-144; b) M.H. Valkenberg, C. deCastro, W.F. Hölderich, Green Chem. 4, 2002, 88-93.
[74] C.P. Mehnert, R.A. Cook, N.C. Dispenziere, M. Afeworki, J. Am. Chem. Soc. 124, 2002, 12932-12933.
[75] W.A. Herrmann, C.W. Kohlpaintner, R.B. Manetsberger, H. Bahrmann, H. Kottmann, J. Mol. Catal. A 97, 1995, 65-72.
[76] a) B. Heinrich, Y. Chen, J. Hjortkjær, J. Mol. Catal. 80, 1993, 365-375; b) A. Riisager, K.M. Eriksen, J. Hjortkjær, R. Fehrmann, J.Mol. Catal. A 193, 2003, 259-272.
[77] A. Riisager, R. Fehrmann, P. Wasserscheid, R. Van Hal, unpublished results.
[78] A. Riisager, K.M. Eriksen, P. Wasserscheid, R. Fehrmann, Catal. Lett. 90, 2003, 149-153.
[79] J.H. Davis Jr., P.A. Fox, Chem. Comm. 11, 2003, 1209-1212.
[80] S.M. Silva, R.P.J. Bronger, Z. Freixa, J. Dupont, P.W.N.M. van Leeuwen, New J. Chem. 27, 2003, 1294-1296.
[81] a) L.A. van der Veen, M.D.K. Boele, F.R. Bregman, P.C.J. Kamer, P.W.N.M. van Leeuwen, K. Goubitz, J. Fraanje, H. Schenk, C. Bo, J. Am. Chem. Soc. 120, 1998, 11616-11626; b) L.A. van der Veen, P.C.J. Kamer, P.W.N.M. van Leeuwen, Organometallics **18**, 1999, 4765-4777; c) L.A. van der Veen, P.H. Keeven, G.C. Schoemaker, J.N.H. Reek, P.C.J. Kamer, P.W.N.M. van Leeuwen, M. Lutz, A.L. Spek, Organometallics **19**, 2000, 872-883.

CHAPTER 8 SUPERCRITICAL FLUIDS

Compressed Gases as Mobile Phase and Catalyst Support

CHARLES M. GORDON and WALTER LEITNER
Institut für Technische Chemie und Makromolekulare Chemie, RWTH Aachen, Worringer Weg 1, 52074 Aachen, Germany

8.1 Introduction to Supercritical Fluids

Supercritical fluids (SCFs) are substances which have been simultaneously heated above their critical temperature T_c and compressed above their critical pressure p_c [1]. At temperatures above T_c, isothermic compression results in a continuous increase in the fluid density, but no condensation to form a liquid occurs. The schematic phase diagram in Figure 8.1 demonstrates this behaviour for carbon dioxide as prototypical SCF. Supercritical fluids fill the entire space available to them like gases, but at the same time can act as solvents for liquids or solids. The solvent properties may be 'tuned' over a wide range by adjustments in the fluid density; this can be achieved by relatively small changes in either temperature or pressure. This unique combination of gas-like and liquid-like tuneable properties offers new approaches for the immobilization of organometallic catalysts and a systematic discussion of these methods is the subject of the present review. General aspects of the use supercritical fluids for organometallic catalysis have been reviewed in detail [1-3]. A short review comparing some of the immobilization techniques described here specifically for hydroformylation as a common benchmark reaction has appeared recently [4].

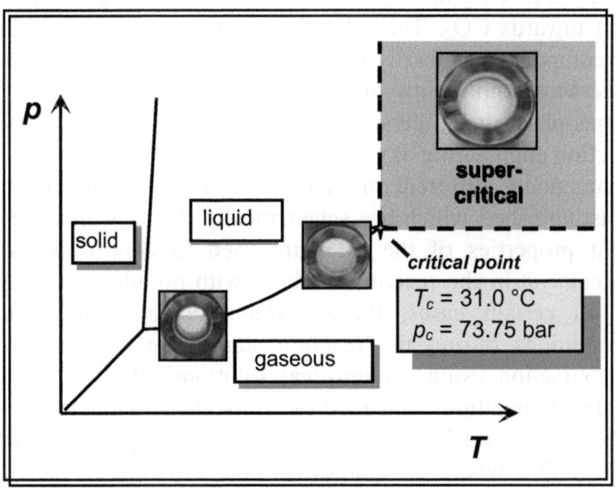

Figure 8.1 Schematic phase diagram of CO_2

D. J. Cole-Hamilton and R. P. Tooze (eds.), Catalyst Separation, Recovery and Recycling, 215–236.

Carbon dioxide ($scCO_2$) is by far the most widely used supercritical fluid. It has mild critical properties (T_c = 31.1 °C, p_c = 73.8 bar, d_c = 0.437 g ml^{-1}), is non-toxic, non-flammable and can be handled safely on laboratory and industrial scales. Unlike classical organic solvents, CO_2 is not classified as a "volatile organic chemical" (VOC) and applications of CO_2 have a GRAS ("generally regarded as safe") status. The economic viability of $scCO_2$ technology has been demonstrated by commercial applications in fields as diverse as natural product extraction [5, 6], wood impregnation [7], wafer production and dry cleaning [8]. A commercial scale multi-purpose plant for heterogeneous catalysis involving $scCO_2$ has recently entered operation [9, 10]. Examples of organometallic catalysis in supercritical fluids are by no means restricted to CO_2 [2, 3, 11], but only this medium has been used to date in applications focusing on catalyst immobilisation. As many of the potential benefits of CO_2 for multiphase catalysis are also retained in the near-critical liquid region, operation at or even below room temperature may be possible in many cases. In certain cases, low density CO_2 at temperatures above T_c can also be used for separation purposes, but generally densities d □ d_c are required to exploit the solvent properties.

Any method for the immobilization of organometallic catalysts faces the paradoxical requirements for intimate contact of reagents and catalyst during the reaction, but a maximum of discrimination at the separation stage. Separation can be facilitated by introducing phase boundaries and confined spaces, but this approach will often create additional mass transfer barriers, thus reducing turnover rates and/or selectivities. This problem is illustrated by the aqueous biphasic hydroformylation system, which cannot be applied to long-chain olefins because of their low aqueous solubility. Furthermore, catalyst leaching and cross contamination between the substrate and catalyst phases are crucial factors for practical implementation of such biphasic approaches. As will be discussed in more detail in the following sections, the properties of $scCO_2$ can be beneficial in this context. The gas-like mass transfer properties, for example, can facilitate exchange between catalyst and substrate phases. Cross contamination of CO_2 into the catalyst phase is not a problem if the catalyst is chemically inert towards CO_2. The relatively poor solvent power of $scCO_2$ for most organometallic compounds generally leads to reduced leaching compared with classical organic solvents. Most importantly, the design of SCF-based continuous flow systems that resemble gas-phase heterogeneous catalytic processes is a very attractive new approach to reaction engineering of organometallic catalysis.

Three fundamentally different approaches to catalyst immobilization involving SCFs can be distinguished which are schematically depicted in Figure 8.2. First, the tuneable solvent properties of the SCF are used to control the solubility of the organometallic catalyst in the reaction medium with no additional support or solvent (Figure 8.2a). In certain cases, these systems operate under truly monophasic supercritical conditions during the reaction stage. We refer to these methods as "catalysis and extraction using supercritical solutions" (CESS). Such processes are reminiscent of the temperature-controlled catalysts in conventional solvents sometimes referred to as "smart catalysts". The second approach is liquid/supercritical multiphase catalysis (Figure 8.2b), where the traditional counterpart is in fact triphasic (liquid-liquid-gas) if gaseous reagents are involved. Finally, the organometallic catalyst can be anchored to a solid organic or inorganic support which is then contacted with the supercritical reaction medium (Figure 8.2c). Again, such a reaction is truly biphasic (solid-supercritical fluid) in nature, whether additional gaseous reagents are present or not. This is in sharp contrast to the situation with supported catalysts in organic solvents, where triphasic (solid-liquid-gas) reaction systems often lead to severe mass transport limitations. Examples of all three approaches will be discussed in more detail in the following sections.

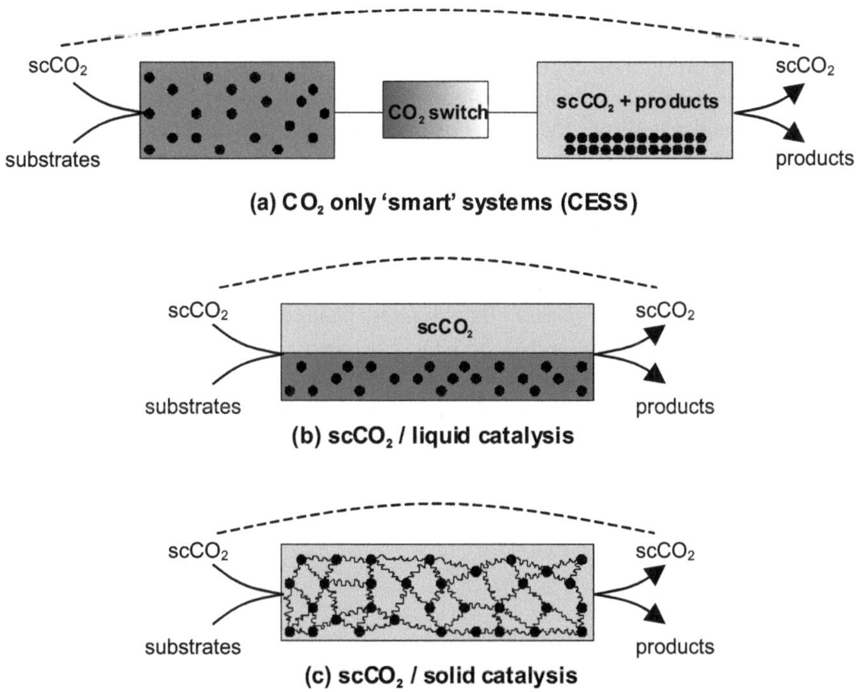

Figure 8.2. Schematic representation of the three approaches discussed in this chapter

Most examples so far have concentrated on $scCO_2$ as the phase containing substrates and/or products, corresponding to the mobile phase in continuous flow operation. More recently, the reverse situation where the catalyst is retained in the $scCO_2$ phase has also found increasing interest. These systems have been referred to as 'inverted' biphasic catalysis. In this chapter we will use the notation 'phase/$scCO_2$' to indicate that the catalyst is contained in the non-CO_2 phase, and '$scCO_2$/phase' when the catalyst is in the $scCO_2$ (*i.e.* 'inverted').

8.2 Applications of $scCO_2$ in Catalyst Immobilisation

8.2.1 CO_2 AS THE ONLY MASS SEPARATING AGENT

The schematic phase behaviour of CO_2 depicted in Figure 8.1 is only valid for the pure compound. The phase behaviour of mixtures is much more complex [6], being a function of composition, and the actual phase diagram can vary considerably even for seemingly similar components. Reaction systems contain at least three substances (substrate, product and catalyst), but in most cases more components are present and a full description of the phase behaviour is a challenging task [12]. On the other hand, this permits operating conditions where a condensed phase is in equilibrium with a compressed CO_2-rich phase at temperatures and pressures beyond the critical point of pure CO_2. Although the whole mixture is then not supercritical, the compressed CO_2 phase will behave like a supercritical fluid in exhibiting solvent properties. Any component in such a mixture will partition between the condensed and the supercritical phase, depending on its molecular structure as well as the pressure and temperature of the system. This rich phase behaviour and the ability to control the partitioning of substrates and catalysts allows the design of integrated reaction/separation schemes that rely on CO_2 as the only mass separating agent.

With traditional solvents, the "solvent power" of a fluid phase is often related to its polarity. Compressed CO_2 has a fairly low dielectric constant under all conditions (ε = 1.2-1.6), but this measure has increasingly been shown to be insufficiently accurate to define solvent effects in many cases [13]. Based on this value however, there is a widespread (yet incorrect!) belief that $scCO_2$ behaves "just like hexane". The Hildebrand solubility parameter (δ) of CO_2 has been determined as a function of pressure, as demonstrated in Figure 8.3. It has been found that the solvent properties of a supercritical fluid depend most importantly on its bulk density, which depends in turn on the pressure and temperature. In general higher density of the SCF corresponds to stronger solvation power, whereas lower density results in a weaker solvent.

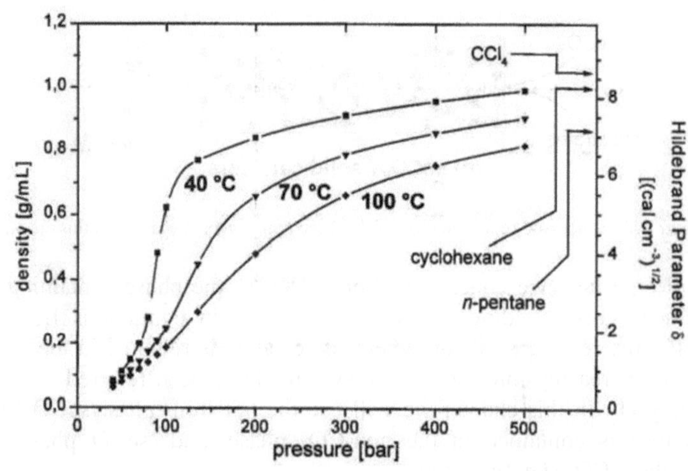

Figure 8.3. Variation of density and δ with temperature and pressure for CO_2. The δ values for three typical non-polar organic solvents are also indicated (reproduced with permission from reference [1])

The solubility of a solute in $scCO_2$ is extremely dependent on its structure, with three features of paramount importance. As expected, compounds of low polarity are more soluble than very polar compounds or salts. However, solubility also increases greatly with increasing vapour pressure of a substrate. To account for the contribution of volatility and solvation to the solubility process, Kurt Zosel coined the term "Destraktion" (from Latin *destillare* and *extrahere*) in his pioneering work on natural product extraction with SCFs [5]. Finally, some specific functional groups like perfluoroalkyl and polysiloxane substituents, or polyether/polycarbonate copolymers are known to give compounds a high affinity to compressed CO_2 that cannot be explained through simple polarity or volatility arguments ([14] and references therein). These "CO_2-philic" substituents can lead to dramatic solubility enhancements, thus allowing control of the phase preference of reaction components at different stages of a reaction/separation process.

Many organic substrates and products have significant solubilities in $scCO_2$, whereas the majority of organometallic catalysts are very poorly soluble in this medium. Attempts have been made to exploit this factor if the main grounds for using $scCO_2$ lie simply in allowing effective catalyst recycling and product isolation. In such cases it is not necessarily essential to work under fully homogeneous conditions during the reaction. In some cases the reaction is carried out in the absence of CO_2, which is used only in the separation step and for downstream processing. An early example of this approach is provided by a BASF patent proposing to regenerate phosphine-modified rhodium hydroformylation catalysts via extraction of the so-called "heavy ends" after the more volatile short chain aldehyde products have been removed by conventional distillation [15]. More recently, the hydroformylation of long chain olefins was carried out using CO_2-insoluble catalysts in a substrate/$scCO_2$ biphasic system. At the end of the reaction, the products were separated by supercritical fluid extraction allowing repeated reuse of the catalyst (Scheme 8.1) [16-18].

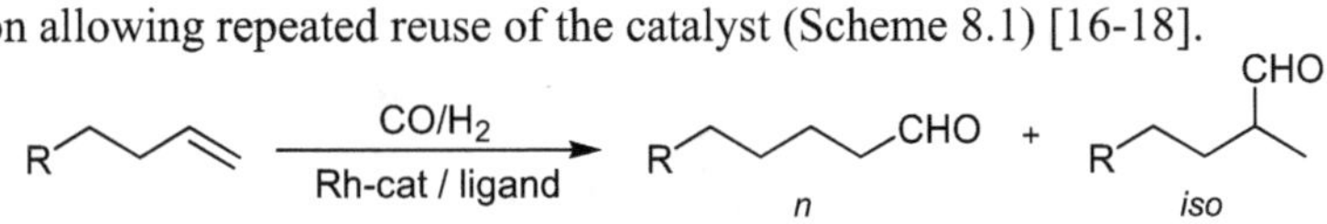

Scheme 8.1. Hydroformylation of long chain olefins

In order to fully exploit the potential benefits of the supercritical state for both reaction and separation, it may be preferable to carry out the reaction in a fully homogeneous single phase, and only afterwards to induce phase separation. In particular, the importance of homogeneity of substrates, catalysts and intermediates in one single phase for high conversion and selectivity in catalytic processes has been noted in several studies [19-21]. For example, the selectivity between intramolecular ring closing metathesis and the competitive intermolecular oligomerisation process can be controlled by pressure control under homogeneous supercritical conditions [22, 23]. Following the reaction step under high density conditions, supercritical extraction of the macrocyclic product at lower density leaves the poorly $scCO_2$ soluble Grubbs catalyst in an active form in the reactor (Scheme 8.2) [24].

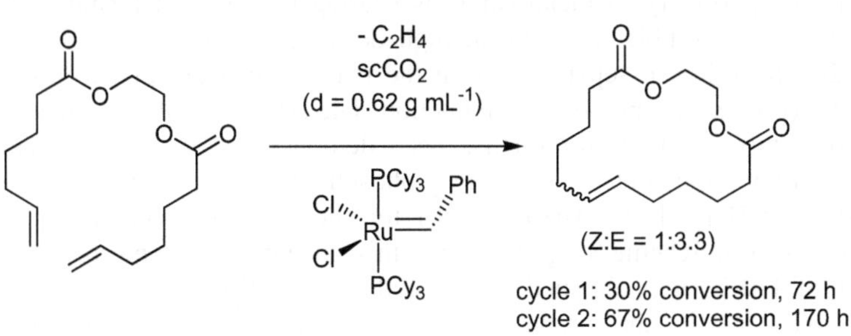

Scheme 8.2. Batch-wise recycling of Grubbs' catalyst for ring closing metathesis

Many transition metal trialkylphosphine complexes have similar solubility properties in $scCO_2$ as the Grubbs catalyst, giving the expectation that similar recycling techniques may be employed in such cases. In fact, there is a number of examples reporting the benefits of employing homogeneous solutions of $scCO_2$ with such catalysts during the reaction step compared to the equivalent reactions in conventional solvents, including the hydrogenation of CO_2 to formic acid and its derivatives using ruthenium trimethylphosphine catalysts [3], the hydroformylation of long chain olefins catalysed by rhodium triethylphosphine complexes [25], and the hydroboration of olefins with rhodium tricyclohexylphosphine catalysts [26]. The remarkable change in chemoselectivity from alcohols to aldehydes in hydroformylation with Rh/PEt_3 catalysts upon passing from biphasic to monophasic conditions is illustrated in Figure 8.4 [25].

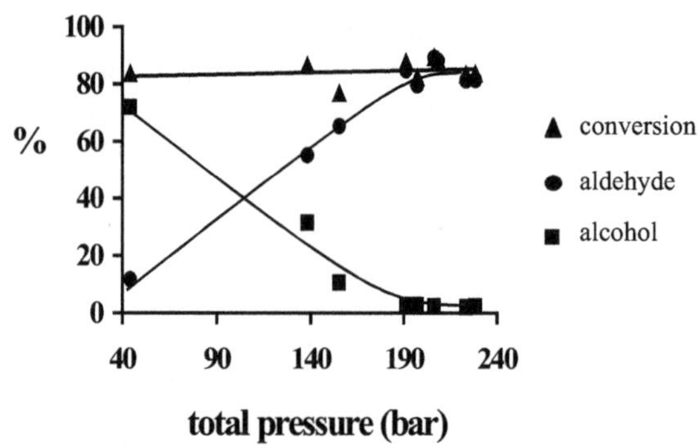

Figure 8.4. Chemoselectivity as a function of system pressure during the hydroformylation of long chain olefins with Rh/PEt3 catalysts. The change in chemoselectivity has been correlated with the transition from biphasic to monophasic reaction conditions

Many, if not most, organometallic complexes exhibit solubilities in $scCO_2$, however, that are too low even for catalytic applications under single phase conditions. This applies particularly to the large class of catalysts bearing aryl phosphine ligands, a structural motif also found frequently in chiral ligands used in enantioselective catalysis. This problem can be overcome by the introduction of perfluoroalkyl groups into the ligand periphery of metal catalysts bearing this type of ligand [27]. Again, these catalysts are soluble in $scCO_2$ at high density, but become insoluble at lower density [28], resulting in separation from the reaction mixture or partitioning into a liquid phase (if present) upon density reduction. The efficiency of this type of regulated system for catalyst immobilization has been demonstrated with a rhodium catalyst bearing fluoro-substituted triphenylphosphine derivatives such as $P[p\text{-}CF_3(CF_2)_6(CH_2)_2C_6H_4]_3$ (3-H^2F^6-TPPP) for the hydroformylation of long chain olefins as shown in Scheme 8.1 [29]. Using this system in a batch process, a turnover frequency of 430 h^{-1} was obtained for the hydroformylation of 1-octene, with a *n*:*iso* ratio of over 5:1 in the products and almost complete suppression of olefin isomerisation. The products were collected after density reduction by careful depressurisation of the reactor, and were found to contain only trace amounts of Rh in the range of 1 ppm under optimised conditions. The catalyst itself could be reused for five successive catalytic runs with no significant change in selectivity or turnover.

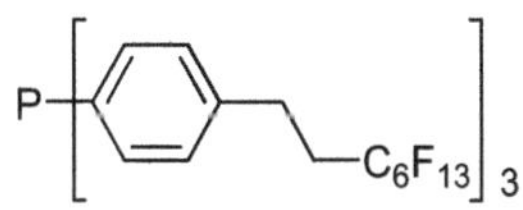

3-H²F⁶-TPPP

Fluorinated groups can be introduced into ligand frameworks via relatively straightforward synthetic routes. Successful examples of their applications in $scCO_2$ have been reported for rhodium catalyzed alkene hydroformylation [30, 31], hydroboration [26], and hydroaminomethylation [32], as well as palladium catalysed C-C bond forming reactions [33-38], oxidation processes [39], and various metal catalyzed polymerization processes [40-43]. Chiral arylphosphine ligands bearing fluoroalkyl groups have been investigated for enantioselective catalytic hydrogenation [44-46] and hydroformylation [45, 47, 48] in $scCO_2$. A representative example will be discussed at greater length below.

Scheme 8.3 exemplifies the CESS approach for batchwise rhodium-catalysed asymmetric hydroformylation of styrene using the CO_2-philic chiral ligand 3-H^2F^6-BINAPHOS [45, 48]. Eight successive runs were performed and the system was still active after a total turnover number of more than 12,000 catalytic cycles per rhodium centre. A noticeable decrease in the enantiomeric excess was encountered after the fourth run, but regioselectivity and conversion remained uniformly high. The decrease of the enantiomeric excess may be attributed partly to racemisation of the product under the experimental conditions [19]. Chemical instability of the phosphite unit in the ligand framework seems to play a more important role in this case, as indicated by the fact that the initial high level of enantioselectivity was restored in the final run by the addition of fresh ligand.

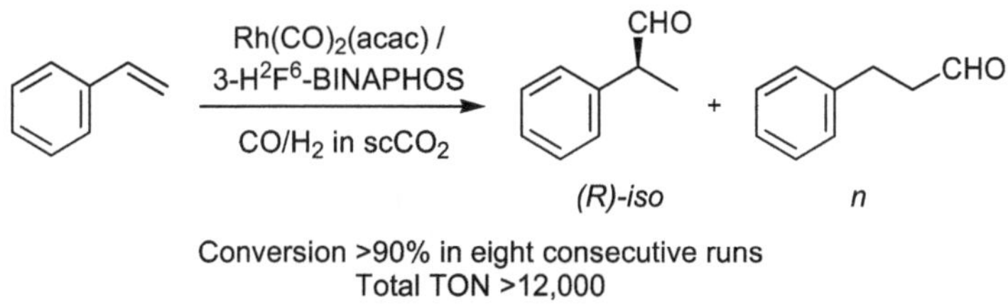

Scheme 8.3. Hydroformylation of styrene with the CO_2-philic *in situ* catalyst 3-H^2F^6-BINAPHOS/[$Rh(acac)(CO)_2$] (2:1) using a batch-wise CESS procedure

Even with the simple laboratory equipment used in these experiments, the CESS procedure allowed quantitative recovery of the product free of solvent, and with rhodium contents ranging from 0.36-1.94 ppm (determined by atomic absorption measurements). Furthermore, using this approach removal of unreacted starting material or side products from the product is possible during extraction from the catalyst, since even small structural differences can result in significant differences in

CO_2 solubility. For example, the desired branched aldehyde formed upon hydroformylation of vinylnaphthalene is considerably more soluble in $scCO_2$ than its linear isomer. Therefore, the ratio of the branched to linear product increases from approximately 90:10 in the condensed phase to up to 98:2 in the CO_2 phase in a single extraction step under the low density conditions of catalyst separation.

The introduction of perfluoroalkyl groups into the ligand periphery of an organometallic complex can lead to a considerable increase in molecular weight and size of the catalytically active species. This has been utilized to separate "CO_2-philic" catalysts from the products in a continuous flow membrane reactor [49]. In the reaction reported, a variant of Wilkinson's catalyst containing the fluorous phosphine ligand $P(C_6H_4\text{-}p\text{-}SiMe_2CH_2CH_2C_8F_{17})_3$ was employed for hydrogenation of 1-butene. In continuous flow operation using a silica membrane, the system gave a turnover number of 1.2×10^5 after 32 hours reaction. ICP-AAS analysis of the products indicated no flux of catalyst or free ligand through the membrane, although a decrease in conversion over the course of the reaction suggested deactivation of the catalyst was occurring. This intriguing methodology (see also Chapter 4 of this book) may be particularly useful where highly fluorinated polymers or copolymers are used to stabilize and solubilise organometallic or colloidal catalysts in $scCO_2$ as the reaction medium.

Many organometallic catalysts and especially many chiral catalysts are cationic and modification of the anion has been found to be very effective for enhancing their solubility in $scCO_2$. Tetrakis-(3,5-bis-trifluoroimethylphenyl)borate (BARF) was an early example of such an anion which has proven extremely useful for this purpose [50], and very pronounced anion effects on the activity and selectivity of the catalysts are observed in many other cases [44, 51, 52]. Using the BARF-modified "CO_2-philic" chiral iridium catalyst shown in Scheme 8.4, catalytic efficiency was greatly enhanced and almost identical *ee* values were observed for the enantioselective hydrogenation of imines upon changing from CH_2Cl_2 to $scCO_2$ as the solvent [51].

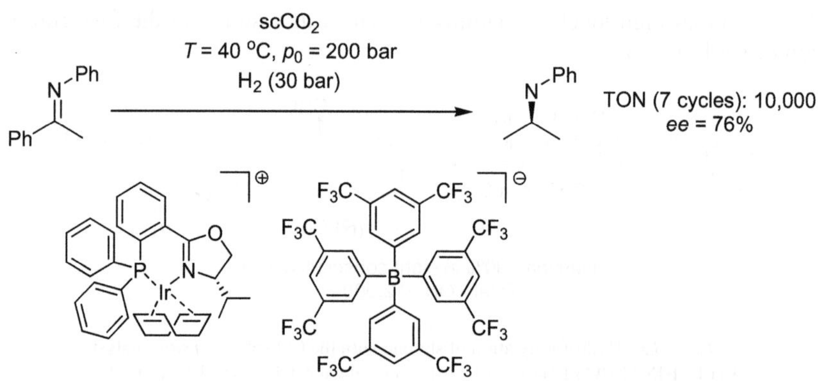

Scheme 8.4. Catalytic hydrogenation of imines using cationic iridium catalysts with the "CO_2-philic" BARF anion

The iridium catalyst was found to be sufficiently soluble for catalysis when in the form of the substrate complex, but precipitated quantitatively once all substrate had been consumed. Supercritical fluid extraction at that stage yielded the solvent- and metal-free product in crystalline form leaving the active and selective catalyst behind for further use (Scheme 8.4). Under batch operation, noticeable deactivation occurred after the fourth cycle which might be avoided by working in a closed system.

8.2.2 BIPHASIC SYSTEMS CONSISTING OF CO_2 AND LIQUID PHASES

In liquid/supercritical systems, the liquid phase acts as support to retain the catalyst rather than using the selective solubility properties of $scCO_2$ as for the CESS approach described above. Thus, there is a permanent phase separation between the mobile phase which is (usually, but not exclusively as we will see below) CO_2 and the stationary liquid phase. The fact that the catalyst is molecularly dispersed in the liquid phase means that the environment resembles the typical situation of homogeneous catalysis. In order to immobilize the catalyst efficiently, the solubility of the organometallic intermediates in the liquid phase must be greatly favoured over their solubility in the supercritical phase. As has been indicated previously, $scCO_2$ is a rather feeble solvent for many organometallic catalysts in the absence of suitable ligands [27]. Therefore, the risk of catalyst leaching can be expected to be reduced compared to similar liquid/liquid multiphase systems such as water/organic, fluorous/organic, or ionic liquid/organic.

No significant cross-contamination of the liquid phase into CO_2 can be tolerated if it is to act as an efficient stationary phase, as this would result in its depletion during continuous flow operation along with contamination of the product. The mixture critical points of most organic solvents are in the vicinity of those of the pure solvents. They would thus be at least partially miscible with the SCF and would not form a stable stationary phase for catalyst immobilization. As has been noted previously, the solubility of a compound in $scCO_2$ is determined not only by its polarity, but also by its vapour pressure. Therefore, highly polar and/or very non-volatile liquids, which do not become miscible with CO_2 until hyperbaric conditions, are the most useful candidates for liquid catalyst phases together with $scCO_2$. Three different approaches that are receiving particular interest are discussed in the following sections.

8.2.2.1 Water as the Liquid Phase

The combination of the two environmentally benign solvents water and CO_2 is a highly attractive approach to "green" solutions for catalyst immobilization. Specific boundary conditions apply for multiphase catalysis with water and CO_2, similarly to classical systems were the catalyst is contained in an aqueous phase. The pH of water is approximately 3 in the presence of the compressed CO_2 phase, requiring that substrates, products and catalysts can tolerate such acidic conditions. If necessary, buffer systems can be added to control the pH even up to slightly basic values [53]. The solubility of water in CO_2 is below 1% under typical conditions [54], but some CO_2 will be lost with the aqueous phase. Cross contamination between the two solvent systems and resulting impurities in the products can be tolerated, however, as both are environmentally and toxicologically benign.

Typical approaches to this biphasic system have involved the immobilization of catalysts in the aqueous phase as colloids [53] or using water-soluble catalysts based on ligands such as the trisulfonated TPPTS [55, 56]. Particularly high reaction rates have been obtained with surfactant-stabilized microemulsions and emulsions that allow for intimate contact of all reagents with the catalyst during the reaction [57]. The emulsions separate readily into two phases by small pressure changes and the CO_2-phase is then vented to isolate the products. The catalyst $RhCl(tppds)_3$ (tppds =

tris(3,5-disulfonatophenyl)phosphine) was employed for hydrogenation of olefins in the presence of anionic, cationic and non-ionic surfactants. For styrene the TOF was found to be 4 h^{-1} in toluene/water, 26 h^{-1} in CO_2/water, and 150-300 h^{-1} in the emulsions. The potential for catalyst recycling was demonstrated by the fact that complete conversion was observed over three reaction cycles using the same aqueous surfactant/catalyst phase. ^{31}P NMR analysis of the catalyst solution showed that the catalyst retained its molecular character.

The techniques discussed up to now use CO_2 as the mobile phase for substrates and products. Naturally, this restricts the applications to relatively non-polar and/or volatile components with sufficient solubility in the supercritical medium. An intriguing alternative for processing highly polar substrates are "inverted" aqueous systems. In this approach, a CO_2-philic catalyst resides in the non-polar CO_2 phase, while water-soluble substrates and products are contained in the aqueous layer [58, 59]. A very attractive and unique feature of the $scCO_2/H_2O$ system is that the stationary supercritical phase is never depressurized and hence the large energy input required for recompression is avoided. Furthermore, the aqueous solution is not contaminated with any organic solvent or catalyst residues, which is particularly important if the product is a fine chemical intended for direct further use in aqueous solution.

As with the CESS system discussed in section 8.2.1, the key prerequisite to immobilize the catalyst in the carbon dioxide phase is the availability of sufficiently "CO_2-philic" organometallic compounds [60]. The catalysts must have a significant absolute solubility in $scCO_2$ as well as an extremely high partitioning preference in the biphasic system, usually attained by the use of highly fluorinated ligands and anions [27]. Gratifyingly, this approach proved successful also for the $scCO_2/H_2O$ system [58, 59]. The fluoroalkyl-substituted derivative of triphenylphosphine, 4-H^2F^6-TPPP, was used to generate a rhodium catalyst for hydroformylation of an unsaturated carboxylic acid (Scheme 8.5).

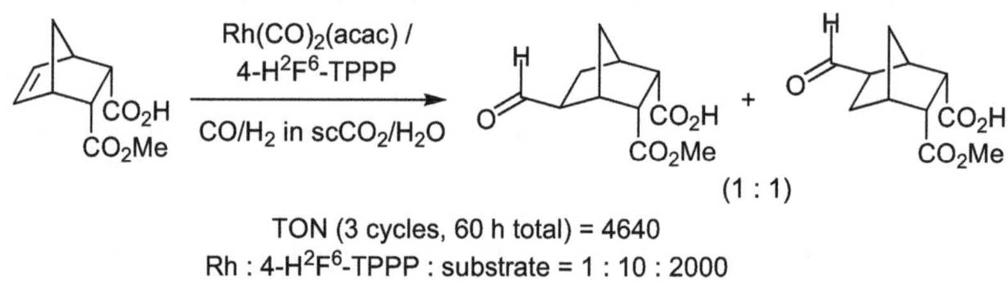

Scheme 8.5. Hydroformylation of polar substrates in an "inverted" biphasic $scCO_2/H_2O$ system

In the inverted system shown in Scheme 8.5, the unmodified rhodium catalyst $[Rh(CO)_2(hfacac)]$ (hfacac = hexafluoroacetylacetonate) gave very inactive and unstable systems. This was somewhat surprising, as unmodified rhodium complexes of this type are known to form very soluble and active hydroformylation catalysts in pure $scCO_2$ [29]. Presumably, both the partitioning in the biphasic system as well as the catalyst stability under acidic conditions is not sufficient for this type of catalyst. The phosphine modified system resulted in reasonable turnover rates and the catalyst could be recycled well at high enough catalyst loadings. A total turnover number of 1500 was achieved over three consecutive runs. This could be increased to more than 4500 at lower catalyst concentrations, albeit at the expense of noticeable deactivation during each recycling step. As catalyst leaching was found to be very low, this loss of activity at lower catalyst loading must again be attributed to catalyst decomposition. Despite these practical shortcomings of the model system, the potential of the inverted aqueous/scCO_2 system for multiphase catalysis with highly polar products is clear to see.

8.2.2.2 Poly(ethyleneglycol) (PEG) as the Liquid Phase

Poly(ethyleneglycol) (PEG) has recently been suggested as non-volatile catalyst phase that can be employed in combination with scCO_2 [61]. Liquid PEG of low to moderate molecular weight serves as a reasonably good solvent for many typical organometallic catalysts, but these systems are still soluble in scCO_2 or form emulsion-type mixtures, which may be interesting media for catalysis in their own right [62]. For multiphase catalysis the use of high molecular weight PEG is preferred, however, owing to its low solubility in scCO_2.

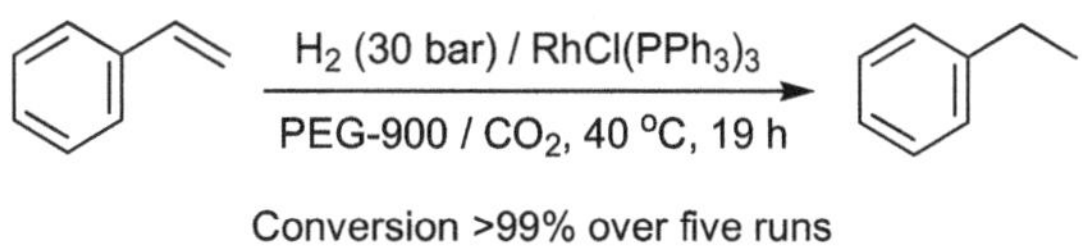

Scheme 8.6. Rhodium-catalysed hydrogenation in the biphasic system PEG/scCO_2

PEGs with average molecular weights above 1000 are waxy solids under ambient conditions, but they melt under CO_2 pressure to become liquids under typical conditions of scCO_2 catalysis [63]. The approach was demonstrated for the rhodium catalysed hydrogenation of styrene as a test reaction using Wilkinson's complex [$(PPh_3)_3RhCl$] as the catalyst (Scheme 8.6) [61].

Batch-wise recycling was successfully achieved by extraction of the ethylbenzene product with the scCO_2 phase. During this process the catalyst remained stable in the PEG environment and could be recycled four times without noticeable loss of activity. Rhodium contamination in the product was below the detection limit (less than one part per million). Using PEG_{900} as the liquid catalyst phase, measurable amounts of PEG were detected among the extracted products, but for PEG_{1500} the contamination was found to be as low as 0.1% by weight.

8.2.2.3 Ionic Liquids as the Liquid Phase

Ionic liquids (ILs) are low melting point (<100°C) salts of organic cations that are finding increasing interest as solvents for organometallic catalysis (for more information see Chapter 7 of this book) [64-67]. Being composed entirely of cations and anions, ILs exhibit no measurable vapour pressure and consequently they generally show no detectable solubility in pure scCO_2. Carbon dioxide, however, has a remarkable affinity for ILs allowing high concentrations of CO_2 in the liquid phase, and thus rapid mass transfer between the two media [68]. Because of this, scCO_2 can be used to extract even high boiling organic substances from the IL phase very efficiently [69]. These properties make IL/scCO_2 biphasic systems highly attractive for application in catalyst immobilization, especially under continuous flow conditions schematically represented in Figure 8.5 (For related examples in biocatalysis, see [70-72]).

A chiral ruthenium catalyst bearing BINAP-type ligands was used in ILs for enantioselective hydrogenation of prochiral α,□-unsaturated acids followed by subsequent CO_2 extraction [73]. A pronounced effect of protic co-solvents such as water or ^{i}PrOH was noted for the reaction stage with certain substrates. The standard substrate tiglic acid (*E*)-H(Me)C=C(Me)COOH was quantitatively hydrogenated under optimized conditions with ee values around 90%. Methylpropionic acid was extracted quantitatively from the IL phase with $scCO_2$ (35°C, 175 bar) and the catalyst/IL solution could be recycled four times without noticeable decrease in conversion or enantioselectivity.

The hydrogenation of CO_2 in the presence of amines to give dialkylformamides has been carried out directly in an IL/$scCO_2$ system. In this case, the ionic liquid was shown to play a dual role [74]. It is an effective solvent for the ruthenium phosphine catalyst and at the same time allows a distinct phase distribution of the polar carbamate intermediates and the less polar products formed during the conversion of CO_2. As a result, the selectivity of the reaction can be increased over conditions where $scCO_2$ is used as the sole reaction medium.

As hydroformylation has been selected as a 'benchmark' reaction for comparison of the different immobilisation techniques discussed in this book, it is instructive to consider this reaction in somewhat more detail. Hydroformylation of long chain olefins has been extensively investigated under biphasic IL/$scCO_2$ conditions. For this reaction it was found that the chemical properties of the IL and catalyst must be carefully matched [75, 76]. Initial investigations using phosphite ligands showed these to be incompatible with ILs based on the $[PF_6]^-$ or $[BF_4]^-$ anions used at first, because small amounts of F^- liberated by anion hydrolysis reacted with the ligands and led to catalyst deactivation. Sulfonated derivatives of triphenylphosphine proved to be chemically more robust, but were only soluble in the ILs in the form of imidazolium salts of monosulfonated triphenylphosphine, *e.g.* [pmim][$Ph_2P(m\text{-}C_6H_4SO_3)$] (pmim = 1-propyl-3-methylimidazolium). An initial flow process based on a catalyst formed *in situ* from [$Rh_2(OAc)_4$] and [pmim][$Ph_2P(m\text{-}C_6H_4SO_3)$] dissolved in the IL [bmim][PF_6] (bmim = 1-butyl-3-methylimidazolium) gave a turnover number of *ca.* 100 after 20 hours continual reaction for the hydroformylation of dodecene. This low level of activity was ascribed to lack of solubility of the substrates in the IL, and optimisation studies showed that significantly higher levels of activity could be attained by increasing the alkyl chain length on the imidazolium ring and changing the anion to $[Tf_2N]^-$. In an extensive optimisation study, other factors considered included substrate flow rate, temperature and gas composition. An optimised system using the IL [omim][Tf_2N] (omim = 1-octyl-3-methylimidazolium) (Scheme 8.7) gave a turnover frequency of 517 h^{-1} for the hydroformylation of 1-octene, with an *n*:*iso* ratio of approximately 3:1.

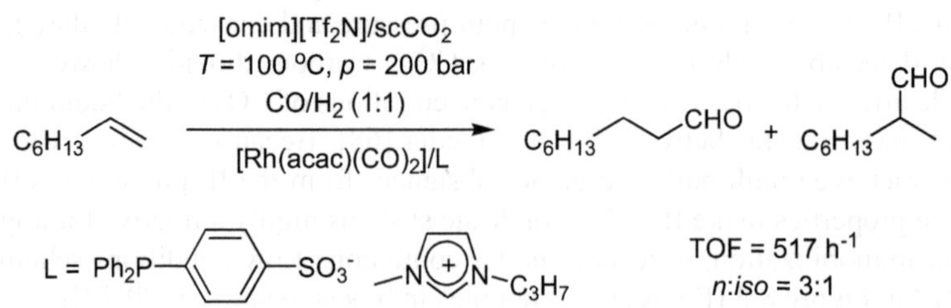

Scheme 8.7. Continuous-flow hydroformylation of long chain olefins in the biphasic system IL/$scCO_2$

The long term stability of the catalyst was also investigated, and found to depend principally on the rate of oxidation of the phosphine. When all the phosphine had been oxidised, the rhodium rapidly started leaching into the CO_2 stream owing to the formation of $Rh_4(CO)_4$. This effect was also demonstrated by an increase in the levels of alkene isomerisation. Under conditions where phosphine oxidation was not a problem, however, the leaching of rhodium in the product stream under steady-state operation could be as low as 0.012 ppm. The authors compared their process with the cobalt catalysed processes used commercially, as well as a rhodium-catalysed process under pilot plant investigation. It was concluded that the optimised system displayed excellent space-time yields in comparison with the established systems, and operated at comparable temperatures. The main problems highlighted for the IL/scCO_2 system were the relatively low *n:iso* selectivity and the relatively high pressures required (200 bar).

Another potential problem common to all IL/scCO_2 systems is the current cost and unknown toxicity of ILs. Although in an optimised system the IL should remain within the reactor, the absence of toxicity data may prove a hindrance to their acceptance for products used in the fine chemicals or pharmaceutical industries. In an attempt to address this problem, Cole-Hamilton and co-workers have recently reported a development of their IL/scCO_2 hydroformylation system discussed above, which uses a similar catalyst system but no IL [18]. In this case, the product nonanal was used as a solvent for the catalyst, which was based on [pent-mim][$Ph_2PC_6H_4SO_3$] (pent-mim = 1-pentyl-3-methylimidazolium) and [Rh(acac)$(CO)_2$]. The longer alkyl chain substituents on the phosphine gave improved solubility in the organic phase (and hence higher activity), but not to the extent that leaching into the CO_2 phase becomes a problem.

A further example of a reaction which may be optimised in IL/scCO_2 by selection of the appropriate anion for the IL is catalytic enantioselective hydrovinylation, a synthetically interesting and truly atom economic C-C bond forming reaction [77-79]. The nickel complex below has been developed by Wilke and co-workers as precursor for a highly active and enantioselective catalyst for this process.

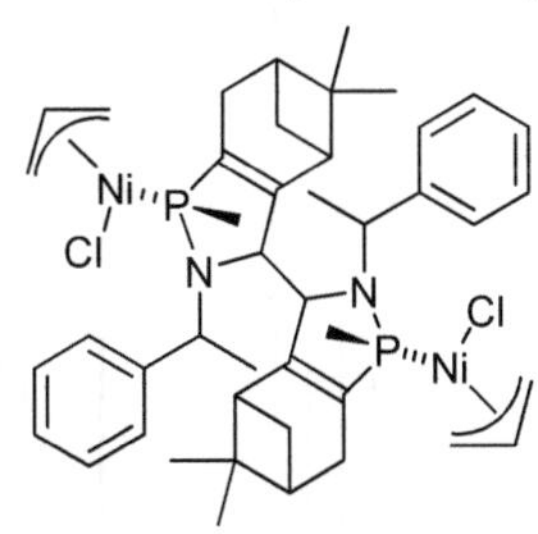

Wilke catalyst

This complex is not the actual catalyst for the hydrovinylation, but needs to be activated in the presence of a suitable co-catalyst. The role of this additive is to abstract the chloride ion from the nickel centre to generate a cationic allyl complex that further converts to the catalytically active nickel hydride species. In conventional solvents this is typically achieved using strong Lewis acids such as Et_2AlCl. Alternatively, sodium or lithium salts of non-coordinating anions such as tetrakis-[3,5-bis(trifluoromethyl)phenyl]borate (BARF) can be used to activate hydrovinylation catalysts [80]. These active intermediates are highly reactive and extremely sensitive so that reactions in conventional solvents need to be carried out at low temperatures, typically -60°C and below. The use of the sodium salt of (BARF) as activator was found to generate an active catalyst from Wilke's catalyst giving high yields and enantioselectivities in liquefied and supercritical CO_2 [42]. The chemo- and enantioselectivities in CO_2 were considerably higher than those obtained in conventional solvents at the same temperatures. Catalyst recycling by selective extraction of the products with low-density CO_2 met with limited success, however, owing to extreme instability of the active species especially in the absence of the substrates.

By transferring this reaction in an IL/scCO_2 system, several significant benefits could be demonstrated. First it was shown that the ionic liquid can serve simultaneously as the catalyst phase and co-catalyst. Furthermore, it was found that dissolving Wilke's catalyst in various ILs resulted in different degrees of activity and selectivity depending on the anion present. The degree of conversion decreases in the order BARF > $Al[OC(CF_3)_2Ph]_4$ > Tf_2N > BF_4, in a manner consistent with the estimated coordination strength of these anions. The high activity in [emim][BARF] (emim = 1-ethyl-3-methylimidazolium) led to significant amounts of consecutive isomerisation and oligomerisation under the non-optimized conditions of the batchwise screening, but could probably be avoided under continuous flow conditions by shorter contact times.

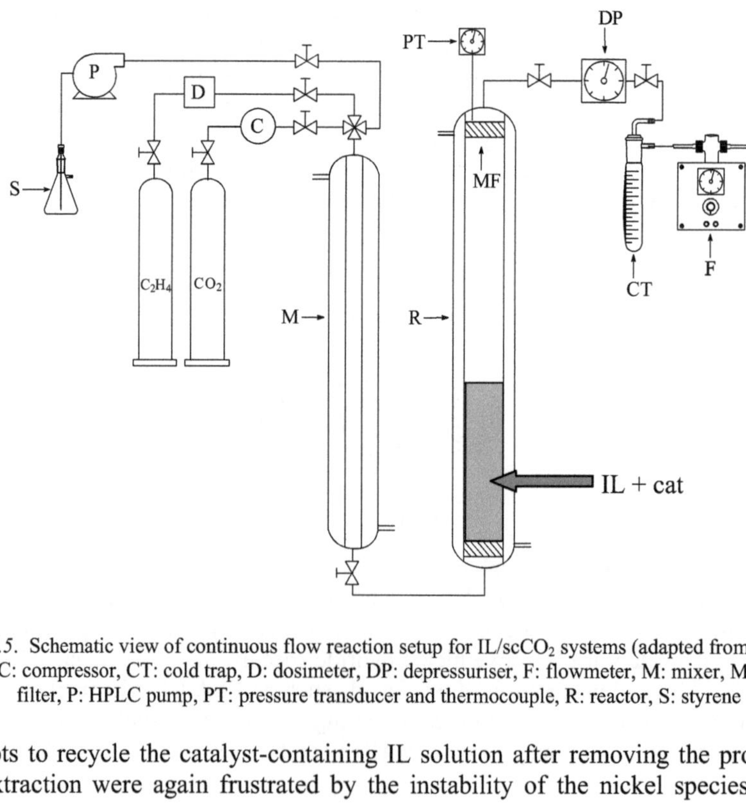

Figure 8.5. Schematic view of continuous flow reaction setup for IL/scCO_2 systems (adapted from reference [81]). C: compressor, CT: cold trap, D: dosimeter, DP: depressuriser, F: flowmeter, M: mixer, MF: metal filter, P: HPLC pump, PT: pressure transducer and thermocouple, R: reactor, S: styrene

Attempts to recycle the catalyst-containing IL solution after removing the product by CO_2 extraction were again frustrated by the instability of the nickel species. It was assumed that the absence of excess substrate was a major reason for the deactivation during batch-wise recycling of this type, so continuous flow operation of the reactions was employed to ameliorate this limitation. Using a setup similar to that shown in Figure 8.5, [emim][Tf_2N] was used as the IL to activate, tune, and immobilize Wilke's catalyst. CO_2, the liquid substrate, and ethylene were delivered from a compressor unit, an HPLC pump, and a gas dosing module, respectively. Liquid CO_2 at 0 °C was used in the long term test study, but similar results could be obtained using $scCO_2$. After passing through a mixer (in its simplest form a four port joint), the substrate flow was fed into the reactor where it bubbled through the ionic liquid solution containing Wilke's catalyst. Styrene was converted to (*R*)-(1-methylallyl)-benzene with high activity and selectivity, and the CO_2 stream transported the product from the reactor into the separation chamber. Initially, some problems were encountered with foaming and consequent loss of the IL, but these were resolved through small modifications to the reactor design. Under optimised conditions, the product separated from the CO_2 stream upon pressure reduction, and was isolated as a colourless liquid in pure form.

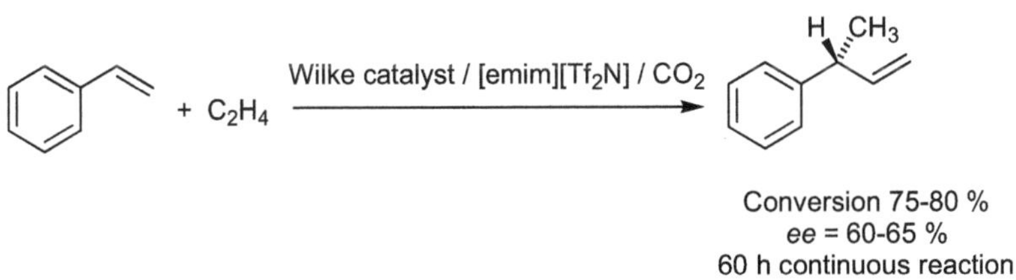

Scheme 8.8. Continuous flow enantioselektive hydrovinylation using the biphasic system IL/$scCO_2$

Scheme 8.8 summarizes the remarkable long term stability of this catalytic system under continuous flow operation. At a reaction temperature of 0°C, conversion remained practically constant over 60 hours. A short induction period was associated with the time required to reach the saturation concentration of substrates and products in the IL. After reaching steady state conditions, the enantiomeric excess showed a marginal decrease from around 65% to 60%. The most exciting feature of this system is the ability to optimise it on both the molecular and the engineering level. For example, based on the results shown in Scheme 8.8, improved matching of the IL and catalyst can be expected to yield products at better levels of conversion and *ee*. At the same time, control of the residence time now becomes an option for homogeneous catalysis, allowing for example the suppression of consecutive reactions.

With the number of successful examples increasing, it is expected that these general advantages of continuous flow multiphase catalysis in IL/$scCO_2$ will be exploited further making this a viable option for fine chemical production. A broad range of reactions can be envisage for the combined IL/$scCO_2$ approach, and it is likely that this list will grow extensively over the coming years. An interesting recent example is the palladium catalyzed dimerisation of methyl acrylate, which was found to proceed smoothly in a biphasic [bmim][BF_4]/$scCO_2$ biphasic system [82]. The catalytic system comprises phosphonium salts and HBF_4 as co-catalysts, making the IL environment particularly attractive. The partitioning of substrates and products between the IL and $scCO_2$ was investigated and the data used for optimisation of the reaction conditions. Turnover numbers up to 560 and turnover frequencies up to 195 h^{-1} were obtained.

8.2.3 BIPHASIC SYSTEMS CONSISTING OF CO_2 AND SOLID PHASES

A third approach for the immobilisation of organometallic catalysts is the use of solid supports. While many successful examples of the use of both inorganic (*e.g.* silica, alumina) and organic (*e.g.* polymer) supports for organometallic complexes have been reported (see Chapter 3 of this book), the activity of such systems can often be limited by the rate of mass transfer between gaseous or liquid phases and the solid phase catalyst. In a similar manner to the situation in heterogeneous catalysis [83], $scCO_2$ offers an intriguing solution to this problem for solid phase supported organometallic complexes.

8.2.3.1 Inorganic Supports

A typical example of organometallic catalysis in $scCO_2$ is the hydrogenation of CO_2 itself to yield formic acid and its derivatives [84, 85]. Ruthenium complexes bearing silylether phosphine ligands (**1**) were anchored on silica supports via sol-gel techniques to yield highly efficient catalysts for hydrogenation of CO_2 under supercritical conditions [86, 87]. In the presence of dimethylamine, turnover numbers of up to 110,800 were observed for the formation of dimethylformamide using these catalysts. The nature of the support has a pronounced effect on the performance of the catalysts, and even higher efficiencies were reported with a mesoporous ruthenium silica hybrid aerogel [88].

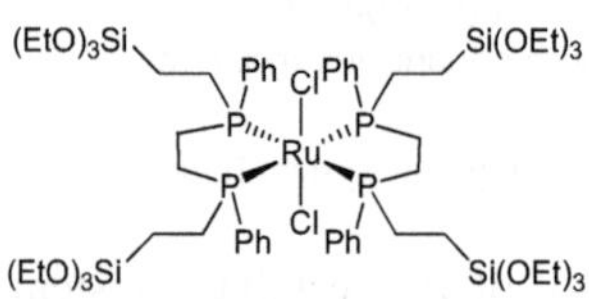

1

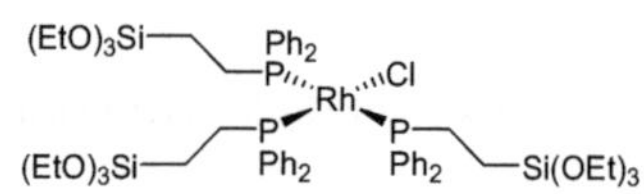

2

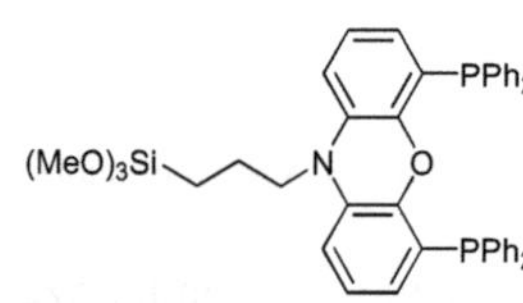

3

The hydroformylation of 1-hexene was carried out using a similar $scCO_2$/solid phase system where a rhodium complex with silylether phosphine ligands (**2**) was anchored on silica [89-91]. Following this approach, an *n/iso* ratio of 4.5 was obtained for the aldehyde product, compared to a value of approximately 3 obtained in toluene solution using the corresponding free homogeneous catalyst. Hydroformylation of the same substrate was also effective with rhodium and platinum/tin phosphine complexes immobilized in the zeolite MCM-41 [92]. A high selectivity for *n*-heptanal was observed using the platinum/tin catalysts, while the rhodium based catalysts gave higher activity but lower selectivity. In another example, silica-supported ligands of the XANTPHOS-type (**3**) were shown to form highly selective rhodium catalysts for continuous-flow hydroformylation of long chain olefins in $scCO_2$ [93]. The structure of the chelating ligand ensured a very high *n/iso* ratio of 40:1 in the conversion of 1-hexene. The silica bound catalyst showed reasonable turnover frequencies under supercritical conditions, comparable to those of its homogeneous counterpart in conventional solvents. Reactions were operated using a fixed-bed arrangement largely identical to that applied with traditional heterogeneous catalysts in $scCO_2$. No apparent rhodium leaching was reported and the catalyst showed stable performance under continuous flow conditions for 6.7 hours.

In a somewhat different approach, supported-aqueous-phase-catalysts (SAPC, see Chapter 5, Section 5.2.5 of this book) have been combined with supercritical CO_2 in catalytic hydrogenation [55]. Ruthenium was supported on silica and combined with the ligand TPPTS in water, after which a $scCO_2/H_2$ phase was applied together with the substrate. Better levels of conversion were obtained using $scCO_2$ than the equivalent system with toluene for the hydrogenation of cinnamaldehyde.

8.2.3.2 Organic Polymer Supports

Organic polymers have also been used as catalyst supports with $scCO_2$ as the substrate phase. Heck and Suzuki cross-coupling reactions were performed with palladium complexes using commercially available amino and phosphino polystyrene resins as scavengers or supports respectively [94]. Yields of greater than 80% were obtained using the amine-resin for the coupling of butyl acrylate and aryl iodides both without added phosphine ligand ("ligandless" reaction), and also with the addition of tri-*tert*-butyl phosphine. The results of the same coupling using palladium complexed to the phosphino resin were very dependent on the base used, those which were CO_2-soluble giving better yields. The use of quaternary phosphonium salts such as tetrabutylphosphonium acetate as the base also resulted in good yields, possibly due to the melting of the salts under the reaction conditions to form an ionic liquid solvent.

In another example, a polymer-supported chromium porphyrin complex was supported on ArgoGel Cl and then employed for the ring-opening polymerization of 1,2-cyclohexene oxide and CO_2 [95]. This complex showed higher activity than a CO_2-soluble equivalent, and the solid nature of the catalyst meant that recycling of the catalyst was much easier.

Amphiphilic resin supported ruthenium(II) complexes similar to those displayed in structure **1** were employed as recyclable catalysts for dimethylformamide production from supercritical CO_2 itself [96]. Tertiary phosphines were attached to crosslinked polystyrene-poly(ethyleneglycol) graft copolymers (PS-PEG resin) with amino groups to form an immobilized chelating phosphine. In this case recycling was not particularly effective as catalytic activity declined with each subsequent cycle, probably due to oxidation of the phosphines and metal leaching.

The chiral BINAPHOS ligand was anchored on an organic resin using a copolymerization strategy. After loading of the polymer with rhodium, the material was packed in a fixed-bed column using commercial equipment for supercritical fluid chromatography. The potential for continuous operation was demonstrated for the asymmetric hydroformylation of styrene to yield the chiral branched aldehydes (81% *iso*-selectivity) with *ee* values up to 86% [97]. Under the optimised reaction conditions, sequential hydroformylation of a small library of structurally diverse olefins was achieved using the same catalyst bed. In all cases excellent regio- and enantioselectivity was achieved using this straightforward technology.

An anionic rhodium iodide carbonyl complex was supported on polyvinylpyrrolidone for the carbonylation of methanol in the presence of $scCO_2$ [98]. Depending on the reaction conditions and method of extraction, less than 0.08% rhodium leaching was observed. Saturation of the support with methyl iodide was found to be vital to enhance the longevity and recyclability of the catalyst.

An "inverted" system based on the CO_2-soluble catalyst $Pd(OAc)_2/P^tBu_3$ has been utilized for Suzuki-coupling of resin-bound substrates [33]. The use of $scCO_2$ with polymer-supported substrates seems highly attractive owing to the known plasticizing and swelling effects of compressed CO_2 on polymeric compounds. Excess reagents and catalysts can readily be stripped off with $scCO_2$, yielding the solid-phase-bound product without the need for subsequent drying or purification steps. Interesting applications of this technique to parallel and/or high-throughput synthesis of biologically active compounds can be envisaged.

8.3. Economic Evaluation and Summary

8.3.1 POTENTIAL FOR SCALE-UP

Supercritical CO_2 offers three fundamentally different procedures, each of which can be carried out as conventional or inverted biphasic systems. Concentrating on the methods using $scCO_2$ as the product stream (*i.e.* conventional systems), the major difference results from the nature of the catalyst phase. Using the CESS procedure the catalyst is homogeneously dissolved in the $scCO_2$ phase during the reaction stage, whereas it is dissolved in a separate liquid phase in liquid/$scCO_2$ systems, or attached to a solid in solid/$scCO_2$ systems. Consequently, three different approaches must be considered for continuous flow operation of these methods. The solid/$scCO_2$ would be accomplished by straightforward fixed bed technology as already implemented on a commercial scale for heterogeneous hydrogenation [9, 10]. A continuously stirred tank reactor would probably be the method of choice for the liquid/$scCO_2$ system. Detailed descriptions of corresponding laboratory setups have been given elsewhere [70, 76].

The CESS approach would require a loop technology to combine the reaction and separation steps into one quasi-continuous process. It is the most complex and expensive of the methods described in this chapter, and can thus be regarded as the most demanding scenario for cost implications of processes employing $scCO_2$. A potential flow sheet for hydroformylation as prototypical catalytic process is illustrated in Figure 8.6, along with process conditions outlined in Table 8.1. The catalyst used for this process is [(cod)Rh(hfacac)]/4-H^2F^6-TPP (cod = cyclooctadiene, hfacac = hexafluoroacetylacetonate) in a molar ratio of 1:10. The 1-octene feed is heated to 65 °C in E101, then pumped into the reactor R101. At the same time the syngas (20 bar) and CO_2 pass through compressors C101 and C102 respectively, are mixed with the recycle stream from V101, heated to 65 °C, and fed into R101 at a total pressure of 200 bar. If necessary (for example at the beginning of the process), a solution containing the catalyst is pumped into R101 by P103 through E104 where it is heated to 65 °C. The product stream from V101 is then heated to 125 °C in E103, after which the pressure is reduced in C103 to 140 bar and the stream passes into separator S101. In this vessel 20% of the product precipitates as a liquid phase containing the catalyst. This stream is pumped back into R101 by P102, at the same time being returned to 65 °C in E104. The remainder of the product stream is then cooled to 25 °C in E105, the pressure adjusted to 50 bar by C104, and then passes into isolation chamber V101. At this stage, the liquid product is separated by gravity, while the remaining gas stream is recompressed to 200 bar by C105, whereupon it rejoins the feed stream. The isolated aldehyde mixture can be subjected directly to a conventional *n*/*iso* distillation without the need for further purification.

TABLE 8.1. Projected process and reactor specifications for hydroformylation of 1-octene in a possible CESS procedure

Reaction pressure (P_1)	200 bar
Reaction temperature (T_1)	65 °C
Separation pressure (P_2)	140 bar
Separation temperature (T_2)	125 °C
Isolation pressure (P_3)	50 bar
Isolation temperature (T_3)	25 °C
Rh:L:substrate (mol)	1:10:5000
1-octene conversion per pass	90%

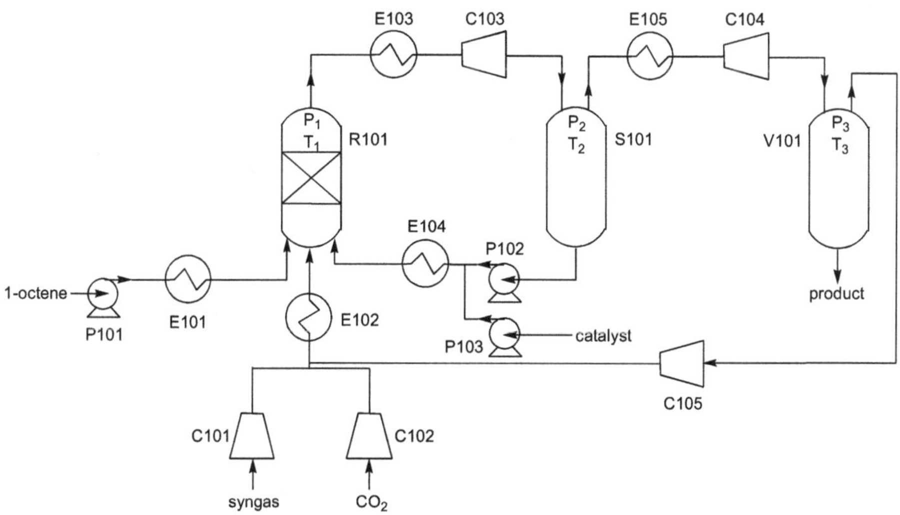

Figure 8.6. Process flow sheet for hydroformylation of 1-octene to 1-nonanal using the CESS approach

Such an approach clearly requires advanced engineering. Large costs are associated with the energy of compression, a feature of all three approaches involving $scCO_2$. This problem can be somewhat ameliorated by the use of temperature to control density, and hence the solubility, thus reducing the need for drastic alterations in pressure. The existing examples of industrial-scale processes employing pressurised CO_2 demonstrate that the cost of generating such pressures need not be prohibitive, and additionally provide a technical framework on which new methods may be developed. Whereas the CESS procedure seems to be particularly applicable to high value added products, the other two approaches described in this chapter could potentially reach the bulk and commodity businesses if suitably optimised. A major advantage of using $scCO_2$ for catalyst immobilisation is the effectively solventless generation of the final products. Thus, all downstream processing related to solvent removal, recycling or disposal is avoided. This can be especially important in cases such as fragrances, food additives or pharmaceuticals where even trace amounts of solvent residues must be avoided.

8.4 Summary

In this chapter we have outlined a number of approaches to catalyst immobilisation in systems employing $scCO_2$ either as mobile phase or catalyst support. Clearly no single approach is ideal for all cases, so parallel development routes are essential. Gratifyingly, $scCO_2$ is compatible with virtually any of the catalyst supports currently known, and further examples are likely to emerge in the near future. Probably the greatest strength of pressurised CO_2 as a solvent is the ability to change its solvent properties simply by altering the density of the medium. The short history of this exciting research area as highlighted in this chapter has already yielded a variety of successful examples of sustainable catalysis on a laboratory scale. While no industrial application has been implemented to date, there seems to be no fundamental technological or economical barrier to prevent such a development.

This chapter illustrates that the potential of $scCO_2$ for reaction engineering of homogeneous catalysis offers a great scope for possible scientific and technical innovation. This newly emerging field of catalysis research lies at the interface of molecular sciences and process engineering and its future development will require truly interdisciplinary efforts from experts in both fields.

8.5 References

[1] P. G. Jessop and W. Leitner, in 'Chemical Synthesis Using Supercritical Fluids', Weinheim, 1999.
[2] W. Leitner, *Acc. Chem. Res.*, 2002, **35**, 746.
[3] P. G. Jessop, T. Ikariya, and R. Noyori, *Chem. Rev.*, 1999, **99**, 475.
[4] D. J. Cole-Hamilton, *Science*, 2003, **299**, 1702.
[5] K. Zosel, *Angew. Chem. Int. Ed.*, 1978, **17**, 702.
[6] M. A. McHugh and V. J. Krukonis, 'Supercritical Fluid Extraction: Principles and Practice', Butterworth-Heinemann, Oxford, 1994.
[7] S. B. Iverson, T. Larsen, O. Henriksen, and K. Felsvang, Proceedings of 6th International Symposium on Supercritical Fluids, Versailles, France, 2003.
[8] S. L. Wells and J. DeSimone, *Angew. Chem. Int. Ed.*, 2001, **40**, 519.
[9] P. Licence, J. Ke, M. Sokolova, S. K. Ross, and M. Poliakoff, *Green Chem.*, 2003, **5**, 99.
[10] http://www.thomas-swan.co.uk.
[11] 'Chemical Synthesis Using Supercritical Fluids', ed. W. Leitner, Wiley VCH, Weinheim, 1999.
[12] J. Ke, B. X. Han, M. W. George, H. K. Yan, and M. Poliakoff, *J. Am. Chem. Soc.*, 2001, **123**, 3661.
[13] T. Sarbu, T. Styranec, and E. J. Beckman, *Nature*, 2000, **405**, 165.
[14] W. Leitner, *Nature*, 2000, **405**, 129.
[15] K. Schwirten, R. Kummer, and W. Richter, in 'Working up of hydroformylation or carbonylation reacion mixtures', USA, 1981.
[16] M. F. Sellin, I. Bach, J. M. Webster, F. Montilla, V. Rosa, T. Aviles, M. Poliakoff, and D. J. Cole-Hamilton, *J. Chem. Soc. Dalton Trans.*, 2002, 4569.
[17] M. F. Sellin and D. J. Cole-Hamilton, *J. Chem. Soc. Dalton Trans.*, 2000, 1681.
[18] P. B. Webb and D. J. Cole-Hamilton, *Chem. Commun.*, 2004, 612.
[19] S. Kainz and W. Leitner, *Catalysis Letters*, 1998, **55**, 223.
[20] G. B. Coombes, F. Dehghani, F. P. Lucien, A. K. Dillow, and N. R. Foster, in 'Reaction Engineering for Pollution Prevention', ed. R. P. Hesketh, Elsevier, Amsterdam, 2000.
[21] Y. Kayaki, Y. Noguchi, and T. Ikariya, *Chem. Commun.*, 2000, 2245.
[22] S. Pandey, M. A. Kane, G. A. Baker, F. V. Bright, A. Fürstner, G. Seidel, and W. Leitner, *J. Phys. Chem. B*, 2002, **106**, 1820.

[23] A. Fürstner, D. Koch, K. Langemann, and W. Leitner, *Angew. Chem. Int. Ed.*, 1997, **36**, 2466.
[24] A. Fürstner, L. Ackermann, H. Beck, H. Hori, D. Koch, K. Langemann, M. Liebl, C. Six, and W. Leitner, *J. Am. Chem. Soc.*, 2001, **123**, 9000.
[25] I. Bach and D. J. Cole-Hamilton, *Chem. Commun.*, 1998, 1463.
[26] C. A. G. Carter, R. T. Baker, S. P. Nolan, and W. Tumas, *Chem. Commun.*, 2000, 347.
[27] S. Kainz, D. Koch, W. Baumann, and W. Leitner, *Angew. Chem. Int. Ed.*, 1997, **36**, 1628.
[28] K. D. Wagner, N. Dahmen, and E. Dinjus, *J. Chem. Eng. Data*, 2000, **45**, 672.
[29] D. Koch and W. Leitner, *J. Am. Chem. Soc.*, 1998, **120**, 13398.
[30] D. R. Palo and C. Erkey, *Organometallics*, 2000, **19**, 81.
[31] A. M. B. Osuna, W. P. Chen, E. G. Hope, R. D. W. Kemmitt, D. R. Paige, A. M. Stuart, J. L. Xiao, and L. J. Xu, *J. Chem. Soc. Dalton Trans.*, 2000, 4052.
[32] K. Wittmann, W. Wisniewski, R. Mynott, W. Leitner, C. L. Kranemann, T. Rische, P. Eilbracht, S. Kluwer, J. M. Ernsting, and C. J. Elsevier, *Chem. Eur. J*, 2001, **7**, 4584.
[33] T. R. Early, R. S. Gordon, M. A. Carroll, A. B. Holmes, R. E. Shute, and I. F. McConvey, *Chem. Commun.*, 2001, 1966.
[34] D. K. Morita, D. R. Pesiri, S. A. David, W. H. Glaze, and W. Tumas, *Chem. Commun.*, 1998, 1397.
[35] M. A. Carroll and A. B. Holmes, *Chem. Commun.*, 1998, 1395.
[36] N. Shezad, R. S. Oakes, A. A. Clifford, and C. T. Rayner, *Tetrahedron Letters*, 1999, **40**, 2221.
[37] N. Shezad, A. A. Clifford, and C. T. Rayner, *Green Chem.*, 2002, **4**, 64.
[38] T. Osswald, S. Schneider, S. Wang, and W. Bannwarth, *Tetrahedron Letters*, 2001, **42**, 2965.
[39] G. Musie, M. Wei, B. Subramaniam, and D. H. Busch, *Coord. Chem. Rev.*, 2001, **219-221**, 789.
[40] H. Hori, C. Six, and W. Leitner, *Macromolecules*, 1999, **32**, 3178.
[41] T. J. de Vries, R. Duchateau, M. A. G. Vorstman, and J. T. F. Keurentjes, *Chem. Commun.*, 2000, 263.
[42] A. Wegner, 'Atomökonomische Synthesen in komprimiertem Kohlendioxid', PhD, Universität Jena, Mülheim an der Ruhr, 2000.
[43] W. Kläui, J. Bongards, and G. J. Reiss, *Angew. Chem. Int. Ed.*, 2000, **39**, 3894.
[44] S. Lange, A. Brinkmann, P. Trautner, K. Woelk, J. Bargon, and W. Leitner, *Chirality*, 2000, **12**, 450.
[45] G. Franciò and W. Leitner, *Chem. Commun.*, 1999, 1663.
[46] D. J. Adams, W. P. Chen, E. G. Hope, S. Lange, A. M. Stuart, A. West, and J. L. Xiao, *Green Chem.*, 2003, **5**, 118.
[47] D. Bonafoux, Z. H. Hua, B. H. Wang, and I. Ojima, *J. Fluorine Chem.*, 2001, **112**, 101.
[48] G. Franciò, K. Wittmann, and W. Leitner, *J. Fluorine Chem.*, 2001, **621**, 130.
[49] E. L. V. Goetheer, A. W. Verkerk, L. J. P. van den Broeke, E. de Wolf, B. J. Deelman, G. van Koten, and J. T. F. Keurentjes, *J. Catal.*, 2003, **219**, 126.
[50] M. J. Burk, S. G. Feng, M. F. Gross, and W. Tumas, *J. Am. Chem. Soc.*, 1995, **117**, 8277.
[51] S. Kainz, A. Brinkmann, and W. Leitner, *J. Am. Chem. Soc.*, 1999, **121**, 6421.
[52] A. Wegner and W. Leitner, *Chem. Commun.*, 1999, 1583.
[53] R. J. Bonilla, B. R. James, and P. G. Jessop, *Chem. Commun.*, 2000, 941.
[54] R. Wiebe and V. L. Gaddy, *J. Am. Chem. Soc.*, 1941, **63**, 475.
[55] B. M. Bhanage, Y. Ikushima, M. Shirai, and M. Arai, *Chem. Commun.*, 1999, 1277.
[56] B. M. Bhanage, Y. Ikushima, M. Shirai, and M. Arai, *Tetrahedron Letters*, 1999, **40**, 6427.
[57] G. B. Jacobsen, C. T. Lee Jr., K. P. Johnston, and W. Tumas, *J. Am. Chem. Soc.*, 1999, **121**, 11092.
[58] D. Hâncu and E. J. Beckman, *Green Chem.*, 2001, **3**, 80.
[59] M. McCarthy, H. Stemmer, and W. Leitner, *Green Chem.*, 2002, **4**, 501.
[60] W. Leitner, *Nature*, 2003, **423**, 930.
[61] D. J. Heldebrant and P. G. Jessop, *J. Am. Chem. Soc.*, 2003, **125**, 5600.
[62] I. Komoto and S. Kobayashi, *Chem. Commun.*, 2001, 1842.
[63] E. Kukova, M. Petermann, and E. Weidner, *Chemie Ingenieur Technik*, 2004, **76**, 280.
[64] C. M. Gordon, *Appl. Catal. A-Gen.*, 2001, **222**, 101.
[65] P. Wasserscheid and W. Keim, *Angew. Chem. Int. Ed.*, 2000, **39**, 3773.
[66] T. Welton, *Chem. Rev.*, 1999, **99**, 2071.
[67] J. D. Holbrey and K. R. Seddon, *Clean. Prod. Proc.*, 1999, **1**, 223.
[68] L. A. Blanchard, D. Hâncu, E. J. Beckman, and J. F. Brennecke, *Nature*, 1999, **399**, 28.
[69] L. A. Blanchard and J. F. Brennecke, *Ind. Eng. Chem. Res.*, 2001, **40**, 287.
[70] M. T. Reetz, W. Wiesenhofer, G. Franciò, and W. Leitner, *Chem. Commun.*, 2002, 992.
[71] P. Lozano, T. De Diego, D. Carrie, M. Vaultier, and J. L. Iborra, *Biotechnol. Prog.*, 2003, **19**, 380.
[72] M. T. Reetz, W. Wiesenhofer, G. Franciò, and W. Leitner, *Adv. Syn. Catal.*, 2003, **345**, 1221.

[73] R. A. Brown, P. Pollet, E. McKoon, C. A. Eckert, C. L. Liotta, and P. G. Jessop, *J. Am. Chem. Soc.*, 2001, **123**, 1254.
[74] F. C. Liu, M. B. Abrams, R. T. Baker, and W. Tumas, *Chem. Commun.*, 2001, 433.
[75] M. F. Sellin, P. B. Webb, and D. J. Cole-Hamilton, *Chem. Commun.*, 2001, 781.
[76] P. B. Webb, M. F. Sellin, T. E. Kunene, S. Williamson, A. M. Z. Slawin, and D. J. Cole-Hamilton, *J. Am. Chem. Soc.*, 2003, **125**, 15577.
[77] P. W. Jolly and G. Wilke, in 'Applied Homogeneous Catalysis with Organometallic Compounds', ed. W. A. Hermann, VCH, Weinheim, 1996.
[78] G. Franciò, F. Faraone, and W. Leitner, *J. Am. Chem. Soc.*, 2002, **124**, 736.
[79] T. V. RajanBabu, *Chem. Rev.*, 2003, **103**, 2845.
[80] M. Nomura, J. Jin, H. Park, and T. V. RajanBabu, *J. Am. Chem. Soc.*, 1998, **120**, 459.
[81] A. Bösmann, G. Franciø, E. Janssen, M. Solinas, W. Leitner, and P. Wasserscheid, *Angew. Chem. Int. Ed.*, 2001, **40**, 2697.
[82] D. Ballivet-Tkatchenko, M. Picquet, M. Solinas, G. Franciò, P. Wasserscheid, and W. Leitner, *Green Chem.*, 2003, **5**, 232.
[83] A. Baiker, *Chem. Rev.*, 1999, **99**, 453.
[84] W. Leitner, *Angew. Chem. Int. Ed.*, 1995, **34**, 2207.
[85] P. G. Jessop, T. Ikariya, and R. Noyori, *Chem. Rev.*, 1995, **95**.
[86] O. Kröcher, R. A. Köppel, and A. Baiker, *Chem. Commun.*, 1996, 1497.
[87] O. Kröcher, R. A. Köppel, M. Froba, and A. Baiker, *J. Catal.*, 1998, **178**, 284.
[88] L. Schmid, O. Kröcher, R. A. Köppel, and A. Baiker, *Microporous and Mesoporous Mater.*, 2000, **35-36**, 181.
[89] O. Hemminger, A. Marteel, M. R. Mason, J. A. Davies, A. R. Tadd, and M. A. Abraham, *Green Chem.*, 2002, **4**, 507.
[90] A. R. Tadd, A. Marteel, M. R. Mason, J. A. Davies, and M. A. Abraham, *Ind. Eng. Chem. Res.*, 2002, **41**, 4514.
[91] A. R. Tadd, A. Marteel, M. R. Mason, J. A. Davies, and M. A. Abraham, *J.Supercritical Fluids*, 2003, **25**, 183.
[92] A. Marteel, J. A. Davies, M. R. Mason, T. Tack, S. Bektesevic, and M. A. Abraham, *Catal. Commun.*, 2003, **4**, 309.
[93] N. J. Meehan, M. Poliakoff, A. J. Sandee, J. N. H. Reek, P. C. J. Kamer, and P. W. N. M. van Leeuwen, *Chem. Commun.*, 2000, 1497.
[94] R. S. Gordon and A. B. Holmes, *Chem. Commun.*, 2002, 640.
[95] L. M. Stamp, S. A. Mang, A. B. Holmes, K. A. Knights, Y. R. de Miguel, and I. F. McConvey, *Chem. Commun.*, 2001, 2502.
[96] Y. Kayaki, Y. Shimokawakoto, and T. Ikariya, *Adv. Syn. Catal.*, 2003, **345**, 175.
[97] F. Shibahara, K. Nozaki, and T. Hiyama, *J. Am. Chem. Soc.*, 2003, **125**, 8555.
[98] R. J. Sowden, M. F. Sellin, N. De Blasio, and D. J. Cole-Hamilton, *Chem. Commun.*, 1999, 2511.

CHAPTER 9 AREAS FOR FURTHER RESEARCH

D. J. COLE-HAMILTON[a] AND R. P. TOOZE[b]
[a] *EaStCHEM, School of Chemistry, Univesity of St. Andrews, St. Andrews, Fife, KY19 9ST, Scotland*
[b] *Sasol Technology UK, Ltd., Purdie Building, North Haugh, St. Andrews, Fife, KY19 9ST, Scotland*

9.1 Introduction

A wide variety of new approaches to the problem of product separation in homogeneous catalysis has been discussed in the preceding chapters. Few of the new approaches has so far been commercialised, with the exceptions of a the use of aqueous biphasic systems for propene hydroformylation (Chapter 5) and the use of a phosphonium based ionic liquid for the Lewis acid catalysed isomerisation of butadiene monoxide to dihydrofuran (see Equation 9.1). This process has been operated by Eastman for the last 8 years without any loss or replenishment of ionic liquid [1]. It has the advantage that the product is sufficiently volatile to be distilled from the reactor at the reaction temperature so the process can be run continuously with built in product catalyst separation. Production of lower volatility products by such a process would be more problematic. A side reaction leads to the conversion of butadiene oxide to high molecular weight oligomers. The ionic liquid has been designed to facilitate their separation from the catalyst (see Section 9.7)

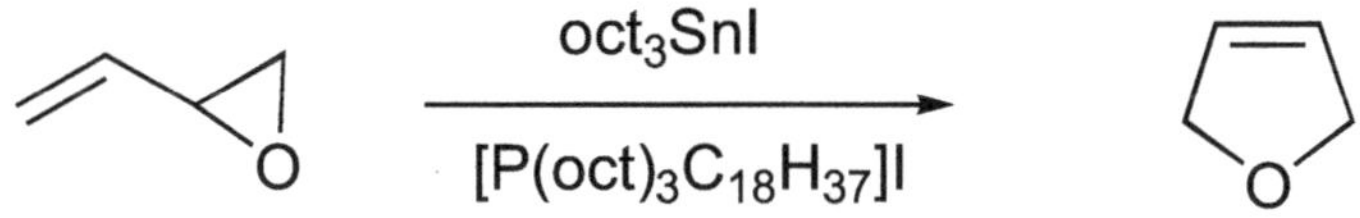

Equation 9.1 Commercialised Lewis acid catalysed production of dihydrofuran in a phosphonium ionic liquid.

In this chapter, we examine the various processes by taking a qualitative look at which parts need to be improved by further research in order to make them commercially attractive for the separation of lower volatility products and especially competitive with low pressure distillation. Once again we focus on the rhodium/tertiary phosphine catalysed hydroformylation of long chain alkenes, specifically 1-octene, since data concerning this reaction is provided in the preceding chapters. A summary of the best results obtained from each of the processes and the problems associated with their implementation appears in Table 9.1. A full economic analysis of each approach to the product separation problem is beyond the scope of this book, so any conclusions as to

D. J. Cole-Hamilton and R. P. Tooze (eds.), Catalyst Separation, Recovery and Recycling, 237–248.

what research is still needed to render any of the processes commercially viable will at least to some extent be subjective.

9.2 Conventional separation methods (See Chapter 2)

Conventional distillation directly from the reaction chamber is not possible for longer chain aldehydes than C_5 because the volatility of the aldehyde is too low for it to be removed at an economically viable rate without raising the temperature above the decomposition temperature of the catalyst. In principle, an alternative to higher temperature is a faster flow of the gas phase through the reactor so that more product is removed per unit time. However, there is a limit in that liquid expansion (foaming) occurs and the reactor would have to be prohibitively large to prevent foam filling it entirely and the liquid catalyst solution also being carried out of the reactor. Once again, this limit is reached at about C_5 aldehydes. The formation of heavy (aldehyde dimer and trimer) products is also enhanced during the distillation process (higher aldehyde concentration and high temperature, see below).

Batch continuous processing, in which part of the catalytic solution is removed to a low pressure distillation unit, on the other hand, has recently been commercialised [2-4]. Very little information is available in the public domain concerning this low pressure distillation process, but the main extra cost will be in generating the reduced pressure required for the distillation. The estimated vapour pressures at 110°C of various long chain linear aldehyde products that are commercially desirable are shown in Figure 9.1. This temperature has been chosen because this is the high temperature limit above which the rhodium triphenylphosphine complex starts to decompose. Any commercial process will require to operate the product distillation step at a pressure no higher than those shown for the individual aldehydes.

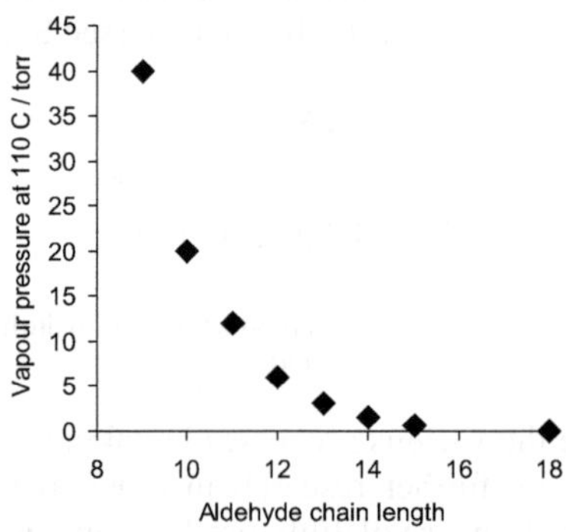

Figure 9.1. Estimated vapour pressures at 110°C of aldehydes of different chain lengths

TABLE 9.1. Rhodium catalysed hydroformylation of 1-octene using a variety of separation methods and catalyst systems compared with results from homogeneous commercial systems (First 4 entries). Modified from [5]

System	Reference	p / bar	T / °C	TOF / h^{-1}	Rate / mol dm^{-3} h^{-1}	l:b	Linear aldehyde / %	Rh loss mg (mol product)$^{-1}$	P loss[a] mg (mol product)$^{-1}$	Problem[a]
Homo / Co	[6]	300	150	38	3.2	2.3	50			
Homo / Co	[7]	80	200	20	1.6	9	80			
Homo[b]	[6]	15	95	770	2.0	8.8	90			
Homo[b]	[6]	16	110	444	0.8	5.3	83			
Supported	[8]	50	80	287	0.12	40	9.5	< 0.1	nr	
Supported / $scCO_2$	[9]	170	90	160		33	94	< 1.2	nr	R, P
Soluble polymer[c]	[10]	30	80	160	0.03	nr	nr	0.3	nr	R, S?, L
Dendrimer	[11]	10	120	1792	7.1	13.9	86	nr	nr	U
Supported Dendrimer[d]	[12]	67	45	25	0.03	0.05	92[e]	20	nr	R,L
Thermo-regulated Aqueous biphasic	[13]	50	100	182	0.3	nr	nr	nr	nr	R, S?
Fluorous biphasic	[14]	10	100	837	0.1	4.5	75	0.12	nr	S, L, E
Fluorous biphasic	[15]	20	70	4400	8.8	6.3	81	0.08	16	E
$ScCO_2$	[16]	200	65	430	14.2	5.5	79	< 0.17	nr	P
$ScCO_2$	[17]	125	100	162	2.5	3.2	75	0.014-0.07	nr	R, L, S
Ionic liquid	[18]	60	100	6200	5.3	44	85	< 0.003	< 0.06	E, I
Supported ionic liquid[c]	[19]	100	100	3600		2.4	71	0.004	nr	S, E?
$ScCO_2$ / Ionic liquid	[20] ([21])	200	100	517 (272)	7.8 (4.1)	3.1 (40)	76 (92)	0.002 (0.03)	nr	P, S, E (P, E)

[a] nr = not recorded, R = low rate, P = high pressure, S = low selectivity to linear aldehyde, L = high catalyst leaching, U = ultrafiltration not attempted, E = expensive ligand, and/or solvent, I high levels of isomerisation; [b] Rh/PPh_3 for propene, which cannot isomerise; [c] 1-hexene; [d] vinyl acetate; [e] branched aldehyde; [f] leaching measured for styrene

The low pressure distillation may also lead to volatilisation of the ligand so that ligand losses may be high. Some of this ligand will be recycled, whilst some must be replaced. Finally, there is a major issue concerning heavies formation in systems manufacturing long chain aldehydes.

High boiling point products (heavies) are dimers or trimers produced from the aldehyde products by aldol condensation reactions. They, thus, have a molecular mass which is approximately twice that of the product aldehyde (or greater) and very much reduced volatility. In the production of butanal, they are removed at about the rate at which they are formed, so once they have built up sufficiently to act as the reaction solvent, they reach a steady state level in the reactor. For longer chain aldehydes, the heavy products are essentially involatile, so they must be removed by a liquid purge, which also removes catalyst. Here the separation problem is extremely severe. One way to reduce this problem is to operate in an inert diluent (reaction solvent). Since aldol condensation is second order in aldehyde (i.e the rate of aldol condensation α $[\text{aldehyde}]^2$), working at lower aldehyde concentration substantially reduces the rate of aldol condensation and the build up of heavies, thus allowing a longer time between liquid purges and a longer lifetime for the catalyst. Processes have been developed for the recovery of some of the catalyst in the liquid purge stream, e. g. by adsorption onto an insoluble support such as an acidic ion exchange resin [4].

Despite these problems, this low pressure distillation process has proven sufficiently economical to be commercialised for the hydroformylation of long chain alkenes and represents the benchmark against which all other processes must be judged.

9.3 Catalysts on insoluble supports (Chapter 3)

Until recently, all attempts that have been made to carry out catalytic reactions using metal complexes anchored to insoluble supports such as solid oxides or polymers have been prone to a degree of leaching that is too large for commercialisation, unless the leached material is recaptured on a guard bed of the same carrier, which in time takes on sufficient catalyst loading that it can be recycled as the active catalyst bed. The recapturing is carried out at lower temperature and pressure than the reaction, so that the stability constant for binding of the catalyst to the resin is maximised. More recently, catalysts have been developed in which the leaching is undetectable and single samples of catalyst have been used in a variety of reactions for over one year [22]. This is an exciting development, which relies upon the use of bidentate binding to the metal, so that the metal is retained with high efficiency, and covalent linking of the ligand into the solid matrix *via* a sol-gel reaction, ensuring that the ligand itself cannot break free from this support. The ligand shown in Figure 9.2 has been used successfully for the hydroformyation of 1-octene with no leaching being detectable (<0.005 ppm Rh, <0.1 ppm P).

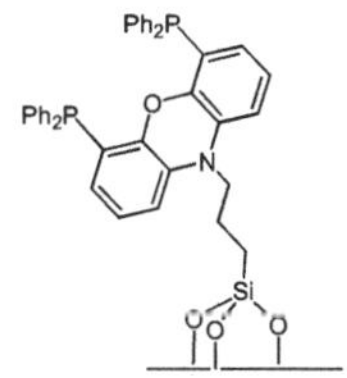

Figure 9.2. Ligand used for supporting a hydroformylation catalyst on a solid support. Leaching of rhodium under the reaction conditions is insignificant.

This is a very attractive process, which does not require any special engineering solutions because the catalyst can be used in a fixed bed, just like a heterogeneous catalyst. The cost of the ligand and the slightly diminished activity mean that cheaper more active catalysts of this kind represent a worthwhile goal and a significant challenge. Multipass operation may overcome the activity problem to some extent, but this has not been demonstrated.

Other factors which affect all new approaches to the catalyst separation problem include ligand degradation and heavies formation, although the latter may not be a serious problem for the supported catalysts as the heavy products should be removed with the organic product. In addition, there are the added complications of catalyst fouling and of mechanical abrasion that can occur with solid phase catalysts. The extent of these problems will only become apparent after continuous long term investigations under the conditions that would be used for a commercial process. The recent advances with supported catalysts suggest that the time is right to carry out these studies.

The same types of catalyst have been employed in 1-octene hydroformylation, but with the substrates and products being transported to and from the reaction zone dissolved in a supercritical fluid (carbon dioxide) [9]. The activity of the catalyst is increased compared with liquid phase operation, probably because of the better mass transport properties of $scCO_2$ than of the liquid. This type of approach may well reduce heavies formation because of the low concentration of aldehyde in the system, but the heavies that do form are likely to be insoluble in $scCO_2$, so may precipitate on and foul the catalyst. The main problem with this process, however, is likely to be the use of high pressure, which is common to all processes where supercritical fluids are used (see Section 9.8).

Supported aqueous phase (Chapter 3, Section 3.6, Chapter 5, Section 5.2.5) and supported ionic liquid phase catalysts, Chapter 7, Section 7.3) are probably not suitable for use with higher alkenes because the liquid feed slowly dissolves some of the water or ionic liquid changing the nature of the catalyst and leading to catalyst leaching.

9.4 Catalysts on soluble supports (Chapter 4)

Dendrimers and soluble polymers provide alternative supports to solids, which have the advantage that access to the catalytically active sites is not restricted. The main problem in these cases is not in the catalysis – reactions with high rates and selectivities have been reported - but rather in the separation which relies on nano- or ultrafiltration. One way of circumventing these problems is to support dendrimers on solid beads so that flow systems or simple filtration can be used. Because of the nature of the dendrimers, all their active sites are on the surface so all are accessible from the liquid phase even if they are supported on an insoluble support. This approach has only been applied in a very restricted number of reactions [12, 23] so increased research in this area could prove to be very productive.

For dendrimers and soluble polymers, there are two main problems: leaching of metal from the catalyst and lack of retention of the dendrimer or polymer by the membrane. Leaching could perhaps be addressed by using bidentate coordination and covalent attachment of the ligand, as in the solid supported catalysts described above, whilst the retention issue is potentially more difficult. The macromolecules could leak through the membrane either because the size of the channels is varied and some are large enough to allow the macromolecules through or because of imperfections such as pinholes.

Although membranes have successfully been developed for ultrafiltration in biological systems in water (removal of whole cells), the research on membrane materials for use with organic solvents and under gas pressure is in its infancy. This is an area where considerable further research could be carried out. Preventing membrane fouling is also an area where more information is required. Cross flow arrangements seem to be much more successful in this regard than dead-end filtration and they also allow continuous (loop) operation. Attention to the detailed shape and design of the channels could well pay large dividends in improving the efficiency of such systems.

Heavies that form in these reactions may be of sufficient size to be excluded by the membrane and hence retained with the catalyst. This would present a major problem as their separation from the catalyst will be non trivial.

9.5 Aqueous biphasic catalysis (Chapter 5)

Water - organic biphasic systems are the most developed of all the alternative technologies for homogeneous catalyst separation. Such a system has been commercialised by Ruhr Chemie/Rhône Poulenc/for the hydroformylation of propene to butanal and has been operating successfully for many years. The major problem with adapting such a process to the hydroformylation of longer chain alkenes does not concern the product separation, but rather the reaction itself. Very low rates are achieved using longer chain alkenes. Despite the fact that this approach was first described 30 years ago, it is still not entirely clear where the reaction takes place – in the aqueous phase or at the water – organic interface. It almost certainly does not occur in the organic phase or catalyst leaching would be a problem. Detailed studies aimed at clarifying this question would be highly beneficial.

As outlined in Chapter 5, Section 5.2.3.2 various approaches to overcoming the low rates of the hydroformylation of long chain alkenes in aqueous biphasic systems have been proposed. Some of these, such as the use of microemulsions [24-26] or pH dependent solubility [27], have provided improvements often at the expense of complicating the separation process. Perhaps the most promising new approaches involve the introduction of new reactor designs where improved mixing allows for much better rates of mass transport between the three phases (water, organic and gas) [28]. The use of thermoregulated systems [13] in which a temperature dependent phase transition makes the catalyst organic phase soluble under the reaction conditions, but water soluble on cooling is a very neat approach worthy of more development. Thermomorphic or latent biphasic polymers may also be important in this regard [29]. Further investment in these types of study would be very useful and could allow the commercialisation of such systems. The very attractive properties of water make this approach one of the most environmentally beneficial and many the problems associated with ligand degradation, long term catalyst stability and spent catalyst recycling have been solved (see Chapter 5, Section 5.3). Heavy aldehyde dimers should be removed with the organic phase and so will have to be separated from the product, probably by simple distillation.

9.6 Fluorous biphasic catalysis (Chapter 6)

It is generally considered that the major obstacle to the commercialisation of reactions employing the fluorous biphasic concept is the cost of the fluorous solvent and the cost of the ligand, which must contain very large amounts of fluorine to retain the catalyst within the fluorous phase. However, preliminary calculations (see Chapter 6, Section 6.14) suggest that careful handling of both the solvent and the ligand could reduce these costs to well below 1 € per tonne of product and that the costs here will be lower than those resulting from rhodium losses. However, this analysis does not take into account ligand degradation to give alkyl phosphines with all the problems, outlined in Chapter 2, Section 2.6 and Chapter 5, Section 5.3, that this brings associated with lower catalyst activity and selectivity as well as poorer catalyst retention. It should also be noted that the environmental persistence and very high global warming potential of fluorinated compounds may make their licensing for wide scale use problematic.

One advantage of the fluorous ligands, which also has implications for reducing the cost, is that the electron withdrawing fluorous ponytails increase the stability constants for complex formation and favour the linear product more than less electron withdrawing ligands. This means that high selectivity towards the linear product can be obtained at low ligand:metal ratio (low ligand loading). Because hydroformylation reactions are inhibited by excess phosphine ligands, the low ligand loadings possible in these systems mean that high rates are obtained and the reaction can be run at lower temperatures, still giving commercially attractive reaction rates [15]. The lower reaction temperature should lead to less degradation of the ligand by all the possible mechanisms, so that ligand losses will be greatly reduced.

Heavy products will be dissolved in the organic phase so that contamination of the catalyst phase and regular purging will not be necessary.

One major problem remains, however, leaching of the ligand and catalyst into the organic phase. The calculations in Chapter 6, Section 6.14, assume that the rhodium leaching can be reduced to 1 ppb and that the ligand losses may be 10 times this. At present, no catalyst has been reported that comes anywhere near this goal, so significant research is required to discover new catalyst designs which retain the ligand much more effectively into the fluorous phase. In addition, the leaching is much reduced if reactions are carried out to high conversion [30]. The reactor design proposed in Chapter 6, Section 6.14, for obtaining high conversion is rather simplistic and research into new reactor design, which would allow high conversions under continuous flow operation would be highly beneficial.

9.7 Reactions involving ionic liquids (Chapter 7)

As indicated in Section 9.1, one of the few examples of the implementation of an alternative strategy for homogeneous catalyst – product separation involves an ionic liquid. In this reaction, the isomerisation of butadiene oxide to dihydrofuran (Equation 9.1), the product is sufficiently volatile to be distilled form the reactor under the reaction conditions in a continuous manner. Where this is not the case, the reaction kinetics require that the substrate and the catalyst both dissolve within the ionic liquid. If the product is much less polar than the starting material, it may phase separate and be removed by decantation. If, however, as in long chain alkene hydroformylation, the product is more polar than the substrate, it is likely that it too will be soluble in the ionic liquid and hence might require extraction. In practice, the engineering solutions described in Chapter 7, Section 7.4 rely upon the ionic liquid becoming saturated with product so that a separate organic phase consisting of unreacted substrate and products forms and can be separated by decantation rather as in the aqueous biphasic process. The main advantage of the using an ionic liquid over using water is that the ionic liquid can be tuned to dissolve the substrate, even if it is of relatively low polarity so that high reaction rates can be achieved. The main disadvantage in a chemical sense is that the ionic liquid may be soluble to some extent in the organic phase, thus leading to the leaching of both ionic liquid and catalyst. The catalyst can be anchored within the ionic liquid phase by making it highly polar, although not so polar that it does not dissolve in the ionic liquid, but rather little has been reported about the solubility of the ionic liquids in the product stream. If the catalyst is retained within the main ionic liquid phase, the extracted ionic liquid can, in principle, be collected from the bottom of the product distillation column. In reality, this could be a major problem as the aldol condensation products, which are also essentially non-distillable, will be almost impossible to separate from this extracted ionic liquid. The cost of the ionic liquid means that such losses must be kept to an absolute minimum, so ionic liquids which dissolve the organic substrate but are insoluble in the product would be the ideal solution to the problem, although whether such liquids represent a realistic possibility remains to be seen. Most of the heavies will be in the organic phase, but they may also build up in the ionic liquid, in which case a purge will be required. Fortunately, on this occasion, the heavies should be extractable from the ionic liquid and catalyst using an organic solvent and the ionic liquid phase could be recycled.

In the Eastman process for 2,5-dihydrofuran production, the situation is different and the problem of heavy products has been tackled in a highly original manner.[31] The oligomers formed in the process are highly polar and insoluble in alkanes. The ionic liquid, $[P(oct)_3C_{18}H_{37}]I$ and the Lewis acid catalyst, $[Sn(oct)_3]I$, which are non toxic ($LD_{50} > 2$ g kg^{-1} for each), non-flammable (flammability 1) and non-corrosive (340 stainless steel is used for the reactor), have been designed to be soluble in heptanes. Part of the ionic liquid containing the oligomers, which is being recycled from the distillation unit to the reactor, is purged to a counter current extractor where the catalyst and the ionic liquid are extracted into heptanes. They are then stripped of the heptanes and returned to te reactor whilst the oligomers, now freed from the catalyst and ionic liquid, are incinerated.

Another solution to the problem of ionic liquid loss to the organic phase is to extract the product from the ionic liquid using a supercritical fluid (See Chapter 8, Section 8.2.2.3). It has been demonstrated that this can be done continuously for a variety of reactions including the hydroformylation of long chain alkenes [20], and that neither the ionic liquid nor the catalyst are leached to significant extents. The only problem here is the high pressures involved (see section 9.8).

Other problems with ionic liquids in addition to their cost, which is falling, are:

- their environmental persistence, especially important if they are lost with the heavies purge
- their lack of flammability, which may mean that destruction by incineration is difficult
- their slight but significant toxicity [32, 33]
- and their corrosive properties [34]

Many of these problems may be overcome by using ionic liquids based on sugars [35] or deep eutectic mixtures. Deep eutectic mixtures such as that derived from choline chloride and urea (m. pt. 12°C [36]) or carboxylic acids [37] can be liquids and have very low vapour pressure. They have been successfully used as electrochemical solvents, but their use in catalysis remains little explored. Urea is a fertiliser and choline chloride (Vitamin B4) is a component of chicken feed so the mixture is environmentally acceptable.

With the myriad of ionic liquids that are available, there is surely ample scope for further research of these important solvents and considerable grounds for optimism that improvements of the current systems to an extent that encourages commercialisation are possible.

9.8 Reactions using supercritical fluids (Chapter 8)

Like water, carbon dioxide is and environmentally very attractive solvent. The ability to control its solvating power by simple swings in temperature and pressure makes it an ideal medium for homogeneous catalysis. The main problem compared with conventional systems is that such swings, especially in pressure, require costly recompression and care must be taken to control the temperature whilst the pressure swing is occurring so that mixed liquid and gas phases do not form. They are very difficult to pump. The results presented in Chapter 8 suggest that rates and selectivities for the hydroformylation of 1-octene and other long chain alkenes can be commercially attractive and that separation of the catalyst can be achieved at least in styrene hydroformylation using the CESS process [38] (see Chapter 8.2). These preliminary experiments on the CESS approach suggest that some rhodium leaching still occurs, so improvements will be required either in the ligand design or in the optimisation of the pressure swings to reduce this still further. The necessity to use costly fluorous derived ligands to render the catalysts soluble in $scCO_2$ adds extra impetus to the search for better catalyst retention. An attractive aspect of the CESS process is that the formation of heavy aldehyde dimers will be reduced because of the low aldehyde concentration and that any that do form will precipitate during the reaction and be easily separated. It is important, however, that they should not extract the catalyst from the $scCO_2$.

Biphasic systems in which the $scCO_2$ simply acts as a transport vector, but the catalyst is dissolved in a separate liquid phase mean not only that catalyst retention can be improved, but also that fewer pressure swings are required. The main disadvantage of these systems is that the solvent (water or polyethyleneglycol) can show some solubility in $scCO_2$, thus complicating the separation. Ionic liquids allow excellent, although not as yet perfect, catalyst retention and no loss of solvent [20]. Heavy products will be reduced because both the ionic liquid and the $scCO_2$ will act as diluents, but they will eventually build up in the ionic liquid phase. This will lead to a requirement to purge the ionic liquid – catalyst – heavies mixture, but once again, the heavies should be extractable with an organic solvent and the catalyst – ionic liquid recyclable. Their main disadvantage, which is shared by other liquid - $scCO_2$ biphasic systems is that the product must be extracted from the liquid phase and hence very high pressures (200 bar) are required. Once again, this increases the costs associated with the recycling of CO_2.

Much lower pressures (125 bar) can be used when the liquid phase, in which the catalyst is dissolved, is the steady state mixture of reactants and products that accumulates during the reaction, because the product simply has to dissolve not be extracted [17]. The lower pressure operation makes this approach particularly attractive, although heavies will build up in the catalyst rich phase and separation of the catalyst from the heavies will require processes similar to those developed for low pressure distillation (see Section 9.2).

Process designs for these types of system have been proposed [21] but once again careful control of temperature and pressure will be essential to avoid phase separation problems especially in the recycling gases (CO, H_2, CO_2). Two alternative approaches to simple decompression and recompression may be worthy of consideration. The first is simply to pass the decompressed gaseous phase over an oxidation catalyst so that any unreacted CO and hydrogen, together with any organic materials are oxidised to water and CO_2 and then to vent the CO_2. This would not increase the overall CO_2 burden very significantly because most of the CO_2 being discharged would simply have been discharged anyway if it had not been involved in the process. This completely gets around the recompression problem but the main losses would be of CO and hydrogen. An alternative to oxidation would be membrane separation, such as is used in reforming. In this way the CO and H_2 could be recycled separately from the CO_2 or the CO_2 could be discharged. Recompressing the CO and H_2 would be much less energy intensive than recompressing CO_2 because the amounts involved would be so much less and there would not be problems with phase separation.

The alternative approach draws form the experience of coffee decaffeination. The currently practiced process is carried out on a massive scale using $scCO_2$ to extract the caffeine from the coffee beans. The caffeine is not precipitated from the $scCO_2$ by decompression, but is extracted into water from which it can be separated. The great advantage of this process is that it can be operated at essentially constant pressure and very little energy is required for recompression of CO_2 in the recycling loop. It is difficult to see how such a process could be applied to truly homogeneous catalytic system, in which the catalyst is dissolved in the $scCO_2$, but it could be applied in systems where the CO_2 is the transport vector. In these cases, it would be necessary to pass the effluent stream from the reactor through a counter current of a solvent which dissolves the product, but is not itself soluble in $scCO_2$. An ionic liquid would be one possibility, especially in the cases where lower pressure operation ensures that the product partitions well into the ionic liquid (e.g. nonanal into 1-octyl-3-methylimidazilium bis(trifluoromethyl)sulphonamide at 125 bar [17]). The high solubility of CO_2 in ionic liquids of this kind will reduce the separation efficiency, but it may be possible to use other solvents which dissolve the product but not CO_2 and are not themselves soluble in $scCO_2$. The removal of the product from this solvent could then be carried out as described in Chapter 7.

9.9 Conclusions

None of the alternative strategies for catalyst/product separation has yet reached the point where it can be commercialised for the rhodium catalysed hydroformyation of long chain alkenes and there are very few examples of commercialisation in any catalytic applications. Batch continuous processing with low pressure product distillation has been commercialised but the complexity of the system suggests that alternatives may be able to compete.

In this book, we have examined the chemistry behind the possible new approaches to solving the separation problem and have discussed process designs that could be used to implement the Chemistry. In this chapter, we have addressed what we believe to be the biggest problems remaining to be solved for the different approaches described in the rest of the book and have attempted to provide pointers towards possible solutions. These should provide impetus for further research in this important area aimed at improving the chemistry, phase behaviour, differential solubility and process design and eventually lead to the commercialisation of many more homogeneous catalytic processes so as to make chemicals production much more environmentally acceptable.

9.10 References

[1] G. Phillips and S. N. Falling, *US Patent*, 1994, 5315019.
[2] *Chem. Eng. News.*, 1999, **77**, 19.
[3] *Chem. Eng. News.*, 2004, **82**, 29.
[4] J. A. Banister and G. E. Harrison, *US Patent*, 2004, 0186323.
[5] D. J. Cole-Hamilton, *Science*, 2003, **299**, 1702.
[6] C. D. Frohling and C. W. Kohlpaintner, in 'Applied Homogeneous Catalysis with Organometallic Compounds', ed. W. A. Herrmann, VCH, Weinheim, 1996.
[7] R. D. Mullineaux and L. H. Slaugh, *US Patent*, 1966, 3.
[8] A. J. Sandee, L. A. van der Veen, J. N. H. Reek, P. C. J. Kamer, M. Lutz, A. L. Spek, and P. van Leeuwen, *Angew. Chem. Int. Edit. Engl.*, 1999, **38**, 3231.
[9] N. J. Meehan, A. J. Sandee, J. N. H. Reek, P. C. J. Kamer, P. van Leeuwen, and M. Poliakoff, *Chem. Commun.*, 2000, 1497.

[10] E. Schwab and S. Mecking, *Organometallics*, 2001, **20**, 5504.
[11] L. Ropartz, K. J. Haxton, D. F. Foster, R. E. Morris, A. M. Z. Slawin, and D. J. Cole-Hamilton, *J. Chem. Soc., Dalton Trans.,*, 2002, 4323.
[12] P. Arya, G. Panda, N. V. Rao, H. Alper, S. C. Bourque, and L. E. Manzer, *J. Am. Chem. Soc.*, 2001, **123**, 2889.
[13] X. L. Zheng, J. Y. Jiang, X. Z. Liu, and Z. L. Jin, *Catal. Today*, 1998, **44**, 175.
[14] I. T. Horvath, G. Kiss, R. A. Cook, J. E. Bond, P. A. Stevens, J. Rabai, and E. J. Mozeleski, *J. Am. Chem. Soc.*, 1998, **120**, 3133.
[15] D. F. Foster, D. Gudmunsen, D. J. Adams, A. M. Stuart, E. G. Hope, D. J. Cole-Hamilton, G. P. Schwarz, and P. Pogorzelec, *Tetrahedron*, 2002, **58**, 3901.
[16] D. Koch and W. Leitner, *J. Am. Chem. Soc.*, 1998, **120**, 13398.
[17] P. B. Webb and D. J. Cole-Hamilton, *Chem. Commun.*, 2004, 612.
[18] R. P. J. Bronger, S. M. Silva, P. C. J. Kamer, and P. van Leeuwen, *Dalton Trans.*, 2004, 1590.
[19] C. P. Mehnert, R. A. Cook, N. C. Dispenziere, and M. Afeworki, *J. Am. Chem. Soc.*, 2002, **124**, 12932.
[20] P. B. Webb, M. F. Sellin, T. E. Kunene, S. Williamson, A. M. Z. Slawin, and D. J. Cole-Hamilton, *J. Am. Chem. Soc.*, 2003, **125**, 15577.
[21] P. Webb, T. E. Kunene, and D. J. Cole-Hamilton, *Green Chem.*, 2005, **7**, 373.
[22] A. J. Sandee, J. N. H. Reek, P. C. J. Kamer, and P. van Leeuwen, *J. Am. Chem. Soc.*, 2001, **123**, 8468.
[23] P. Arya, N. V. Rao, J. Singkhonrat, H. Alper, S. C. Bourque, and L. E. Manzer, *J. Org. Chem.*, 2000, **65**, 1881.
[24] M. Haumann, H. Koch, P. Hugo, and R. Schomäcker, *Appl. Catal. A-Gen.*, 2002, **225**, 239.
[25] M. Haumann, H. Yidiz, H. Koch, and R. Schomäcker, *Appl. Catal. A-Gen.*, 2002, **236**, 173.
[26] M. Haumann, H. Koch, and R. Schomäcker, *Catal. Today*, 2003, **79**, 43.
[27] J. K. Lee, T. J. Yoon, and Y. K. Chung, *Chem. Commun.*, 2001, 1164.
[28] K. D. Wiese, O. Möller, G. Protzmann, and M. Trocha, *Catal. Today*, 2003, **79**, 97.
[29] D. E. Bergbreiter, in 'Immobilized Catalysts', 2004.
[30] E. Perperi, Y. L. Huang, P. Angeli, G. Manos, C. R. Mathison, D. J. Cole-Hamilton, D. J. Adams, and E. G. Hope, *Chem. Eng. Sci*, 2004, In the press.
[31] S. N. Falling and S. A. Godleski, *US Patent*, 1993, 5238889.
[32] B. Jastorff, R. Stormann, J. Ranke, K. Molter, F. Stock, B. Oberheitmann, W. Hoffmann, J. Hoffmann, M. Nuchter, B. Ondruschka, and J. Filser, *Green Chem.*, 2003, **5**, 136.
[33] R. P. Swatloski, J. D. Holbrey, S. B. Memon, G. A. Caldwell, K. A. Caldwell, and R. D. Rogers, *Chem. Commun.*, 2004, 668.
[34] M. Uerdingen, C. Treber, M. Balser, G. Schmitt, and C. Werner, *Green Chem.*, 2005, **7**, 321.
[35] E. B. Carter, S. L. Culver, P. A. Fox, R. D. Goode, I. Ntai, M. D. Tickell, R. K. Traylor, N. W. Hoffman, and J. H. Davis, *Chem. Commun.*, 2004, 630.
[36] A. P. Abbott, G. Capper, D. L. Davies, R. K. Rasheed, and V. Tambyrajah, *Chem. Commun.*, 2003, 70.
[37] A. P. Abbott, D. Boothby, G. Capper, D. L. Davies, and R. K. Rasheed, *J. Am. Chem. Soc.*, 2004, **126**, 9142.
[38] G. Francio and W. Leitner, *Chem. Commun.*, 1999, 1663.

Catalysis by Metal Complexes

Series Editors:
R. Ugo, *University of Milan, Milan, Italy*
B.R. James, *University of British Colombia, Vancouver, Canada*

1. F.J. McQuillin: *Homogeneous Hydrogenation in Organic Chemistry.* 1976 ISBN 90-277-0646-8
2. P.M. Henry: *Palladium Catalyzed Oxidation of Hydrocarbons.* 1980 ISBN 90-277-0986-6
3. R.A. Sheldon: *Chemicals from Synthesis Gas.* Catalytic Reactions of CO and H_2. 1983 ISBN 90-277-1489-4
4. W. Keim (ed.): *Catalysis in C_1 Chemistry.* 1983 ISBN 90-277-1527-0
5. A.E. Shilov: *Activation of Saturated Hydrocarbons by Transition Metal Complexes.* 1984 ISBN 90-277-1628-5
6. F.R. Hartley: *Supported Metal Complexes.* A New Generation of Catalysts. 1985 ISBN 90-277-1855-5
7. Y. Iwasawa (ed.): *Tailored Metal Catalysts.* 1986 ISBN 90-277-1866-0
8. R.S. Dickson: *Homogeneous Catalysis with Compounds of Rhodium and Iridium.* 1985 ISBN 90-277-1880-6
9. G. Strukul (ed.): *Catalytic Oxidations with Hydrogen Peroxide as Oxidant.* 1993 ISBN 0-7923-1771-8
10. A. Mortreux and F. Petit (eds.): *Industrial Applications of Homogeneous Catalaysis.* 1988 ISBN 90-2772-2520-9
11. N. Farrell: *Transition Metal Complexes as Drugs and Chemotherapeutic Agents.* 1989 ISBN 90-2772-2828-3
12. A.F. Noels, M. Graziani and A.J. Hubert (eds.): *Metal Promoted Selectivity in Organic Synthesis.* 1991 ISBN 0-7923-1184-1
13. L.I. Simándi (ed.): *Catalytic Activation of Dioxygen by Metal Complexes.* 1992 ISBN 0-7923-1896-X
14. K. Kalyanasundaram and M. Grätzel (eds.): *Photosensitization and Photocatalysis Using Inorganic and Organometalic Compounds.* 1993 ISBN 0-7923-2261-4
15. P.A. Chaloner, M.A. Esteruelas, F. Joó and L.A. Oro: *Homogeneous Hydrogenation.* 1994 ISBN 0-7923-2474-9
16. G. Braca (ed.): *Oxygenates by Homologation or CO Hydrogenation with Metal Complexes.* 1994 ISBN 0-7923-2628-8
17. F. Montanari and L. Casella (eds.): *Metalloporphyrins Catalyzed Oxidations.* 1994 ISBN 0-7923-2657-1

Catalysis by Metal Complexes

18. P.W.N.M. van Leeuwen, K. Morokuma and J.H. van Lenthe (eds.): *Theoretical Aspects of Homogeneous Catalisis.* Applications of *Ab Initio* Molecular Orbital Theory. 1995 ISBN 0-7923-3107-9
19. T. Funabiki (ed.): *Oxygenases and Model Systems.* 1997 ISBN 0-7923-4240-2
20. S. Cenini and F. Ragaini: *Catalytic Reductive Carbonylation of Organic Nitro Compounds.* 1997 ISBN 0-7923-4307-7
21. A.E. Shilov and G.P. Shul'pin: *Activation and Catalytic Reactions of Saturated Hydrocarbons in the Presence of Metal Complexes.* 2000 ISBN 0-7923-6101-6
22. P.W.N.M. van Leeuwen and C. Claver (eds.): *Rhodium Catalyzed Hydroformylation.* 2000 ISBN 0-7923-6551-8
23. F. Joó: *Aqueous Organometallic Catalysis.* 2001 ISBN 1-4020-0195-9
24. R.A. Sánchez-Delgado: *Organometallic Modeling of the Hydrodesulfurization and Hydrodenitrogenation Reactions.* 2002 ISBN 1-4020-0535-0
25. F. Maseras and A. Lledós (eds.): *Computational Modeling of Homogeneous Catalysis.* 2002 ISBN 1-4020-0933-X
26. L.I. Simándi (ed.): *Advances in Catalytic Activation of Dioxygen by Metal Complexes.* 2003 ISBN 1-4020-1074-5
27. A. Sen (ed.): *Catalytic Synthesis of Alkene-Carbon Monoxide Copolymers and Cooligomers.* 2003 ISBN 1-4020-1129-6
28. A. Kotschy and G. Timári: *Heterocycles from Transition Metal Catalysis.* Formation and Functionalization. 2005 ISBN 1-4020-3624-8
29. P.J. Dyson and T.J. Geldbach: *Metal Catalysed Reactions in Ionic Liquids.* 2005 ISBN 1-4020-3914-X
30. D.J. Cole-Hamilton and R.P. Tooze (eds): *Catalyst Separation, Recovery and Recycling.* Chemistry and Process Design. 2006 ISBN 1-4020-4086-5

** Volume 1 is previously published under the Series Title: Homogeneous Catalysis in Organic and iNorganic Chemistry*

MIX
Papier aus verantwortungsvollen Quellen
Paper from responsible sources
FSC® C105338

If you have any concerns about our products,
you can contact us on
ProductSafety@springernature.com

In case Publisher is established outside the EU,
the EU authorized representative is:
Springer Nature Customer Service Center GmbH
Europaplatz 3, 69115 Heidelberg, Germany

Printed by Libri Plureos GmbH
in Hamburg, Germany